Emerging Protein Biotherapeutics

Emerging Protein Biotherapeutics

Edited by Iqbal S. Grewal

CRC Press
Taylor & Francis Group
Boca Raton London New York

CRC Press is an imprint of the
Taylor & Francis Group, an **informa** business

CRC Press
Taylor & Francis Group
6000 Broken Sound Parkway NW, Suite 300
Boca Raton, FL 33487-2742

Printed in the United States of America on acid-free paper
10 9 8 7 6 5 4 3 2 1

International Standard Book Number-13: 978-1-4200-6320-2 (Hardcover)

Library of Congress Cataloging-in-Publication Data

Emerging protein biotherapeutics / editor, Iqbal S. Grewal.
p. ; cm.
Includes bibliographical references and index.
ISBN-13: 978-1-4200-6320-2 (hardcover : alk. paper)
ISBN-10: 1-4200-6320-0 (hardcover : alk. paper)
1. Protein drugs. I. Grewal, Iqbal S. II. Title.
[DNLM: 1. Proteins--therapeutic use. 2. Autoimmune Diseases--drug therapy. 3. Biological Products--therapeutic use. 4. Biological Therapy. 5. Neoplasms--drug therapy. QU 55 E525 2009]

RM666.P87E44 2009
615'.3--dc22 2009003643

Visit the Taylor & Francis Web site at
http://www.taylorandfrancis.com

and the CRC Press Web site at
http://www.crcpress.com

Contents

Preface

Over the last two decades, development of protein biotherapeutics has dramatically increased and has become an essential part of modern medical treatments for cancers and autoimmune diseases. Protein biotherapeutics are expected to be the major part of the pharmaceutical market in the next few years. The current estimates of different genes in the human genome indicate that there are 25,000 to 40,000 functional genes. Together with alternative splicing of these genes and posttranslational modification of proteins, human genes have the potential to code for a large number of functionally distinct protein molecules. As more and more novel protein molecules are being linked with the underlying disease mechanisms, these estimates pose an immense challenge to modern medicine. On the other hand, these estimates represent a tremendous opportunity in terms of harnessing the potential of targeting these proteins to develop protein biotherapeutics to alleviate disease.

Protein biotherapeutics have several distinct advantages over small-molecule drugs, and serve highly specific and complex biologic functions that cannot be mimicked by simple chemical compounds. Because the action of protein biotherapeutics is highly specific, their interference with normal biologic processes and induction of adverse effects is very limited and usually well tolerated. In addition, protein biotherapeutics have the potential for far-reaching patent protection due to the unique nature of functions of these molecules. This aspect is particularly attractive for biopharmaceutical and biotechnology companies because it may provide them with assurance for return on the large amount of investment typically needed to develop protein biotherapeutics. Together, these advantages make protein biotherapeutics not only very attractive when compared with small-molecule drugs but also help to propel the field further.

Targeting protein antigens has been the most highly investigated area of basic medical research for over two decades. These investigations have benefited from the enormous growth in our understanding of mechanisms of disease progression and the explosion in the knowledge involved in regulation of normal and pathological immune response. In addition, much has been learned about the molecular mechanisms of tumor progression, escape of tumor cells from host surveillance mechanisms, and from discovery of the novel protein molecules important in tumor growth. These new findings are shifting currently established paradigms and providing a foundation for novel drug design efforts. This exemplary growth in our knowledge has provided us with several new protein biotherapeutics that have advanced to human clinical trials, while many more are still being tested in preclinical settings.

As in other rapidly evolving fields, protein biotherapeutics is still an area of intensive research. Progress made in this area is not necessarily congruent and is often difficult to simulate into a cogent whole. A concise account of advances in this area in a readily available format is urgently needed. The aim of *Emerging Protein Biotherapeutics* is to make readily available the major research important in the exploitation of protein targets for developing therapeutic strategies for human diseases, in a single volume. Under the auspices of Taylor & Francis Group/CRC Press, I have undertaken the task to concisely consolidate current knowledge of key protein targets important in autoimmunity and cancers focusing on both basic aspects and their clinical application.

In this volume, a number of leading scientists in the autoimmunity and cancer fields have attempted to cover many aspects of biology of protein targets, ranging from the *in vivo* role in the disease process to various strategies to exploit the development of these targets for therapeutic use. Relevant background information, a discussion of the clinical implications of biology regulated by these protein targets, an account of preclinical and clinical testing of various candidates, and a useful bibliography is provided in each chapter. This consolidated information will help the reader to make more detailed analysis and will make the study of protein biotherapeutics accessible at all levels of expertise. This book offers researchers an efficient and rational way to organize the theoretical and experimental knowledge currently available and presents a framework that is readily applicable to develop strategies for targeting proteins for clinical applications. This book will also be highly valuable for basic researchers to understand the disease process.

I express my sincere thanks to all of my contributors for their excellent effort and undertaking this project with such enthusiasm, to Judith E. Spiegel for commissioning me to edit this volume, and to Patricia Roberson and the staff of Taylor & Francis Group for help with publication coordination.

I anticipate that many researchers will be inspired by the comprehensive work presented here and will contribute to this field by further investigations. Looking ahead, I predict that this field will gain much more importance in the future because of the rising awareness of use of protein biotherapeutics and their effect on the disease process. In coming years researchers studying protein biotherapeutics will play an important role in the development of therapeutic strategies to target many human diseases including cancers, autoimmune disease, and others. The information provided in this book has broad implications and will be of value to biologists, immunologists, cancer biologists, biotechnology and pharmaceutical scientists, researchers in basic medical sciences, and clinicians as well.

Iqbal S. Grewal

About the Editor

Iqbal S. Grewal, PhD., FRCPath, is a well-established scientist and biotechnologist with many years of research in immunology and oncology. He has extensive experience of discovering and developing innovative protein-based biotherapeutics in many disease areas. Some of Dr. Grewal's findings have already helped to translate research into key drug candidates for the treatment of autoimmune diseases and cancers. Currently, he is serving as Vice President of Preclinical Therapeutics at Seattle Genetics and as a leading preclinical researcher for the development of monoclonal antibodies and antibody–drug conjugates as therapeutics in the areas of autoimmunity and oncology. Dr Grewal's career also includes leading discovery research and preclinical development of biologics at Genentech, where he identified and validated several novel molecules as therapeutic candidates. Prior to working in the biotech industry, Dr. Grewal held various research positions at Yale University School of Medicine and at the University of California, Los Angeles (UCLA). Dr. Grewal has presented his research work at both national and international meetings, as well as authored more than 100 scientific publications and numerous worldwide patent applications. Dr Grewal is also serving on editorial boards of research journals and has edited two books. He is a fellow of the Royal College of Pathologists, London and member of several distinguished societies. Dr. Grewal holds a PhD in immunology from UCLA and completed his postdoctoral fellowship at Howard Hughes Medical Institute at Yale University School of Medicine.

Contributors

Alessandro Alessandrini
Cellular and Molecular Immunology Laboratory
Department of Surgery, Transplantation Unit
Massachusetts General Hospital and Harvard Medical School
Boston, Massachusetts

Stephen M. Ansell
Division of Hematology
Mayo Clinic
Rochester, Minnesota

Jürgen C. Becker
Department of Dermatology
University of Wuerzburg
Wuerzburg, Germany

Celeste Bello
Department of Malignant Hematology
H. Lee Moffitt Cancer and Research Institute
Tampa, Florida

Gilles Benichou
Cellular and Molecular Immunology Laboratory
Department of Surgery, Transplantation Unit
Massachusetts General Hospital and Harvard Medical School
Boston, Massachusetts

Dennis R. Benjamin
Department of Chemistry
Seattle Genetics, Inc.
Bothell, Washington

Sergio Chimenti
Department of Dermatology
University of Rome 'Tor Vergata'-PTV
Rome, Italy

A. Benedict Cosimi
Cellular and Molecular Immunology Laboratory
Department of Surgery, Transplantation Unit
Massachusetts General Hospital and Harvard Medical School
Boston, Massachusetts

Robert Eisenberg
University of Pennsylvania School of Medicine
Philadelphia, Pennsylvania

Paulo Fontoura
Department of Immunology
Faculty of Medical Sciences
New University of Lisbon
Lisbon, Portugal

Hideki Garren
Bayhill Therapeutics, Inc.
Palo Alto, California
and
Department of Neurology and Neurological Sciences
Stanford University
Stanford, California

Hans-Peter Gerber
Department of Preclinical Therapeutics
Seattle Genetics, Inc.
Bothell, Washington

Iqbal S. Grewal
Department of Preclinical Therapeutics
Seattle Genetics, Inc.
Bothell, Washington

Stephanie de Haseth
Cellular and Molecular Immunology Laboratory
Department of Surgery, Transplantation Unit
Massachusetts General Hospital and Harvard Medical School
Boston, Massachusetts

Martin H. Holtmann
1st Department of Medicine
Johannes Gutenberg-University
Mainz, Germany

Mark J. Houseman
Musculoskeletal Research Group
Institute of Cellular Medicine
Newcastle University
Newcastle-upon-Tyne, United Kingdom

John D. Isaacs
Musculoskeletal Research Group
Institute of Cellular Medicine
Newcastle University
Newcastle-upon-Tyne, United Kingdom

Mohamed A. Kharfan-Dabaja
Department of Blood and Marrow Transplantation
H. Lee Moffitt Cancer Center and Research Institute
Tampa, Florida

Jiyoun Kim
Department of Pathology and Laboratory Medicine
Boston University School of Medicine
Boston, Massachusetts

Sudha Natarajan
Department of Pathology and Laboratory Medicine
Boston University School of Medicine
Boston, Massachusetts

Markus F. Neurath
1st Department of Medicine
Johannes Gutenberg-University
Mainz, Germany

Ezogelin Oflazoglu
Department of Preclinical Therapeutics
Seattle Genetics, Inc.
Bothell, Washington

Shuji Ozaki
Division of Hematology and Transfusion Medicine
Tokushima University Hospital
Tokushima, Japan

Javier Pinilla-Ibarz
Department of Malignant Hematology
H. Lee Moffitt Cancer and Research Institute
Tampa, Florida

Daniel G. Remick
Department of Pathology and Laboratory Medicine
Boston University School of Medicine
Boston, Massachusetts

Edgardo S. Santos
Division of Hematology–Medical Oncology
University of Miami Miller School of Medicine
and
Sylvester Comprehensive Cancer Center
Miami, Florida

Rosita Saraceno
Department of Dermatology
University of Rome 'Tor Vergata'-PTV
Rome, Italy

David Schrama
Department of Dermatology
University of Wuerzburg
Wuerzburg, Germany

Peter D. Senter
Department of Chemistry
Seattle Genetics, Inc.
Bothell, Washington

Louis J. Vaickus
Department of Pathology and Laboratory Medicine
Boston University School of Medicine
Boston, Massachusetts

Tom L. Van Belle
La Jolla Institute for Allergy and Immunology
La Jolla, California

Matthias G. von Herrath
La Jolla Institute for Allergy and Immunology
La Jolla, California

Section I

Introduction

1 Overview of Antibody-Based Therapeutics

Present and Future Promise

Iqbal S. Grewal

CONTENTS

KEY WORDS: mAb, therapeutic antibodies, immunotherapy, antibody–drug conjugates, cancer, autoimmunity

ABSTRACT

Antibodies are an essential part of the human immune system. They are exquisitely specific, naturally evolved molecules that recognize and eliminate disease antigens and pathogenic cells. With the landmark discovery of hybridoma technology by Kohler and Milstein about 30 years ago, intense antibody research has offered a promise of new versatile limitless therapeutics to fight human diseases. Since the approval of anti-CD3 monoclonal antibody (mAb), muromonab-CD3, for suppressing allograft rejection in 1986, mAbs have become a major part of the therapeutic strategy for a variety of indications including transplantation, autoimmunity, cancer, infections, and cardiovascular diseases. However, the early development of mAbs and their present acceptance as viable therapeutics by the pharmaceutical industry met with many challenges. One of the major limitations of the mAbs produced in rodents was their inherent immunogenicity. This issue was resolved by the replacement of murine amino acid sequences with human counterparts, leading to chimeric and humanized mAbs. Remarkable advances in *in vivo* gene targeting and display technologies have now enabled us to produce fully human mAbs with refined antibody properties such as size, affinity, specificity, and effector functions. Continuous advances in antibody engineering technologies are being applied to improve antibodies further to generate lead clinical molecules with highly optimized properties. Several optimization platforms for mAbs are now firmly established within the pharmaceutical industry and have altered our view of how mAbs might be used for therapeutic purposes in the future. The second generations of genetically engineered therapeutic mAbs entering the clinic today are greatly optimized for high specificity and functionality, and in some cases are engineered to serve as vehicles for delivery of cytotoxic drugs. The safety issues with small molecule therapeutics and remaining unmet medical needs, especially in oncology and autoimmune and inflammatory diseases, have prompted a need for safer and more effective alternatives. Although mAbs are filling this void, their complex biology, the variety of recombinant methods used to produce them, and unwanted exposure of normal tissues as a result of systemic use all pose new safety concerns. Additional clinical success of mAbs will depend on meeting these new challenges, and their market success will depend on clearly establishing their clinical and cost benefits. In this review, aspects of antibodies relevant to their use as therapeutics are discussed.

INTRODUCTION

About a century ago, Paul Ehrlich coined the term "magic bullets" to refer to chemicals with the ability to target and destroy microorganisms and spare normal tissues of the host harboring these microorganisms.[1] Ehrlich's concepts and seminal ideas established the foundation for targeted therapies. These specific targeting approaches include the use of antibodies and other biologics that have affinity for binding to particular antigens. With a remarkable breakthrough in mouse hybridoma technology in

1975 when Köhler and Milstein developed a method to generate murine monoclonal antibodies (mAbs),[2] mAbs became extremely useful tools for specific targeting in basic and clinical research.[3–5]

Unlike small molecule therapies that only offer short-term therapeutic benefit and usually cause serious side effects, mAbs provide effective treatment with relatively high efficacy and tolerability. One of the key advantages of antibody-based therapeutics is the ability to target specific antigens present in cells, tissues, and organs involved in the pathology of disease, while reducing the toxicity in the normal tissues. Antibodies usually have a much longer half-life than small molecules, which adds to their efficacy and avoids the need for frequent or continuous infusions.[6] Antibodies achieve superior efficacy by binding to membrane expressed targets and recruiting host immune system to mediate powerful effector functions such as antibody-dependent cellular cytotoxicity (ADCC), complement-dependent cytotoxicity (CDC), and antibody-dependent cellular phagocytosis (ADCP) that ultimately lyse the target cell. In addition, mAbs can also be used to block receptor–ligand interactions, neutralize unwanted soluble molecules, or send a productive signal to the target cell to regulate its status. Antibody therapeutics also offer a relatively faster track to the clinic compared to small molecule inhibitors.[7,8] As antibodies share common structural and functional features and work through similar mechanisms, advances made in the development of antibody therapeutics can also be applied as a general rule to new antibodies having similar mechanism of action. Interestingly, with the advancement of antibody engineering technology, the potency of the antibody can readily be altered to suit one's requirement.[9,10] Thus, antibodies can be exploited to harness multiple mechanisms of action to develop superior therapeutics such as effector functions, their blocking and neutralizing properties, and their ability to directly target pathogenic cells.[10] Besides utilizing unconjugated antibodies, antibodies can also be used as delivery vehicles for cytotoxic drugs, cytokines, toxins, or radionucleotides.[11]

Although antibody-based therapies have proved successful in the clinic and are a viable option for many autoimmune diseases and cancers, there are still some limitations with this approach. For example, antibodies can target only readily accessible antigens usually expressed on the target cell surface or a pathogenic soluble cytokine and, unlike small molecules, antibodies cannot access intracellular targets. As antibodies are proteins and need to be produced by cells, they require special controlled handling and production methods, and their production is usually associated with high cost. One of the biggest challenges of developing antibody-based therapeutics is understanding the biology of the target and its relationship to the disease process. Human use of the antibodies without thorough understanding of basic biology of the target in some rare cases has led to serious undesirable effects. A case of the use of super agonistic anti-CD28 (TGN1412) antibody in healthy volunteers resulted in life-threatening side effects, highlighting this risk.[12]

With remarkable progress in antibody engineering, Erlich's concepts now have been successfully applied to the development of innovative treatment strategies, which have become a practical and successful means of treating cancers, autoimmune and inflammatory diseases. Indeed, in 1986 the first hybridoma-based mAb, muromonab-CD3, and in 1997 the first engineered chimeric therapeutic mAb, rituximab, were approved by the U.S. Food and Drug Administration (FDA) for human use.[13,14] For the past two decades, therapeutic mAbs have emerged as the fastest growing area of the biotechnology industry, and have become a major therapeutic strategy for cancers and autoimmune disease. Today, large portions of the biologics in biotechnology product pipelines are mAbs. There are currently 21 monoclonal antibodies that are approved by the FDA for the treatment of cancers and autoimmunity (Table 1.1), and many more are in clinical testing.[3]

TABLE 1.1
Approved Antibodies for Therapeutic Use

Approval Date	Name	Company	Antibody	Target	Antibody Type	Indication
1986 (US)	OKT3	Johnson & Johnson	Muromonab-CD3	CD3	Murine, IgG_2a	Autoimmune
1984 (US)	ReoPro	Centocor	Abciximab	PIIb/IIIa	Chimeric, IgG_1, Fab	Homeostasis
1997 (US) 1998 (EU)	Rituxan	Genentech	Rituximab	CD20	Chimeric, IgG_1	Cancer
1997 (US) 1999 (EU)	Zenapax	Roche	Daclizumab	CD25	Humanized, IgG_1	Autoimmune
1998 (US) 1998 (EU)	Simulect	Novartis	Basiliximab	CD25	Chimeric, IgG_1	Autoimmune
1998 (US) 1999 (EU)	Synagis	MedImmune	Palivizumab	RSV	Humanized, IgG_1	Infections
1998 (US) 1999 (EU)	Remicade	Centocor	Infliximab	TNFα	Chimeric, IgG_1	Autoimmune
1998 (US) 2000 (EU)	Herceptin	Genentech	Trastuzumab	HER2	Humanized, IgG_1	Cancer
2000 (US)	Mylotarg	Wyeth	Gemtuzumab ozogamicin	CD33	Humanized, IgG_4, immunotoxin	Cancer
2001 (US) 2001 (EU)	Campath	Genzyme	Alemtuzumab	CD52	Humanized, IgG_1	Cancer
2002 (US) 2004 (EU)	Zevalin	Biogen Idec	Ibritumomab tiuxetan	CD20	Murine, IgG_1, radiolabeled (Yttrium 90)	Cancer
2002 (US) 2003 (EU)	Humira	Abbott	Adalimumab	TNFα	Human, IgG_1	Autoimmune
2003 (US)	Xolair	Genentech	Omalizumab	IgE	Humanized, IgG_1	Autoimmune
2003 (US)	Bexxar	Corixa/GSK	Tositumomab-I-131	CD20	Murine, IgG_2a, radiolabeled (Iodine-131)	Cancer
2003 (US) 2004 (EU)	Raptiva	Genentech	Efalizumab	CD11a	Humanized, IgG_1	Autoimmune

TABLE 1.1 (CONTINUED)
Approved Antibodies for Therapeutic Use

Approval Date	Name	Company	Antibody	Target	Antibody Type	Indication
2004 (US) 2004 (EU)	Erbitux	Imclone	Cetuximab	EGFR	Chimeric, IgG_1	Cancer
2004 (US) 2005 (EU)	Avastin	Genentech	Bevacizumab	VEGF	Humanized, IgG_1	Cancer
2004 (US)	Tysabri	Biogen Idec	Natalizumab	α4-ntergrin	Humanized, IgG_4	Autoimmune
2005 (JP)	Actemra	Chugai/ Roche	Tocilizumab	Anti-IL-6R	Humanized, IgG_1	Autoimmune
2006 (US)	Vectibix	Amgen	Panitumumab	EGFR	Human, IgG_2	Cancer
2007 (US)	Soliris	Alexion	Eculizumab	C5	Humanized $IgG_{2/4}$	Blood disorders
2008 (US)	Cimzia	UCB	Certolizumab	TNFα	Humanized, pegylated Fab fragment	Autoimmune

STRUCTURAL FEATURES OF THE ANTIBODIES

Antibodies are large heterodimeric proteins composed of four polypeptide chains, two identical heavy (H) chains and two identical light (L) chains linked together with disulfide bonds to form a stable molecule with a flexible hinge region.[15,16] The amino acid sequence in the amino-terminal ends of both H and L chains varies greatly among different antibodies. These variable regions form a pocket for binding antigens. Recombination events during B-cell maturation rearranges variable (V), diversity (D), and junctional (J) gene segments of the H chain and the VJ gene segments of the L chain to generate the DNA that encodes unique antibody combining sites that confer specificity to the variable regions.[16] These variable regions also give the antibody diversity and affinity of antigen binding. The amino acid sequence variability in the variable regions is especially pronounced in three hypervariable domains also called complementarity determining regions (CDRs). The CDRs of immunoglobulin G (IgG) are responsible for the strength of binding (affinity) of the antibody. Treating the antibody with proteases can cleave the binding regions producing fragments called fragment antigen binding (Fab) and $F(ab')_2$.[16] The carboxyl terminals of H and L chains are called the constant regions because only few differences in the sequences of amino acid residues are found in this region among different antibodies within the same class. The constant region of the H chains of antibodies is called fragment crystallizable (Fc), and determines the effector mechanism and catabolism of the antibodies. Based on the structural properties, each chain of the antibody is divided into variable and constant domains (Figure 1.1). Fc region constant domains (CH2 and CH3) in the heavy chain of the IgG antibody contain binding sites for complement components and for Fc receptors (FcRs) found on immune cells.

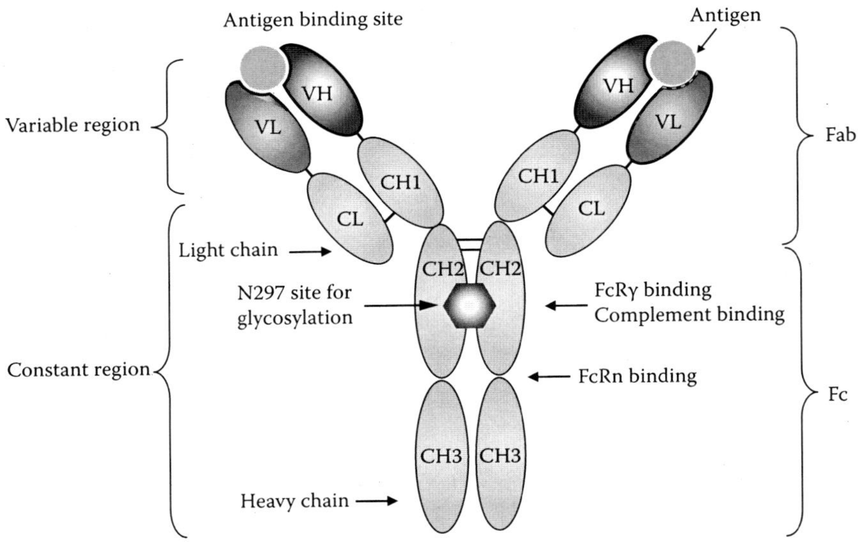

FIGURE 1.1 ***A color version of this figure follows page 176.*** Typical structure of IgG. IgG is composed of two identical heavy and two identical light chains. Each heavy chain and light chain contains a variable region and a constant region. The antigen-binding sites comprise the variable regions of the heavy and light chains. The heavy constant regions (CH2 and CH3) interact with complement component C1q and Fcγ receptors, and are responsible for the effector functions and pharmacokinetics of the IgG. IgG is also glycosylated via carbohydrate moieties attached at conserved amino acid residue, asparagine at position 297 (N297) in CH2 domain, which is important for Fc interactions with FcRγ. FcRn binds to both CH2 and CH2 domains in the Fc region of the mAb. VH, heavy chain variable domain; VL, light chain variable domain; CH1, CH2, and CH3, heavy chain constant domains; CL, light chain constant domain; Fab, antigen-binding fragment; Fc, constant region fragment; FcRγ, Fcγ receptor; FcRn, neonatal Fc receptor.

The binding of the Fc region with complement initiates CDC, and with FcRγ results in ADCC and ADCP. Fc regions of the antibodies also bind to FcRn, the salvage receptor, which controls the catabolism of IgG. The Fc region of each H chain of IgG also contains an *N*-linked glycan chain attached to the aspargine residue at position 297 (N297).[17–20] When IgG is produced in human serum or by mammalian cells in culture, a heterogeneous population of glycoforms with varying oligosaccharide structures such as G_0, G_1, G_2 forms is produced (Figure 1.2). In fully processed forms of human IgG, the oligosaccharide core is a bi-antennary type which consists of N297 linked to *N*-acetyl glucosamine (GlcNAc)-(Fucose)-GlcNAc-Mannose-(Mannose-GlcNAc)$_2$.[15] Partially processed IgG with varying terminal glactose, sialic acids at the terminal GlcNAc, presence of bisecting GlcNAc, or fucose residues attached to GlcNAc that is linked to N297 are also produced (Figure 1.2). Fc region glycosylation is important in the interaction of Fc with FcRγ[17–20] and is sensitive to the changes in the oligosaccharide structures attached to the Fc.[21,22]

Based on the constant region structure and immune function, antibodies are divided into five major classes, IgM, IgG, IgA, IgD, and IgE, with either κ or λ L

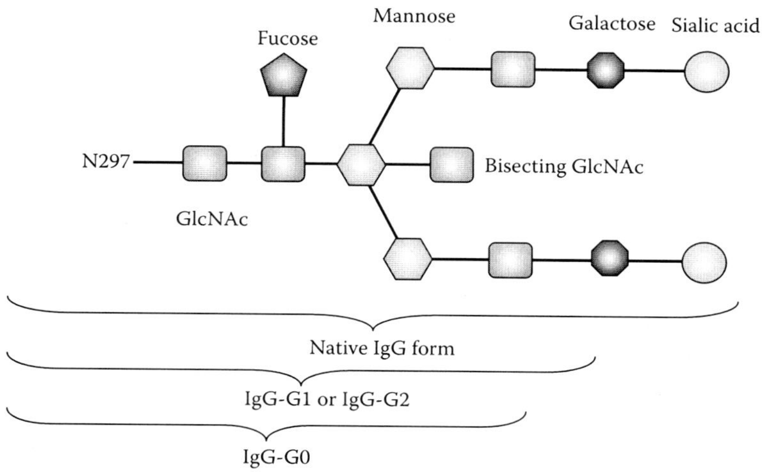

FIGURE 1.2 ***A color version of this figure follows page 176.*** Schematic representation of glycosylation of IgG_1. IgG carbohydrate moiety attached to N297 of each CH2 domain of IgG_1 in the fully processed (native form) with all potential branching and terminal sugar residues are shown in the figure. When IgG_1 is produced *in vivo*, only a minor fraction of IgG_1 molecules have this carbohydrate side chain configuration. More frequently, partially processed moieties are identified, such as IgG_1 G_0 variants that lack all terminal sialic-acid and galactose residues and G_1 or G_2 glycoforms with one or two galactose residues present in the carbohydrate side chain. GlcNAc, N-acetylglucosamine, N297, asparagine residue at position 297.

chains within any single antibody molecule. The majority of therapeutic antibodies are of IgG subclass because of their *in vivo* properties, including high stability and bioavailability, relatively longer half-life, and excellent effector functions. IgG subclasses are also relatively easy to produce and commercial manufacturing methods to produce them are well established.[23]

TYPES OF THERAPEUTIC ANTIBODIES

The advent of mouse hybridoma technology for production of mAbs coupled with the discovery of a large number of surface expressed antigens on pathogenic cells fueled an intensive interest in the development of mAbs for therapeutic use. Although mouse hybridoma-produced mAbs were the source of many early therapeutic modalities, development of mAbs has progressed to the generation of four major antibody types, including murine, chimeric, humanized, and fully human mAbs (Figure 1.3). The types of antibodies are determined by the percentage of mouse and human parts making up the antibody molecules. The amino acid sequence of murine antibodies contains 100 percent mouse sequences, chimeric antibodies are composed of approximately 35 percent mouse sequence, humanized antibodies contain approximately 5 percent mouse amino acid residues, and fully human mAbs contain 100 percent human amino acid sequences.[15]

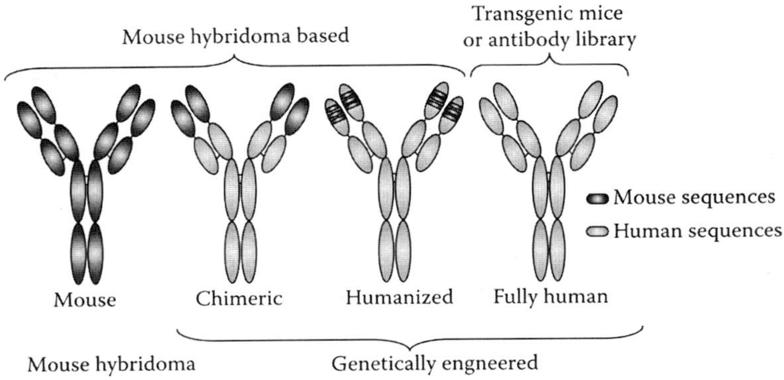

FIGURE 1.3 ***A color version of this figure follows page 176.*** Different forms of therapeutic mAbs. Mouse hybridoma technology is used to produce mouse mAbs. Genetic engineering techniques are used to develop chimeric and humanized therapeutic mAbs. Chimeric mAbs are constructed by cloning variable region genes derived from a murine source and constant region genes derived from a human source and are produced in transfected cells. Humanized therapeutic mAbs are predominantly derived from a human source except for the CDRs, which are murine and are generated by insertion of murine CDRs to human constant and variable framework regions. Fully human mAbs are generated either by selection of *in vitro* antibody libraries or are produced in transgenic animals by generating human hybridomas.

Murine Monoclonal Antibodies

Initially, antibodies against human antigens were generated in mice and murine mAbs obtained by hybridoma technology were tested in the clinic for human therapeutic use.[23] Although hybridoma-derived mouse mAbs offered a reliable source of large quantities of antibodies for clinical testing, their clinical use met with many disappointments.[24] Because of dissimilarity between murine and human immune systems, initial murine therapeutic antibodies largely failed in clinical trials. The major factors responsible for failure of these antibodies included their immunogenicity, resulting in neutralization and rapid clearance, and short *in vivo* half-life and an inability to efficiently recruit host effector functions.[23] In many cases, formation of drug/antidrug antibody complexes after repeated administration resulted in mild allergic reactions and sometimes in a major anaphylactic shock.[25] Many technical advances in antibody generation in the 1980s helped to circumvent some of these issues. To overcome these problems, part of the amino acid sequences of mouse antibodies was replaced with human sequences resulting in chimeric mAbs, which were considered as viable therapeutic candidates.[15,23] Understanding the key structural and functional features of the antibodies was extremely important to resolve some of these issues and to improve the efficacy of the therapeutic antibodies.

Chimeric Monoclonal Antibodies

Chimeric mAbs are molecularly engineered antibody molecules that contain mouse variable domains that are fused with human constant domains from both H and L chains, and are produced by transfected cells as recombinant proteins.[15] Chimeric antibodies engineered by taking gene sequences from the kappa light chain and the IgG_1 H chain of human antibodies and fusing them with mouse variable region sequences are approximately 65 percent human. Removal of mouse content from the antibody reduces murine antibody immunogenicity and increases serum half-life while retaining the same specificity as the parent murine antibody.[25] In addition, chimeric mAbs also interact more efficiently with FcRγ on human immune cells, which results in increased potency.[25] Indeed, antibodies such as infliximab, rituximab, basiliximab, and cetuximab are all chimeric FDA-approved mAbs that have significantly reduced immunogenicity, relatively longer half-life in human patients, and have demonstrated clinical efficacy.[23]

Humanized Monoclonal Antibodies

Humanization of the murine antibodies was initially achieved by grafting murine hypervariable domains into human antibodies, a technique called CDR grafting. The amino acid sequence of the resulting molecule is approximately 95 percent human origin. Although CDR grafting produces a near human antibody, several studies have shown that humanized antibodies generated this way have lower antigen binding affinity than the parent murine monoclonal antibodies. In many cases antigen binding affinities have been reported to decrease up to several hundred fold.[26,27] In order for humanized antibodies to be effective, high antigen binding affinity is believed to be required. Many approaches have been developed to improve the antigen binding affinity by introducing mutations into the CDRs of these antibodies.[28] These techniques include chain-shuffling, randomization of CDRs, and generation of antibody libraries with mutations within the variable regions using error-prone polymerase chain reaction (PCR) and site-specific mutagenesis.[29] Chimeric and humanized antibodies have now generally replaced murine antibodies in modern therapeutic antibody applications. Progress in production of antibodies also has been made to supplement hybridoma technology with recombinant DNA technology to produce antibodies in mammalian and nonmammalian cells.[30] Several humanized mAbs have been approved by the FDA for autoimmune diseases and cancers and have shown acceptable clinical efficacy (Table 1.1).

Fully Human Monoclonal Antibodies

Fully human and humanized antibodies are now the most common protein therapeutics in development. These antibodies interact very well with human immune cells. Differences in fully human and humanized antibodies are negligible and, as such, both types of antibodies have equal efficacy and safety. Fully human antibodies are routinely produced in transgenic mice carrying a repertoire of human immunoglobulin germline gene segments.[15,30,31] These transgenic mice, when immunized with a desired human antigen, make fully human antibodies against the antigen, and spleen cells from these mice can be used to produce antigen-specific monoclonal antibodies.[30] These mice have been successfully used to develop a large number of clinical molecules.[31] Fully human antibodies can also be generated from phage libraries.[32] By using randomized combinations of human antibody variable regions, synthetic phage libraries have been created. These libraries are used to screen for antigen specific sequences for making fully human antibodies. Antibodies made this way have the variable regions that are very humanlike in nature.[32] Two such antibodies, panitumumab and adalimumab, are approved by the FDA for cancers and autoimmune diseases, respectively, and have proved to be successful therapeutic agents.[23]

FUNCTIONAL FEATURES OF THE ANTIBODIES

As most commonly used therapeutic antibodies belong to the IgG subclass, their clinical activity is usually mediated by multiple mechanisms. For example, some therapeutic antibodies induce their effect by blocking receptor–ligand interactions or by neutralization of activity of the soluble ligand involved in the disease pathology. On the other hand, some antibodies work through their agonistic properties to send a productive signal directly to the target cell, resulting in profound changes in the status of the target cells such as induction of apoptosis or inhibition of the growth of the pathogenic cell. Agonistic antibodies can also be used to activate immune effector cells for a beneficial host immune response. One other major mechanism of action of therapeutic antibodies is the delivery of Fc-mediated effector function such as ADCC, ADCP, and CDC, especially in the area of oncology where elimination of the tumor's cells by virtue of effector functions proved to be a successful strategy to fight cancers.[3,5,10,33]

NEUTRALIZATION OF CYTOKINES AND EFFECTOR MOLECULES

In the last few decades, considerable progress has been made in the identification of key cytokines and effector molecules that are important in the development of cancers, autoimmunity, and inflammatory disease. For example, tumor necrosis factor alpha (TNFα) was found to be a key player in the pathogenesis of multiple autoimmune and inflammatory diseases.[34,35] Accordingly, biologic molecules to neutralize this molecule have been developed and successfully launched in the clinic. These include two recombinant therapeutic antibodies, infliximab and adalimumab. Infliximab is an anti-TNFα chimeric mAb, whereas Adalimumab is a recombinant human IgG_1 mAb specific for human TNFα. Adalimumab was generated by using phage display technology resulting in an antibody with human-derived H and L chain variable regions and human IgG_1 and kappa chain constant regions. Both of these antibodies neutralize TNFα and inhibit TNFα action. In addition, adalimumab also lyses surface TNF expressing cells. Infliximab has been approved for active Crohn's disease and rheumatoid arthritis,[34,35] and adalimumab is approved for rheumatoid arthritis, psoriatic arthritis, ankylosing spondylitis, and Crohn's disease.[36]

Several angiogenic cytokines and immunosuppressive agents secreted by tumor cells also have been identified recently. These include molecules like transforming growth factor beta (TGFβ) and vascular endothelial growth factor A (VEGF-A), which are important for the growth of tumors. Antibodies to neutralize these factors are being developed for therapeutic use. For example, *in vivo* neutralization of TGFβ by anti-TGFβ mAbs have shown efficacy in abdominal tumor, lung metastases, and breast carcinomas models in mice.[37] Likewise, neutralization of VEGF-A with bevacizumab, a humanized IgG_1, which binds with high affinity to all human VEGF-A isoforms and their bioactive proteolytic fragments, has been shown to inhibit tumor-associated angiogenesis in tumor sites and is approved for colorectal cancer and breast tumors and being tested for many other carcinomas.[38]

Blocking Receptor–Ligand Interactions

Apart from their ability to neutralize and inactivate soluble factors, mAbs can also be used to bind cell surface expressed receptors to block binding and thereby antagonize the ligand functions. For example, anti-CD25 mAbs, daclizumab and basiliximab, bind to interleukin-2 (IL-2) receptor and block the actions of IL-2 resulting in efficacious suppression of the graft rejection.[39,40] Similarly anti-CD11a mAb, efalizumab, binds α-subunit of leukocyte function antigen-1 (LFA-1) and inhibits the binding of LFA-1 to intercellular adhesion molecule-1 (ICAM-1), thereby inhibiting the adhesion of leukocytes to other cell types.[41] As interactions between LFA-1 and ICAM-1 are important for activation and adhesion of T cells to endothelial cells, and migration of T cells to sites of inflammation including psoriatic skin, efalizumab has been approved for the treatment of psoriasis.[41]

Some growth factor receptor family antigens associated with tumors are known to be important for activation and maintenance of cellular growth and survival of tumors. Antibodies targeting these antigens can either induce cell cycle arrest and/or apoptosis or inhibit the function of the targeted receptor. Monoclonal antibodies such as trastuzumab, cetuximab, and panitumumab work through inhibiting the tumor growth in multiple tumor types.[42,43] Likewise, anti-VEGF-A mAb, bevacizumab, inhibits angiogenesis, and anti-type 1 insulin-like growth factor (IGF) receptor mAb blocks the IGF signaling pathway, after binding to their specific receptors on target cells.[44,45] Apart from interfering with tumor cell growth and induction of apoptosis, some mAbs can also sensitize cells to chemotherapeutic agents, such as chimeric anti-Her-1 mAb, cetuximab, which blocks epidermal growth factor receptor (EGFR) signaling.[43]

Signaling

Antibodies can also send direct signals to the cells that express the antigens on their surface.[46–48] Certain receptors belonging to TNF superfamily such as CD95, DR4, and DR5 are death-inducing receptors; their activation initiates a series of signals that can either induce cell cycle arrest or apoptosis.[49] Agonistic antibodies can trigger these receptors, induce apoptosis and, in some cases, sensitize cells to chemotherapeutic agents. For example, agonistic mAb against CD95 has been shown to induce apoptosis in CD95 expressing cells via activation of CD95 associated pathways.[50] Similarly, binding of rituximab to CD20 on B cells also leads to apoptosis of the B cells, which is mediated by activation of the Src kinase and of mitogen-activated protein kinases family members and upregulation of activator protein-1. Rituximab also promotes chemosensitization of various B cell lines by inhibiting survival signals such as extracellular signal-regulated kinases1/2 and the nuclear factor-κB pathway.[51]

Apart from directly targeting tumor cells, signaling antibodies can also be used to activate host immune cells to mount a protective host antitumor immune response. Thus, agonistic antibodies to key receptors in the immune system have been considered for antitumor therapy. These include anti-CTLA-4,[52] anti-4-1BB,[53] anti-OX40,[54] anti-CD40,[55] and anti-BAT,[56] which bind to signaling receptors on T cells or antigen-presenting cells (APCs). These antibodies have been shown to enhance antigen presentation capabilities of the APCs and activate tumor specific cytotoxic T cells (CTLs). For example, anti-CTLA-4 antibody binds to an inhibitory receptor, CTLA-4, on T cells and selectively blocks the effects of B7.1/B.2, the ligands for this receptor, thereby allowing the T cell to receive a co-stimulatory signal through activating receptors in the absence of inhibitory signal.[57] Both anti-4-1BB and anti-OX40 directly activate T cells and anti-CD40 activates APC for enhanced tumor antigen presentation.[54,58,59] Although the mechanisms of action of these antibodies are different from each other, they all lead to enhancement of antitumor CTL responses. Many of these antibodies are in clinical testing and are showing promising results.[54,60,61]

Antibodies as Vehicles for Payload Delivery

An ever expanding area of research, especially for the development of mAbs for therapeutic use in cancers, is to arm the antibodies to carry payloads such as radionucleotides, prodrugs, toxic chemical drugs, potent toxins, enzymes, chemokines, or cytokines and selectively deliver them to cancer cells.[5] This promising approach has resulted in the approval of toxin- and radionucleotide-conjugated antibodies for the treatment of cancers. For example, two radionucleotide-conjugated anti-CD20 antibodies, ibritumomab tiuxetan and tositumomab-I-131, have shown excellent efficacy in non-Hodgkin's lymphoma (NHL). Their superior efficacy is attributed to the synergy between signals induced with anti-CD20 and delivery of radioisotope payload delivered by the antibodies.[62]

The first generation of antibody–drug conjugates composed of anticancer drugs such as methotrexate, doxorubicin, cisplatin, and vinca alkaloids was not optimal for clinical use, because the agents were only marginally efficacious and were less potent than their corresponding drugs in unconjugated form. However, considerable progress has been made to optimize antibody–drug conjugates for the clinical utility.[63,64] Currently, many different types of cytotoxic drugs are being exploited for conjugation with antibodies, including enediyne antibiotics, dolstatin analogues, maytansinoids, auristatins, CC-1065 analogs, anthracyclines, and many other novel cytotoxic natural and synthetic compounds. These antibody conjugates show promising pre-clinical data and are being advanced as potential clinical candidates.[63,65] One such drug conjugate, gemtuzumab ozogamicin, an anti-CD33 calicheomicin conjugate, is already approved as a drug conjugate.[66,67] Many other drug conjugates are being tested in the clinical trials and early clinical data from these trials are encouraging.[63,65] For example, phase I data from clinical testing of SGN-35, a CD30-targeting antibody–drug conjugate consisting of auristatin drug, MMAE, linked to anti-CD30 antibody via protease-cleavable linker, has demonstrated efficacy in patients with Hodgkin's lymphoma. Preliminary data from 22 evaluable patients who were treated with SGN-35 at escalating doses ranging from 0.1 to 1.2 mg/kg intravenously every 3 weeks show that SGN-35 was well tolerated and induced tumor regressions in over 86 percent of patients with 45 percent objective responses and 23 percent complete responses.[68]

One major reason for tumor progression in humans is considered to be the inability of the host to successfully activate antitumor immune response. In addition to activation of antitumor T cells using unconjugated antibodies such as anti-CTLA-4 mAb as discussed earlier, alternative approaches using conjugated mAbs to activate immune response against tumors are also increasingly being considered. One such approach is to use mAbs as tools to recruit and activate host immune cells. In this respect, antibody–cytokine conjugates have been constructed and biologic activity of these conjugates is demonstrated. The best-studied antibody conjugate is an antibody-IL-2 construct that has been shown to eradicate established tumors in syngeneic mice.[69] These encouraging preclinical data have resulted in clinical testing of IL-2 conjugated humanized anti-GD2 and humanized anti-EpCAM antibodies for metastatic melanoma and prostate cancer, respectively.[70,71] Other cytokines such as granulocyte–macrophage colony-stimulating factor (GM-CSF), IL-12, TNFα, interferon gamma (IFNγ), and lymphotoxin alpha (LTα) conjugated to antibodies are under development and have shown superior *in vivo* antitumor activity.[72–75] Thus, these cytokine–antibody conjugates offer an efficient and specific targeting approach without causing systemic toxic effects of the cytokines.

FC-MEDIATED ACTIVITIES OF THE ANTIBODIES

Binding of an antibody to an antigen expressed on the surface of cells in many cases only results in a limited direct activity. The significant biologic effects of the antibodies are shown to be the consequence of secondary effector functions. Antibodies mediate a variety of these effector functions through the interaction of their Fc portion with the FcRs expressed on a variety of effector cells. This has been clearly demonstrated in animal studies where lack of effective Fc interaction with FcR-expressing effector cells results in poor efficacy of the antibodies.[76] In addition, antibodies also fix complement, which results in lysis of cells and release of biologically active molecules.[16] Fc-mediated effector functions of mAbs are discussed in the following sections.

RECRUITMENT OF EFFECTOR CELLS VIA FC–FCRγ INTERACTIONS

Phagocytic cells, lymphocytes, platelets, mast cells, and basophils have FcRs that bind antibodies through their Fc portion.[76] Effector cells can express both activating receptors, such as FcRγI (CD64) and FcRγIII (CD16), and an inhibitory receptor, such as FcRγIIB (CD32b). Binding of antibodies to FcRs can either activate or inhibit functions of the effector cells. The outcome of the effector response is regulated by the balance between activating and inhibitory signals delivered through these receptors. Each IgG subclass has a different affinity for different FcRs, and differentially activates cells to perform effector functions.[6] Binding of Fc to FcR also induces endocytosis of antigen/antibody immune complexes resulting in antigen presentation for the antigen. Activation of FcR via Fc binding can also lead to the production and release of cytokines and proinflammatory molecules from the cell it binds.[76,77]

Two of the main mechanisms mediated through Fc–FcR interactions of the therapeutic antibodies against cancer targets are believed to be ADCC and ADCP (Figure 1.4). The role of FcRs in these effector functions had been demonstrated in mice deficient in FcRγI and FcRγIII, where IgG_1 antibodies exhibit impaired *in vivo* ADCC and ADCP.[78] By using a molecular biology approach, antibodies have been constructed to selectively abolish Fc–FcRγIIIA interactions by making mutations in the Fc domain of the antibodies. For example, an anti-Her-1 mAb containing mutation in the Fc portion induced by changing amino acid residue 265 from asparagine to alanine reduced *in vitro* ADCC activity.[79] Antitumor activity of rituximab is also dramatically reduced in a common FcRγ-deficient athymic mouse (lacking activating FcRγs) model for human lymphoma.[79] Likewise, trastuzumab also shows impaired antitumor activity in an FcRγ-deficient athymic mouse

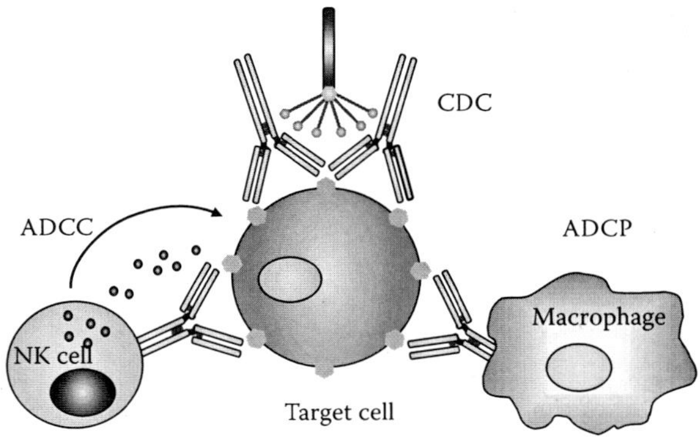

FIGURE 1.4 ***A color version of this figure follows page 176.*** Typical antibody Fc-dependent effector functions of the mAbs. Following the binding of IgG to a specific target cell, C1q complement factor interacts with the CH2 constant region of the mAb, which leads to the activation of a proteolytic cascade of the classical complement pathway and consequently induces the formation of a membrane-attack complex for the lysis of target cells. This activity is termed complement-dependent cytotoxicity (CDC). Binding of IgG to target cells can also induce antibody-dependent cellular cytotoxicity (ADCC) mediated by immune-effector cells such as macrophages and NK cells which interact with the CH3 domain in the Fc region of the IgG through FcRγIIIA expressed by effector cells. Target cells coated with IgG can also be phagocytosed by macrophages via binding of Fc of IgG to FcRγIIIA expressed on macrophages by an active process called antibody-dependent cellular phagocytosis (ADCP). IgGs can also negatively regulate the cytotoxic response against target cells through FcRIIB, which is expressed on the cell surface of macrophages. IgG_1 and IgG_3 interact with Fc receptors more effectively than IgG_2 or IgG_4, and therefore mediate potent effector functions.

model of human breast cancer.[79] Rituximab efficacy in human patients has been also influenced by mutations in patients' FcRγ to which the antibody binds on leukocyte effector cells.[80,81] Genes responsible for an allelic polymorphism that results in allo-types with different receptor properties have been identified.[80,81] Two such genes, FCGR3A, encoding FcRγIIIA (CD16a) with either amino acid phenylalanine (F) or valine (V) at amino acid position 158, and FCGR2A gene for Fc gamma RIIa (CD32a) encoding amino acid histidine (H) or arginine (R) at position 131, have been identified. These changes in a single amino acid significantly affect the affinity of the FcRγIIIA or FcRγIIA for human immunoglobulin G_1 and the efficacy of antibodies mediated through ADCC. Now it is well documented that human IgG_1 has higher affinity for homozygous FcRγIIIA-158 valine/valine (V/V) than homozygous FcRγIIIA-158 phenylalanine/phenylalanine (F/F) or heterozygous (F/V) carriers.[80] Indeed, the therapeutic response to rituximab was found to be superior in NHL patients with homozygous FcRγIIIA-158V/V than those with FcRγIIIA-158F/F genotype.[80,81] Likewise, therapeutic response to rituximab was found to be superior in NHL patients with homozygous FcRγIIA-131H/H than those of 131R/R genotype.[81] However, in patients with chronic lymphocytic leukemia, the similar correlation between FcRγIIIA 158V/V or FcRγIIA 131H/H polymorphism and responses to rituximab has not been observed.[82] In any case, these findings provided the rationale for many laboratories to engineer the Fc portion of mAbs to increase binding to low-affinity receptors and to enhance ADCC to improve the efficacy of antibody therapy. This is discussed in more detail in the following sections.

Activation of Complement

Activation of the classical pathway of complement activation is mediated by binding of IgG complexed with antigen expressed on the target cell to the first component of the complement, C1q, with CH2 domain of the IgG molecule.[16] This step initiates the activation of complement components C5-C9, also known as the membrane attack complex, which leads to the lysis of the target cell. Several therapeutic antibodies have the potential to activate complement, which may contribute to the mechanisms of activity *in vivo*. Rituximab is one such therapeutic antibody which displays effective CDC activity. This has been demonstrated by showing lack of *in vivo* efficacy of rituximab in $CD20^+$ lymphoma xenograft models in mice that are deficient in C1q.[83] *In vivo* efficacy of rituximab in mice harboring human NHL was severely impaired and when complement was depleted by treatment with cobra venom factor.[84] However, it remains to be established whether CDC is a dominant mechanism of action of rituximab in human patients. As CDC activity is considered one of the potential mechanisms of action of therapeutic antibodies, antibodies that show potent CDC activities are being increasingly considered for development for clinical application in cancers. In addition, approaches are being developed to increase the CDC activity of mAbs by making mutations in the Fc portion of the antibodies and are discussed in the following sections.

APPROACHES TO IMPROVE THE POTENCY OF ANTIBODIES

Activities mediated by the interactions of target antigen with the Fab portion of the antibody and often effector functions of the Fc portion of the antibody are considered to be a major mechanism of action of therapeutic antibodies. The contribution of each of these activities is dependent on antibody isotype, its pharmacokinetic (PK) behavior, and the nature of the target antigen. In addition, other properties of the antibodies, such as antigen binding affinity, immunogenicity, and ability to internalize upon binding with cell surface antigen also contribute to the activities of the antibodies.[9,85–87] As multiple aspects of the activities associated with the antibodies may contribute to their efficacy *in vivo,* at present the complete details of the mechanism of action of therapeutic antibodies are not understood. As such, it is very challenging to design an ideal therapeutic antibody with the right properties to harness the maximum potential. Future progress will depend on our knowledge of structural and functional relationships of the antibodies. Many molecular engineering approaches along with refined approaches to select targets for antibodies are being developed to improve the efficacy of clinical antibody candidates.

ANTIBODY AFFINITY

For antibodies to be effective targeted therapy, they must selectively bind to their target antigens. There is a common belief that high affinity of mAbs for binding to their antigen target translates into high efficacy and selectivity. Thus, improving antigen-binding affinity of the antibodies is thought to improve potency, PK, and reduce the dose that will allow selective targeting and reduced toxicity associated with the mAbs[32,87] Indeed, improving antibody binding affinity has in many cases translated into improved *in vitro* and *in vivo* efficacy of the antibodies. For example, improvement in the binding affinity of motavizumab, a respiratory syncytial virus (RSV) neutralizing antibody, has resulted in a second-generation mAb, palivizumab, which shows much improved efficacy in *in vivo* models of RSV infections and is now in clinical testing.[88,89] Many attributes of the antibodies including the nature of the antigen, antigen density, and biology of the antigen are all important for the efficacy of the antibodies. Increasing binding affinity of the antibodies does not always result in improvement in the efficacy. However, in some cases improving binding affinity has translated to enhanced toxicity, notably, when mAbs are designed to target tumor antigens. In these cases, higher affinity antibodies have not improved tumor targeting but have resulted in increased toxicity due to better binding of the antibodies to target antigens expressed at low levels on normal tissues. Although reducing the dose may have an economic benefit, several studies have suggested a linear correlation between the affinity and potency only up to the threshold level.[90] Antibodies generated via conventional hybridoma technology or using transgenic animals usually have high affinity due to the affinity maturation process *in vivo,* and in many cases need not to be improved. In contrast, mAbs generated via phage display technologies usually have lower affinity. Two common approaches are used to improve affinity of the antibodies. In the first approach, a large library of CDRs or entire variable domains containing random mutations are created and high-affinity variants from this library are selected and further optimized.[91–94] The second approach is based on the principle of *in vivo* affinity maturation and relies on creation of a much smaller library with mutations in the hotspots or using focused mutagenesis. In this procedure, every amino acid in the CDRs and hotspot areas is randomly mutagenized and high-affinity variants are selected.[95–100] By using this approach, the binding affinity of many antibodies has been increased. For example, affinity of human immunodeficiency virus (HIV) glycoprotein gp-120 specific antibody and ERBB2-specific antibody fragment has been increased from nanomolar to picomolar range.[92,93]

Improving Fc-Mediated Effector Functions

Some antibodies exert their protective effects by simply blocking receptor–ligand interactions. For example, two anti-TNFα mAbs, infliximab and adalimumab (approved for autoimmune indications), in part work by neutralizing soluble TNFα resulting in the blockade of TNFα-mediated signals.[34,35] Similarly, trastuzumab and anti-EGFR antibodies block growth signals induced by ligands on the tumor cells and inhibit their growth.[5] However, in some cases, particularly in cancers, it is desirable to eliminate tumor cells altogether. To accomplish efficient depletion of the target cell, therapeutic antibodies recruit effector cell or fix complement to destroy the tumor cells.[5] As the Fc portion of the antibodies is critical in mediation of effector functions such as ADCC, CDC, and ADCP, it is no surprise that approaches are being developed to understand the nature of interaction of Fc with FcRγs or C1q.[5,9,10,15]

ADCC

By using alanine scanning of the solvent-accessible residues in the Fc portion of the IgG, the amino acid residues of IgG_1 that are critical for binding to FcRγs have been mapped.[17] Based on this approach many residues that either enhance or reduce the binding of IgG_1 to FcRγI, II, or III have been identified. Likewise, amino acid residues that improve the binding to activation receptors, such as FcRγII and FcRγIIIA, and reduce the binding to inhibitory FcR, such as FcRγIIB, have been identified. The functional consequences of alteration of Fc–FcR interactions and their impact on *in vitro* ADCC activity have been studied.[17] For example, these mutations S298A/E333A/K334A in Fc portion have been demonstrated to improve the ADCC activity of alemtuzumab, rituximab, cetuximab, and trastuzumab by improving binding to FcRγIIIA and decreasing binding to FcRγIIB.[17] Thus, several mutations in Fc portion of the antibodies containing one or combination of more than one amino acid residue that show improvement in ADCC activity have been identified.[17] By using phage display technology, additional amino acid residues in the Fc portion have been identified that enhance binding to FcRγIIIA and improve the ADCC activity.[101] On the other hand, under certain circumstances, it may be desirable to abrogate the ADCC activity of the antibodies to offset toxicity resulting from the effector functions. Accordingly, amino acid residues that abolish binding of Fc to FcRγ have also been identified.[10] Several antibodies using Fc engineering to alter effector functions are in clinical testing now.[9,10] However, it remains to be seen if making alterations in the Fc portion of therapeutic antibodies leads to the enhanced immunogenicity.

A second approach to enhance the ADCC activity has been to modify the carbohydrate moieties in the Fc portion of the antibodies. Elimination of fucose moiety from the core of Fc *N*-linked glycans has been demonstrated to alter binding of the Fc to FcRγ.[102–104] Modifications leading to hypo-fucosylated IgG have been documented to increase binding of Fc to FcRγIIIA to 42-fold in (F158) polymorphic form and 19-fold to (V158) form.[10] Removal of the fucose moiety attached to innermost GlcNAc in Fc has also been shown to profoundly improve the ADCC activity of the IgG while no effect on CDC activity was noticed.[20,102,105] For example, ADCC activity of hypo-fucosylated humanized IgG_1 anti-IL-5 mAb and chimeric IgG_1 anti-CD20 mAb were found to be at least 50-fold higher than their fully fucosylated IgG_1 forms.[102] Based on this concept, genetic approaches to produce large quantities of hypo-fucosylated mAbs in mammalian cells have been recently developed.[105–109] The first approach involves production of high levels of nonfucosylated glycoform of IgG molecules in transgenic Chinese hamster ovary (CHO) cells that are transfected with β-(1-4)-*N*-acetylglucosaminyl transferase III (GnT-III).[107] These antibodies are shown to have 100-fold higher ADCC activity. A second approach to produce hypo-fucosylated glycoforms of IgG is to downregulate α-1,6-fucosyltransferase (FUT8) in host cells by using FUT8-knockout CHO cells where core-fucosyl transferase activity is lacking[105] or by expression of sRNAi to inhibit the FUT8 expression in CHO cells.[108] A third approach involves downregulation of the expression of guanosine diphosphate (GDP)-mannose 4,6-dehydratase (GMD) by using GMD-knockout CHO cells to produce hypo-fucosylated IgG_1.[109] Several antibodies produced using these approaches are being tested in the clinic. For example, a defucosylated antibody specific for CCR4, which has shown robust ADCC and antitumor activity in mouse models, is being tested in clinical trials.[110,111]

CDC Activity

CDC activity of IgG is initiated by binding of two C1q complement components to the hinge region of CH2 domain of IgG antibodies.[112–115] The amino acid residues present in the CH2 domains that are critical for the binding of C1q and initiation of complement activation pathway have been mapped.[116] Thus, an approach to increase the binding of CH2 domain of IgG to C1q by making mutations in these amino acid residues has been developed, and large numbers of mutant IgGs with enhanced binding C1q have been identified.[116,117] Because many mAbs approved by the FDA for oncology indications have the potential to activate complement and demonstrate *in vitro* CDC activity, the design of mAbs with enhanced CDC activity for therapeutic use is being considered. At present, however, no *in vivo* data are available to validate this concept.

Immunogenicity

As with other protein antigens, the host can mount an immune response against mAbs, which could lead to rapid clearance of the mAbs from the system and in some cases lead to serious toxicities including hypersensitivity reactions and lethal anaphylactic response.[84,118] For example, muromonab-CD3 induces an

IgE anaphylaxis response, gemtuzumab ozogamicin and rituximab induce infusion reactions, and infliximab administration is documented to result in serum sickness. To make mAbs less immunogenic, murine mAbs have been genetically engineered to replace murine parts with human counterparts. This progression from murine to chimeric and from chimeric to humanized or fully human mAbs not only has provided better effector functions but also has provided therapeutic molecules with dramatically reduced immunogenicity.[119] As variable regions of mouse and human idotype are potentially immunogenic regardless of their origin, both mouse and human mAbs could still stimulate production of anti-Id antibodies. Interestingly, even fully human mAbs produced by phase display technology (such as human IgG_1 anti-TNFα, adalimumab) could induce an immune response in some patients. In addition, several other factors, including physiochemical properties, dose and route of administration, and status of the host all play a profound role in determining immunogenicity of the mAbs.[119] A recent survey of antiantibody immune response of murine, chimeric, and humanized mAbs in humans has indicated that all forms of antibodies are capable of inducing an immune response, but the magnitude of the response varies among different types.[120] Chimerization of murine mAbs was found to reduce their immunogenicity and humanization of the mAbs improves the antibodies further. Surprisingly, adalimumab, a fully human antibody, is found to be immunogenic in 12 percent of patients after repeat doses.[119] As a limited set of data, with only two fully human mAb (adalimumab and panitumumab) is available, at present it is difficult to determine if fully human antibodies are less immunogenic than humanized antibodies.

Pharmacokinetic Properties of mAbs

The terminal half-life of mAbs in plasma is one important aspect of an antibody that determines clinical utility. The PK properties of mAbs in the blood can be altered by a variety of ways to meet the needs of the patients. Half-lives of the mAbs can be increased where it is desirable to increase the efficacy and reduce the frequency of the dosing. On the other hand, the half-life of an mAb can be significantly reduced to lower the systemic exposure and to improve the target-to-normal tissue ratio. One of the mechanisms to alter the half-life of the mAbs is to alter the binding of the Fc portion of the mAbs to FcRn, the salvage receptor. FcRns are expressed on multiple cell types such as vascular endothelial cells, APCs, and epithelial cells.[121,122] These receptors are involved in the transport of the IgG across the placenta and also play a critical role in maintaining the serum concentration of antibodies. Antibodies can bind to FcRns under acidic conditions at pH 6.5 with nanomolar affinity and not at a physiological pH (7.0 to 7.4).[123,124] IgGs are taken up by the cells via nonspecific micropinocytosis and transported to endosomes where they bind to FcRn under acidic conditions. These FcRn/IgG complexes are transcytosed to cell surface where IgG is released under pH 7.0 to 7.4 of plasma.[125] IgGs that do not bind to FcRn are degraded. Binding sites for FcRn binding on the Fc region that are important for pH-dependant interaction have been mapped. These binding sites are located in the IgG in the CH2–CH3 domains of IgG and are distinct from binding sites for FcRγ binding.[17,125,126] Because binding of IgG to FcRn regulates IgG half-life and serum concentration, one of the major areas of research to improve the serum half-life is to regulate affinity of mAbs to FcRn.[10]Accordingly, many amino acid residues in the Fc portion have been identified either to reduce or to increase binding to FcRn; however, altering the binding to FcRn does not always result in the changes in serum half-life.[10,127,128] For example, making mutations in the Fc portion of anti-RSV IgG_1 mAb led to a 10-fold increase in binding to FcRn in cynomolgus monkey FcRn at pH 6.0, but only a 4-fold increase in the half-life in cynomolgus monkey and no increase in half-life in mice.[129,130] Similarly, introducing mutations in the Fc portion of anti-human hepatitis B mAb could increase the binding of both human and Rhesus monkey FcRn up to 30-fold with only a 2.5-fold increase in the serum half-life in Rhesus monkey.[131] Likewise, increasing the binding affinity of humanized anti-TNFα mAb 40-fold to FcRn of cynomolgus monkey at pH 6.0 has not resulted in prolongation of half-life in cynomolgus monkey.[132] Although altering the binding affinity to FcRn has resulted in changes in the half-life of some of the antibodies, the benefit of increasing FcRn binding to improve PK properties of the antibodies still remains unknown.

ANTIBODY FRAGMENTS

One of the most important factors for clinical efficacy of mAbs is effective targeting and the ability to effectively penetrate tissues. Whole antibody molecules with high affinity have been shown to diffuse poorly from vascular space into solid tissues. For example, high-affinity antibodies are retained at the periphery of the tumors whereas low- to medium-affinity antibodies penetrate deep in the tumor mass.[90] To improve penetration and selective targeting, a wide variety of antibody fragments can potentially be developed. Antibody fragments lacking Fc can also be engineered, which can do part of the job that whole antibodies do, for example, blocking soluble cytokines binding to their receptors. Antibody fragments that are engineered to have better tumor/tissue penetration are good selective targeting agents while they can be rapidly cleared from the systemic circulation. Thus, antibody fragments such as Fab, scFv, diabodies, triabodies, and minibodies ranging from molecular weight (MW) 15 kD to more than 150 kD can be engineered (Figure 1.5). These fragments are routinely used for imaging studies and are also increasingly considered for antibody drug conjugates.[133–136] Furthermore, these molecules can also be produced in large quantities in bacteria, bypassing some of the issues associated with the mammalian culture systems.

The smallest antibody fragment that retains the antigen-binding specificity of a whole antibody molecule is Fv, which is composed of the VH and VL domains of the antibodies.[137,138] Due to its instability at low concentrations, the VH and VL domains are covalently linked via a stable peptide linker to make a scFv.[139] Fv can also be engineered to form a disulfide bond between VH and VH domains resulting in a molecule called dsFv.[140] Homodimers of scFv linked via a short peptide forcing the formation of a molecular complex of cognate VH and VL pairing is called a diabody.[141] Minibodies consisting of homodimers of scFv linked to CH3 can also be engineered.[142] These antibody fragments have several properties that can be harnessed to improve their utility as effective clinical molecules. For example, antibody

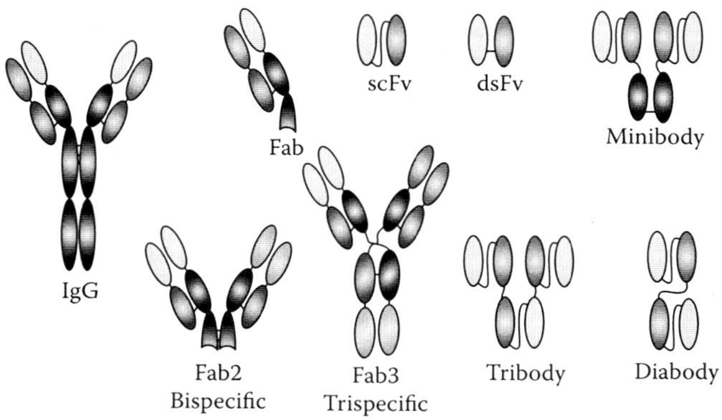

FIGURE 1.5 ***A color version of this figure follows page 176.*** Different forms of most common antibody fragment formats. Schematic representation of most studied antibody formats including intact IgG molecule along with various fragments are shown in the figure. A variety of antibody fragments, including Fab, scFv, dsFv, and multimeric formats, such as minibodies, diabodies, triabodies, and chemically conjugated Fab multimers, are shown in the figure. Sizes of these fragments vary between approximately 15 to 150 kD.

fragments lack effector functions such as ADCC or CDC, which in some cases are associated with toxicities of whole antibodies. Under certain circumstances it is desirable to have molecules lacking effector functions to avoid toxicities. Although antibody fragments have a relatively short half-life, an advantage for tumor targeting and for imaging studies, these can be modified by conjugation to polyethylene glycol (PEG) to improve the efficacy if desired.[143] The advantage of PEGylated molecules over IgG is that they lack any undesirable Fc-mediated effector functions while they maintain a relatively longer life. Indeed, several antibody fragments are being tested in clinical trials and a few of them have already been approved for humans, including Fab abciximab, ranibizumab, and a humanized-TNF-specific Fab conjugated to PEG, certolizumab pegol.[144,145]

NEW EMERGING ANTIBODY FORMATS

One of the most popular ways to improve the efficacy of mAbs is to arm the antibody with payloads or engineer improvements in their functions as discussed in the previous sections. Several other novel approaches to harness the power of the antibodies are being developed industrywide. These approaches are mostly targeted for the treatment of cancer, where mAbs have limited activity as single agent immunotherapy. One such approach is based on the power of the CTLs and their exploitation for elimination of tumors. Antibody-based molecules are thus designed to recruit CTL to the site of the tumors. A bispecific molecule that engages CTLs and also binds to the target antigen on tumor cells is currently in clinical trials.[146] Activity of a bispecific antibody construct called blinatumomab, specific for anti-CD19 and CD3, which engages and activates CTLs via CD3 and binds to the NHL tumor cells expressing CD19, has been demonstrated. Doses as low as 0.005 mg/m^2/day in NHL patients have been reported to eliminate tumor cells in patients. Thus, this novel approach using T cell-engaging antibodies seems to have therapeutic potential for the treatment of malignant diseases.[146] Additional molecules targeting solid tumors and engaging CTLs are showing promise in the preclinical studies and are awaiting clinical trials.[147]

This approach is also applied to harness the power of mononuclear cells where bispecific antibodies are designed to bind CD16 and tumor antigens of choice to recruit effector cells such as natural killer (NK) cells and macrophages to combat tumors (Figure 1.6).[148] The activation and recruitment of CTLs for tumor lysis can also be accomplished by delivering two signals to T cells and bringing these activated T cells close to the tumors. This has been achieved by making tribodies consisting of three scFvs specific for CD28, CD3, and tumor antigens and demonstrating *in vitro* activities.[149] Bispecific and trispecific small molecule antibody-derived constructs offer a promising approach for tumor targeting, particularly, to achieve better penetration and retention in the solid tumor environment with rapid clearance from the circulation.[149] However, several technical challenges, such as mis-pairing of H and L chain, stability, and large-scale production of clinical-grade material, need to be met in order for these formats to become part of the therapeutic strategy for cancers.

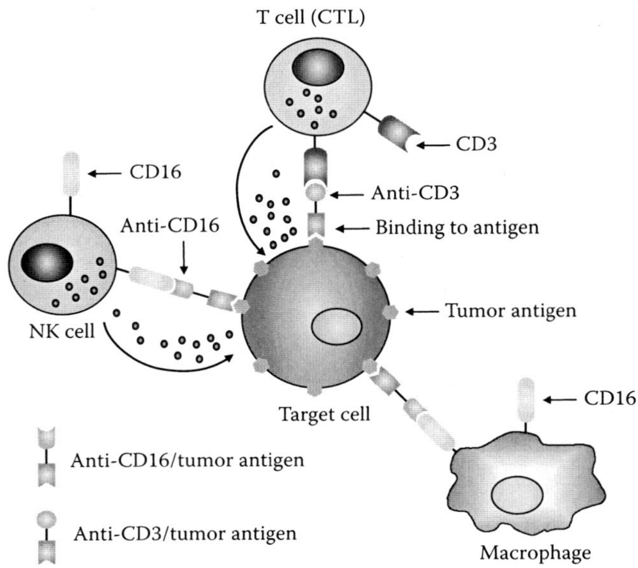

FIGURE 1.6 ***A color version of this figure follows page 176.*** Mechanism of action of bispecific antibody fusion molecules. Bispecific antibody fragments allow targeting of the target cell by recruiting effector immune cells, such as NK cells and cytotoxic T cells. Bispecifc molecules with binding specificity for the antigen expressed on target cells such as CD19 bind to target cells and at the same time recruit effector cells by binding to receptors expressed on the effector cells. Two bispecific molecules, CD3/CD19 and CD16/CD19, are shown in the figure.

FUTURE PROSPECTIVE

The phenomenal success of therapeutic antibodies in the clinic has established these molecules as important therapeutic agents for several different human diseases. However, a number of challenges in developing antibody therapeutics still need to be met. Our future success developing antibody therapeutics will depend on multiple factors including advances in molecular engineering of mAbs, innovative methods to reduce development times, and effectively creating and protecting intellectual property. In addition, higher success rates will be aided by the development of cost-effective manufacturing.

The past 30 years of experience in developing antibody therapeutics has taught us that success of antibody therapeutics depends on the choice of antigen targets. However, clear criteria of antibody target selection are not yet established. As most of the obvious antibody targets have already been identified, it will be difficult to discover and develop novel antibody targets with the same remarkable success we have seen in the past. Understanding the basic biology of the targets and their potential role in the disease pathogenesis will be crucial to contend with the few remaining challenging antibody targets. As the number of novel antibody targets will decrease over time, increased competition is definitely expected, which we have already seen with some antibody therapeutics where multiple companies are targeting the same antigen.

Together with advances in target selection, antibody generation, protein engineering, glycoengineering, and new insights in molecular mechanisms of Fc–FcR interactions have allowed us to optimally design second generation therapeutic antibodies entering the clinic today. The ability to generate mAbs that are highly optimized for effector functions, PK properties, and are fully human with remarkable *in vivo* activity in preclinical models in many cases has provided us with clinical candidates to access challenging targets. One of the major steps in antibody technology was the creation of fully human therapeutic mAbs by the application of display technology, which has already resulted in FDA approval of two fully human mAb, adalimumab and panitumumab, for human use. Although we do not have much clinical experience with second-generation molecules at the present time, they are expected to emerge as more potent and effective clinical candidates. Many factors will aid in their success such as commitment to finding better treatments for patients, making full use of the emerging new technologies to harness the power of mAbs, and use of creative approaches to compete for the same target.

As a result of the high cost of present antibody development and production methods in the mammalian systems, most companies have adopted a premium pricing approach for mAb products. Antibody-based therapies are usually more expensive than small molecule treatments, especially, when mAbs are used in combination with conventional drugs, or when continuous treatment is required. These high costs may be offset in the future as competition increases and development procedures are refined, particularly, the development of alternative antibody production technologies. Furthermore, for mAbs to become less restricted in usage, biopharmaceutical companies need to provide the benefits of the mAbs and pharmocoeconomic analysis in terms of overall cost savings and effectiveness. This will facilitate granting reimbursement to additional patient groups and enhancement in the rate of uptake.

Earlier, the scope of clinical application of therapeutic mAbs was restricted to the approved target population. Now with the increased understanding of the disease process, a number of mAbs in clinical development are being investigated for the treatment of more than one disease or for different forms of the same disease. For example, infliximab, which was initially approved for the treatment of Crohn's disease, was later approved for treatment of rheumatoid arthritis. Likewise, rituximab was initially approved for NHL and now is approved for rheumatoid arthritis and is being tested in multiple other indications. The emerging trend now is that most mAb therapeutics initially approved for a specific form of a disease will likely be investigated for other indications to expand the patient population that can be treated and to increase revenue potential.

Antibody therapeutics have been very effective in hematological and lymphocyte-mediated diseases, probably due to easy access of the mAb to the antigenic cell and effective recruitment of effector cells to the site of disease. In contrast, for mAbs to be effective in solid tumors many additional challenges exist including (1) lack of easy access and barriers posed by tumor vasculature, stromal cells. and epithelial cells, (2) high interstitial pressure within the tumor microenvironment, (3) heterogeneity of the target antigen expression, and (4) high level of expression of inhibitory molecules that prevent functioning of the effector cells. One approach to effectively target solid tumors may be to use small molecular weight antibody fragments for better penetration of the solid tumors. However, small size molecules are disadvantageous in that they have a shorter half-life in the serum. Another reasonable approach is to target tumor vasculature because of the relative genetic stability of the endothelial antigens, and a plethora of available tumor vasculature targets. The activation of host antitumor immune response could also potentially help in elimination of solid tumors. In this regard, mAbs with immunostimulatory properties may offer a viable approach to treat solid tumors. Promising combinations of mAbs with a variety of small molecule inhibitors are also being exploited to treat solid tumors and could become routine in developing modalities for solid tumors. This approach could overcome issues of tumor heterogeneity but still offer selectivity of the treatment. Other emerging approaches include targeting solid tumors with antibody–drug conjugates and targeting stem cells using mAbs. All the approaches have shown good preclinical data and in many cases are awaiting clinical data. Extensive research in this area is urgently needed to harness the power of mAbs to effectively target solid tumors.

Although use of mAbs in humans is usually well tolerated, some unexpected toxicity associated with some mAbs has been seen. These toxicities in some cases resulted from the target antigen expression on normal tissues and activation of the antigen-associated pathway by the therapeutic mAbs. Toxicities can also result from effector functions of mAbs and their binding to effector cells. Antibodies with immunomodulatory properties pose additional risks for toxicities and could result in unexpected severe toxicities. An increased number of second-generation antibodies that are highly modified from their natural structure by engineering for improved efficacy and are produced as recombinant proteins by a variety of cell lines are entering in clinical testing today, but whether their toxicities in humans will be within the manageable range still remains to be seen. In particular, alterations in the antibody structure introduced by changed amino acid sequences and carbohydrate moieties could affect immunogenicity, PK, or biodistribution, all of which could pose new safety risks. Furthermore, escalating trends in antibody technologies, novel approaches of antibody drug conjugates, and delivery of immune stimulatory molecules in pursuit of developing highly potent drugs may pose additional safety challenges. A thorough understanding of the target biology coupled with detailed knowledge of antibody structure–function relationship and a well-thought-out preclinical safety evaluation is required for successful development of future therapeutic mAbs.

As a growing number of patents for approved mAbs are expiring, manufacturing and marketing of second-generation "biosimilar" mAbs with reduced costs are being considered by many companies now. However, the conventional approach applied for the development and marketing of generic small-molecule drugs is not suitable for mAbs. Thus, new methods to produce mAbs in a variety of systems, including insect cells, bacteria, plants, yeast, or whole organisms, are being developed. As these systems are inherently different from the current manufacturing practices for therapeutic mAbs, it is likely that product quality and potency could also be significantly different from the first-generation molecules. In addition, mAbs are very complex molecules and undergo posttranslational modifications that determine their physicochemical and their pharmacological properties. Such variability in mAbs may influence the biologic activity, stability, pharmacokinetics, immunogenicity, and clinical efficacy of biosimilar mAbs. Thus, there is an urgent need to develop new systems to deal with biosimilar mAbs. Full comparability studies between the original product and the new biosimilar mAbs on both preclinical and clinical levels together with vigorous quality evaluation of these molecules must be performed. In addition, a clear regulatory strategy with appropriate guidelines from various agencies on biosimilar mAbs is critical to facilitate the development and approval of high-quality biosimilar mAbs.

A view into the future, inspired by the remarkable success of the mAbs in the clinic as exquisitely specific and versatile therapeutic molecules, coupled with evolution of their basic understanding, their targets, and technology to produce them in large quantities for the past 30 years, predicts that researchers studying mAbs will continue to make rapid progress in the development of mAbs for multiple indications. A growing number of second-generation mAbs including fully optimized human antibodies, antibody–drug conjugates, and antibody fragments are likely to be a major part of therapeutic strategies for clinical development of mAbs in the future. Extensive research in industry and academia alike to identify novel ways to harness the power of mAbs and advancements in antibody technologies and cost-effective production methods will no doubt continue in the future.

ACKNOWLEDGMENTS

I thank Maureen Ryan and Julie McEarchern for critically reading the manuscript.

REFERENCES

1. Schwartz, R. S. 2004. Paul Ehrlich's magic bullets. *N Engl J Med.* 350:1079–80.
2. Köhler, G., Milstein, C. 2005. Continuous cultures of fused cells secreting antibody of predefined specificity. 1975. *J Immunol.* 174:2453–55.
3. Brekke, O. H., Sandlie, I. 2003. Therapeutic antibodies for human diseases at the dawn of the twenty-first century. *Nat Rev Drug Discov.* 2:52–62.
4. Glennie, M. J., Johnson, P. W. 2000. Clinical trials of antibody therapy. *Immunol Today.* 21:403–10.
5. Schrama, D., Reisfeld, R. A., Becker, J. C. 2006. Antibody targeted drugs as cancer therapeutics. *Nat Rev Drug Discov.* 5:147–59.
6. Salfeld, J. G. 2007. Isotype selection in antibody engineering. *Nat Biotechnol.* 25:1369–72.
7. Kola, I., Landis, J. 2004. Can the pharmaceutical industry reduce attrition rates? *Nat Rev Drug Discov.* 3:711–15.
8. Baker, M. 2005. Upping the ante on antibodies. *Nat Biotechnol.* 23:1065–72.
9. Carter, P. J. 2006. Potent antibody therapeutics by design. *Nat Rev Immunol.* 6:343–57.
10. Liu, X. Y., Pop, L. M., Vitetta, E. S. 2008. Engineering therapeutic monoclonal antibodies. *Immunol Rev.* 222:9–27.
11. Milenic, D. E., Brady, E. D., Brechbiel, M. W. 2004. Antibody-targeted radiation cancer therapy. *Nat Rev Drug Discov.* 3:488–99.
12. Suntharalingam, G., Perry, M. R., Ward, S., et al. 2006. Cytokine storm in a phase 1 trial of the anti-CD28 Monoclonal antibody TGN1412. *N Engl J Med.* 355:1018–28.
13. Grillo-López, A. J., Hedrick, E., Rashford, M., Benyunes, M. 2002. Rituximab: ongoing and future clinical development. *Semin Oncol.* 29:105–12.
14. Cosimi, A. B., Burton, R. C, Colvin R. B., et al. 1981. Treatment of acute renal allograft rejection with OKT3 monoclonal antibody. *Transplantation.* 32:535–39.
15. Kim, S. J., Park, Y., Hong, H. J. 2005. Antibody engineering for the development of therapeutic antibodies. *Mol Cells.* 20:17–29.
16. Murphy, K. M., Travers, P., Walport, M. 2007. *Janeway's Immunobiology*, 7th ed. Garland Sciences, New York.
17. Shields, R. L., Namenuk, A. K., Hong, K. et al. 2001. High resolution mapping of the binding site on human IgG1 for Fc gamma RI, Fc gamma RII, Fc gamma RIII, and FcRn and design of IgG1 variants with improved binding to the Fc gamma R. *J Biol Chem.* 276:6591–604.
18. Leatherbarrow, R. J., Rademacher, T. W., Dwek, R. A., et al. 1985. Effector functions of a monoclonal aglycosylated mouse IgG2a: Binding and activation of complement component C1 and interaction with human monocyte Fc receptor. *Mol Immunol.* 22:407–15.
19. Takahashi, N., Ishii, I., Ishihara, H. et al. 1987. Comparative structural study of the N-linked oligosaccharides of human normal and pathological immunoglobulin G. *Biochemistry.* 26:1137–44.
20. Tao, M. H., Morrison, S. L. 1989. Studies of aglycosylated chimeric mouse-human IgG. Role of carbohydrate in the structure and effector functions mediated by the human IgG constant region. *J Immunol.* 143:2595–601.
21. Lund, J., Takahashi, N., Pound, J. D., Goodall, M., Jefferis, R. 1996. Multiple interactions of IgG with its core oligosaccharide can modulate recognition by complement and human Fc gamma receptor I and influence the synthesis of its oligosaccharide chains. *J Immunol.* 157:4963–69.
22. Wright, A., Morrison, S. L. 1998. Effect of C2-associated carbohydrate structure on Ig effector function: studies with chimeric mouse-human IgG1 antibodies in glycosylation mutants of Chinese hamster ovary cells. *J Immunol.* 160:3393–402.

23. Reichert, J. M., Rosensweig, C. J., Faden, L. B., Dewitz , M. C. 2005. Monoclonal antibody successes in the clinic. *Nat Biotechnol.* 23:1073–78.
24. Shawler, D. L., Bartholomew, R. M., Smith, L. M., Dillman, R. O. 1985. Human immune response to multiple injections of murine monoclonal IgG. *J Immunol.* 135:1530–35.
25. Stern, M., Herrmann, R. 2005. Overview of monoclonal antibodies in cancer therapy: Present and promise. *Crit Rev Oncol Hematol.* 54:11–29.
26. Carter, P., Presta, L., Gorman, C. M., et al. 1992. Humanization of an anti-p185HER2 antibody for human cancer therapy. *Proc Natl Acad Sci U S A.* 89:4285–89.
27. Presta, L. G., Lahr, S. J., Shields, R. L., et al. 1993. Humanization of an antibody directed against IgE. *J Immunol.* 151:2623–32.
28. Chothia, C., Lesk, A. M., Tramontano, A., et al. 1989. Conformations of immunoglobulin hypervariable regions. *Nature.* 342:877–83.
29. Waldmann, T. A. 2003. Immunotherapy: Past, present and future. *Nat Med.* 9:269–77.
30. Hudson, P. J, Souriau, C. 2003. Engineered antibodies. *Nat Med.* 9:129–34.
31. Lonberg, N. 2005. Human antibodies from transgenic animals. *Nat Biotechnol.* 23:1117–25.
32. Hoogenboom, H. R. 2005. Selecting and screening recombinant antibody libraries. *Nat Biotechnol.* 23:1105–16.
33. Imai, K, Takaoka, A. 2006. Comparing antibody and small-molecule therapies for cancer. *Nat Rev Cancer.* 6:714–27.
34. Maini, R., St Clair, E. W., Breedveld, F., et al. 1999. Infliximab (chimeric anti-tumour necrosis factor alpha monoclonal antibody) versus placebo in rheumatoid arthritis patients receiving concomitant methotrexate: A randomised phase III trial. ATTRACT Study Group. *Lancet.* 354:1932–39.
35. Sandborn, W. J., Hanauer, S. B. 1999. Antitumor necrosis factor therapy for inflammatory bowel disease: A review of agents, pharmacology, clinical results, and safety. *Inflamm Bowel Dis.* 5:119–33.
36. Devlin, S. M, Panaccione, R. 2008. Adalimumab for the treatment of Crohn's disease. *Expert Opin Biol Ther.* 8:1011–19.
37. Arteaga, C. L., Hurd, S. D., Winnier, A. R., Johnson, M. D., Fendly, B. M., Forbes, J. T. 1993. Anti-transforming growth factor (TGF)-beta antibodies inhibit breast cancer cell tumorigenicity and increase mouse spleen natural killer cell activity. Implications for a possible role of tumor cell/host TGF–beta interactions in human breast cancer progression. *J Clin Invest.* 92:2569–76.
38. Gerber, H. P., Wu, X., Yu, L., et al. 2007. Mice expressing a humanized form of VEGF-A may provide insights into the safety and efficacy of anti-VEGF antibodies. *Proc Natl Acad Sci U S A.* 104:3478–83.
39. Mottershead, M., Neuberger, J. 2007. Daclizumab. *Expert Opin Biol Ther.* 7:1583–96.
40. Ramirez, C. B., Marino, I. R. 2007. The role of basiliximab induction therapy in organ transplantation. *Expert Opin Biol Ther.* 7:137–48.
41. Hodulik, S., Hadi, S. 2006. Efalizumab: A biological agent for the treatment of psoriasis. *Rev Recent Clin Trials.* 1:165–68.
42. Hudis, C. A. 2007. Trastuzumab—Mechanism of action and use in clinical practice. *N Engl J M*ed. 357:39–51.
43. Galizia, G., Lieto, E., De Vita, F., et al. 2007. Cetuximab, a chimeric human mouse anti-epidermal growth factor receptor monoclonal antibody, in the treatment of human colorectal cancer. *Oncogene.* 26:3654–60.
44. Lainakis, G., Bamias, A. 2008. Targeting angiogenesis in renal cell carcinoma. *Curr Cancer Drug Targets.* 8:349–58.
45. Keyhanfar, M., Forbes, B. E., Cosgrove, L. J., Wallace, J. C., Booker, G. W. 2006. Production and characterization of monoclonal antibodies against insulin-like growth factor type 1 receptor. *Hybridoma* (Larchmt). 25:230–37.

46. Ghetie, M. A., Picker, L. J., Richardson, J. A., Tucker, K., Uhr, J. W., Vitetta, E. S. 1994. Anti-CD19 inhibits the growth of human B-cell tumor lines in vitro and of Daudi cells in SCID mice by inducing cell cycle arrest. *Blood.* 83:1329–36.
47. Chaouchi, N., Vazquez, A., Galanaud, P., Leprince, C. 1995. B cell antigen receptor-mediated apoptosis. Importance of accessory molecules CD19 and CD22, and of surface IgM cross-linking. *J Immunol.* 154:3096–104.
48. Cragg, M. S., French, R. R., Glennie, M. J . 1999. Signaling antibodies in cancer therapy. *Curr Opin Immunol.* 11:541–47.
49. Ashkenazi, A. 2002. Targeting death and decoy receptors of the tumour-necrosis factor superfamily. *Nat. Rev. Cancer.* 2:420–30.
50. Krammer, P. H., Galle, P. R., Möller, P., Debatin, K. M. 1998. CD95(APO-1/Fas)-mediated apoptosis in normal and malignant liver, colon, and hematopoietic cells. *Adv Cancer Res.* 75:251–73.
51. Glennie, M. J., French, R. R., Cragg, M. S., Taylor, R. P. 2007. Mechanisms of killing by anti-CD20 monoclonal antibodies. *Mol Immunol.* 44:3823–37.
52. Leach, D. R., Krummel, M. F., Allison, J. P. 1996. Enhancement of antitumor immunity by CTLA-4 blockade. *Science.* 271:1734–36.
53. Li, Q., Iuchi, T., Jure-Kunkel, M. N., Chang, A. E. 2007. Adjuvant effect of anti-4-1BB mAb administration in adoptive T cell therapy of cancer. *Int J Biol Sci.* 3:455–62.
54. Weinberg, A. D., Thalhofer, C., Morris, N., et al. 2006. Anti-OX40 (CD134) administration to nonhuman primates: Immunostimulatory effects and toxicokinetic study. *J Immunother.* 29:575–85.
55. French, R. R., Chan, H. T., Tutt, A. L., Glennie, M. J. 1999. CD40 antibody evokes a cytotoxic T-cell response that eradicates lymphoma and bypasses T-cell help. *Nat Med.* 5:548–53.
56. Hardy, B., Niv, Y., Fadaeev, L., Raiter, A. 2005. BAT mAb induces lymphopoiesis in nude mice. *Int Immunol.* 17:615–19.
57. Korman, A. J., Peggs, K. S., Allison, J. P. 2006. Checkpoint blockade in cancer immunotherapy. *Adv Immunol.* 90:297–339.
58. Rudolf, D., Silberzahn, T., Walter, S., et al. 2008. Potent costimulation of human CD8 T cells by anti-4-1BB and anti-CD28 on synthetic artificial antigen presenting cells. *Cancer Immunol Immunother.* 57:175–83.
59. Hunter, T. B., Alsarraj, M., Gladue, R. P., Bedian, V., Antonia, S. J. 2007. An agonist antibody specific for CD40 induces dendritic cell maturation and promotes autologous anti-tumour T-cell responses in an in vitro mixed autologous tumour cell/lymph node cell model. *Scand J Immunol.* 65:479–86.
60. O'Day, S. J., Hamid, O., Urba, W. J. 2007. Targeting cytotoxic T-lymphocyte antigen-4 (CTLA-4): a novel strategy for the treatment of melanoma and other malignancies. *Cancer.* 110:2614–27.
61. Vonderheide, R. H., Flaherty, K. T., Khalil, M., et al. 2007. Clinical activity and immune modulation in cancer patients treated with CP-870,893, a novel CD40 agonist monoclonal antibody. *J Clin Oncol.* 25:876–83.
62. Illidge, T. M., Cragg, M. S., McBride, H. M., French, R. R., Glennie, M. J. 1999. The importance of antibody-specificity in determining successful radioimmunotherapy of B-cell lymphoma. *Blood.* 94:233–43.
63. Lambert, J. M. 2005. Drug-conjugated monoclonal antibodies for the treatment of cancer. *Curr Opin Pharmacol.* 5:543–49.
64. Kovtun, Y. V., Goldmacher, V. S. 2007. Cell killing by antibody-drug conjugates. *Cancer Lett.* 255:232–40.
65. Carter, P. J, Senter, P. D. 2008. Antibody-drug conjugates for cancer therapy. *Cancer J.* 14:154–69.

66. Bross, P. F., Beitz, J., Chen, G., et al. 2001. Approval summary: gemtuzumab ozogamicin in relapsed acute myeloid leukemia. *Clin Cancer Res.* 7:1490–96.
67. Sievers, E. L. 2001. Efficacy and safety of gemtuzumab ozogamicin in patients with CD33-positive acute myeloid leukaemia in first relapse. *Expert Opin Biol Ther.* 1:893–901.
68. Younes, A., Forero-Torres, A., Bartlett, N. L., et al. 2008. Objective responses in a phase I dose-escalation study of SGN-35, a novel antibody-drug conjugate (ADC) targeting CD30, in patients with relapsed or refractory Hodgkin lymphoma. *J Clin Oncol* 26: 8526 Abstract.
69. Dela Cruz, J. S., Trinh, K. R., Chen, H. W., et al. 2006. Anti-HER2/neu IgG3-(IL-2) and anti-HER2/neu IgG3-(GM-CSF) promote HER2/neu processing and presentation by dendritic cells: Implications in immunotherapy and vaccination strategies. *Mol Immunol.* 43:667–76.
70. King, D. M., Albertini, M. R., Schalch, H., et al. 2004. Phase I clinical trial of the immunocytokine EMD 273063 in melanoma patients. *J Clin Oncol.* 22:4463–73.
71. Ko, Y. J., Bubley, G. J., Weber, R., et al. 2004. Safety, pharmacokinetics, and biological pharmacodynamics of the immunocytokine EMD 273066 (huKS-IL2): Results of a phase I trial in patients with prostate cancer. *J Immunother.* 27:232–39.
72. Reisfeld, R. A., Gillies, S. D., Mendelsohn, J., Varki, N. M., Becker, J. C. 1996. Involvement of B lymphocytes in the growth inhibition of human pulmonary melanoma metastases in athymic nu/nu mice by an antibody-lymphotoxin fusion protein. *Cancer Res.* 56:1707–12.
73. Borsi, L., Balza, E., Carnemolla, B., et al. 2003. Selective targeted delivery of TNFalpha to tumor blood vessels. *Blood.* 102:4384–92.
74. Ebbinghaus, C., Ronca, R., Kaspar, M., et al. 2005. Engineered vascular-targeting antibody-interferon-gamma fusion protein for cancer therapy. *Int J Cancer.* 116:304–13.
75. Hombach, A., Heuser, C., Abken, H. 2005. Simultaneous targeting of IL2 and IL12 to Hodgkin's lymphoma cells enhances activation of resting NK cells and tumor cell lysis. *Int J Cancer.* 115:241–47.
76. Nimmerjahn, F., Ravetch, J. V. 2008. Fcgamma receptors as regulators of immune responses. *Nat Rev Immunol.* 8:34–47.
77. Nimmerjahn, F., Ravetch, J. V. 2007. Fc-receptors as regulators of immunity. *Adv Immunol.* 96:179–204.
78. Takai, T., Li, M., Sylvestre, D., Clynes, R., Ravetch, J. V. 1994. FcR gamma chain deletion results in pleiotrophic effector cell defects. *Cell.* 76:519–29.
79. Clynes, R. A., Towers, T. L., Presta, L. G., Ravetch, J. V. 2000. Inhibitory Fc receptors modulate in vivo cytoxicity against tumor targets. *Nat Med.* 6:443–46.
80. Cartron, G., Dacheux, L., Salles, G., et al. 2002. Therapeutic activity of humanized anti-CD20 monoclonal antibody and polymorphism in IgG Fc receptor FcgammaRIIIa gene. *Blood.* 99:754–58.
81. Weng, W. K., Levy, R. 2003. Two immunoglobulin G fragment C receptor polymorphisms independently predict response to rituximab in patients with follicular lymphoma. *J Clin Oncol.* 21:3940–47.
82. Farag, S. S., Flinn, I. W., Modali, R., Lehman, T. A., Young, D., Byrd, J. C. 2004. Fc gamma RIIIa and Fc gamma RIIa polymorphisms do not predict response to rituximab in B-cell chronic lymphocytic leukemia. *Blood.* 103:1472–74.
83. Di Gaetano, N., Cittera, E., Nota, R., et al. 2003. Complement activation determines the therapeutic activity of rituximab in vivo. *J Immunol.* 171:1581–87.
84. Cragg, M. S., Glennie, M. J. 2004. Antibody specificity controls in vivo effector mechanisms of anti-CD20 reagents. *Blood.* 103:2738–43.
85. Carter, P. 2001. Improving the efficacy of antibody-based cancer therapies. *Nat Rev Cancer.* 1:118–29.

86. Presta, L. G. 2002. Engineering antibodies for therapy. *Curr Pharm Biotechnol.* 3:237–56.
87. Chowdhury, P. S., Wu, H. 2005. Tailor-made antibody therapeutics. *Methods.* 36:11–24.
88. Wu, H., Pfarr, D. S, Tang, Y., et al. 2005. Ultra-potent antibodies against respiratory syncytial virus: Effects of binding kinetics and binding valence on viral neutralization. *J Mol Biol.* 350:126–44.
89. Mejías, A., Chávez-Bueno, S., Ríos, A. M., et al. 2005. Comparative effects of two neutralizing anti-respiratory syncytial virus (RSV) monoclonal antibodies in the RSV murine model: time versus potency. *Antimicrob Agents Chemother.* 49:4700–7.
90. Adams, G. P., Schier, R., McCall, A. M., et al. 2001. High affinity restricts the localization and tumor penetration of single-chain fv antibody molecules. *Cancer Res.* 61:4750–55.
91. Adams, G. P, Schier, R., Marshall, K., et al. 1998. Increased affinity leads to improved selective tumor delivery of single-chain Fv antibodies. *Cancer Res.* 58:485–90.
92. Schier, R., McCall, A., Adams, G. P., et al. 1996. Isolation of picomolar affinity anti-c-erbB-2 single-chain Fv by molecular evolution of the complementarity determining regions in the center of the antibody binding site. *J Mol Biol.* 263:551–67.
93. Yang, W. P., Green, K., Pinz-Sweeney, S., Briones, A. T., Burton, D. R., Barbas, C. F. 3rd. 1995. CDR walking mutagenesis for the affinity maturation of a potent human anti-HIV-1 antibody into the picomolar range. *J Mol Biol.* 254:392–403.
94. Zhang, M. Y., Shu, Y., Rudolph, D., et al. 2004. Improved breadth and potency of an HIV-1-neutralizing human single-chain antibody by random mutagenesis and sequential antigen panning. *J Mol Biol.* 335:209–19.
95. Chowdhury, P. S. 2003. Engineering hot spots for affinity enhancement of antibodies. *Methods Mol Biol.* 207:179–96.
96. Chowdhury, P. S., Pastan, I. 1999. Improving antibody affinity by mimicking somatic hypermutation in vitro. *Nat Biotechnol.* 17:568–72.
97. Glaser, S. M., Yelton, D. E., Huse, W. D. 1992. Antibody engineering by codon-based mutagenesis in a filamentous phage vector system. *J Immunol.* 149:3903–13.
98. Ho, M., Kreitman, R. J., Onda, M., Pastan, I. 2005. In vitro antibody evolution targeting germline hot spots to increase activity of an anti-CD22 immunotoxin. *J Biol Chem.* 280:607–17.
99. Wu, H., Beuerlein, G., Nie, Y., et al. 1998. Stepwise in vitro affinity maturation of Vitaxin, an alphav beta3-specific humanized mAb. *Proc Natl Acad Sci U S A.* 95:6037–42.
100. Wu, H., Nie, Y., Huse, W. D., Watkins, J. D. 1999. Humanization of a murine monoclonal antibody by simultaneous optimization of framework and CDR residues. *J Mol Biol.* 294:151–62.
101. Stavenhagen, J. B., Gorlatov, S., Tuaillon, N., et al. 2007. Fc optimization of therapeutic antibodies enhances their ability to kill tumor cells in vitro and controls tumor expansion in vivo via low-affinity activating Fcgamma receptors. *Cancer Res.* 67:8882–90.
102. Shields, R. L., Lai, J., Keck, R., et al. 2002. Lack of fucose on human IgG1 N-linked oligosaccharide improves binding to human Fcgamma RIII and antibody-dependent cellular toxicity. *J Biol Chem.* 277:26733–40.
103. Shinkawa, T., Nakamura, K., Yamane, N., et al. 2003. The absence of fucose but not the presence of galactose or bisecting *N*-acetylglucosamine of human IgG1 complex-type oligosaccharides shows the critical role of enhancing antibody-dependent cellular cytotoxicity. *J Biol Chem.* 278:3466–73.
104. Okazaki, A., Shoji-Hosaka, E., Nakamura, K., et al. 2004. Fucose depletion from human IgG1 oligosaccharide enhances binding enthalpy and association rate between IgG1 and FcgammaRIIIa. *J Mol Biol.* 336:1239–49.

105. Yamane-Ohnuki, N., Kinoshita, S., Inoue-Urakubo, M., et al. 2004. Establishment of FUT8 knockout Chinese hamster ovary cells: An ideal host cell line for producing completely defucosylated antibodies with enhanced antibody-dependent cellular cytotoxicity. *Biotechnol Bioeng.* 87:614–22.
106. Niwa, R., Natsume, A., Uehara, A., et al. 2005. IgG subclass-independent improvement of antibody-dependent cellular cytotoxicity by fucose removal from Asn297-linked oligosaccharides. *J Immunol Methods.* 306:151–60.
107. Umaña, P., Jean-Mairet, J., Moudry, R., Amstutz, H., Bailey, J. E. 1999. Engineered glycoforms of an antineuroblastoma IgG1 with optimized antibody-dependent cellular cytotoxic activity. *Nat Biotechnol.* 7:176–80.
108. Mori, K., Kuni-Kamochi, R., Yamane-Ohnuki, N., et al. 2004. Engineering Chinese hamster ovary cells to maximize effector function of produced antibodies using FUT8 siRNA. *Biotechnol Bioeng.* 88:901–8.
109. Kanda, Y., Imai-Nishiya, H., Kuni-Kamochi, R., et al. 2007. Establishment of a GDP-mannose 4,6-dehydratase (GMD) knockout host cell line: a new strategy for generating completely non-fucosylated recombinant therapeutics. *J Biotechnol.* 130:300–10.
110. Yano, H., Ishida, T., Inagaki, A., et al. 2007. Defucosylated anti CC chemokine receptor 4 monoclonal antibody combined with immunomodulatory cytokines: A novel immunotherapy for aggressive/refractory *Mycosis fungoides* and Sezary syndrome. *Clin Cancer Res.* 13:6494–500.
111. Yano, H., Ishida, T., Imada, K., et al. 2008. Augmentation of antitumour activity of defucosylated chimeric anti-CCR4 monoclonal antibody in SCID mouse model of adult T-cell leukaemia/lymphoma using G-CSF. *Br J Haematol.* 140:586–89.
112. Klein, M., Haeffner-Cavaillon, N., Isenman, D. E., 1981. Expression of biological effector functions by immunoglobulin G molecules lacking the hinge region. *Proc Natl Acad Sci U S A.* 78:524–28.
113. Sensel, M. G., Kane, L. M., Morrison, S. L. 1997. Amino acid differences in the N-terminus of C(H)2 influence the relative abilities of IgG2 and IgG3 to activate complement. *Mol Immunol.* 34:1019–29.
114. Tao, M. H., Canfield, S. M., Morrison, S. L. 1991. The differential ability of human IgG1 and IgG4 to activate complement is determined by the COOH-terminal sequence of the CH2 domain. *J Exp Med.* 173:1025–28.
115. Tao, M. H., Smith, R. I., Morrison, S. L. 1993. Structural features of human immunoglobulin G that determine isotype-specific differences in complement activation. *J Exp Med.* 178:661–67.
116. Idusogie, E. E., Wong, P. Y., Presta, L. G. et al. 2001. Engineered antibodies with increased activity to recruit complement. *J Immunol.* 166:2571–75.
117. Dall'Acqua, W. F., Cook, K. E., Damschroder, M. M., Woods, R. M., Wu, H. 2006. Modulation of the effector functions of a human IgG1 through engineering of its hinge region. *J Immunol.* 177:1129–38.
118. Pendley, C., Schantz, A., Wagner, C. 2003. Immunogenicity of therapeutic monoclonal antibodies. *Curr Opin Mol Ther.* 5:172–79.
119. Bender, N. K., Heilig, C. E., Dröll, B., Wohlgemuth, J., Armbruster, F. P., Heilig, B. 2007. Immunogenicity, efficacy and adverse events of adalimumab in RA patients. *Rheumatol Int.* 27:269–74.
120. Hwang, W. Y., Foote, J. 2005. Immunogenicity of engineered antibodies. *Methods.* 36:3–10.
121. Zhu, X., Meng, G., Dickinson, B. L., et al. 2001. MHC class I-related neonatal Fc receptor for IgG is functionally expressed in monocytes, intestinal macrophages, and dendritic cells. *J Immunol.* 166:3266–76.

122. Akilesh, S., Christianson, G. J., Roopenian, D. C., Shaw, A. S. 2007. Neonatal FcR expression in bone marrow-derived cells functions to protect serum IgG from catabolism. *J Immunol.* 179:4580–88.
123. Raghavan, M., Gastinel, L. N., Bjorkman, P. J. 1993. The class I major istocompatibility complex related Fc receptor shows pH-dependent stability differences correlating with immunoglobulin binding and release. *Biochemistry.* 32:8654–60.
124. Raghavan, M., Bonagura, V. R., Morrison, S. L., Bjorkman, P. J. 1995. Analysis of the pH dependence of the neonatal Fc receptor/immunoglobulin G interaction using antibody and receptor variants. *Biochemistry.* 34:14649–57.
125. Wines, B. D., Powell, M. S., Parren, P. W., Barnes, N., Hogarth, P. M. 2000. The IgG Fc contains distinct Fc receptor (FcR) binding sites: The leukocyte receptors Fc gamma RI and Fc gamma RIIa bind to a region in the Fc distinct from that recognized by neonatal FcR and protein A. *J Immunol.* 164:5313–18.
126. Martin, W. L., West, A. P. Jr., Gan, L., Bjorkman, P. J. 2001. Crystal structure at 2.8 A of an FcRn/heterodimeric Fc complex: mechanism of pH-dependent binding. *Mol Cell.* 7:867–77.
127. Vaccaro, C., Zhou, J., Ober, R. J., Ward, E. S. 2005. Engineering the Fc region of immunoglobulin G to modulate in vivo antibody levels. *Nat Biotechnol.* 23:1283–88.
128. Kenanova, V., Olafsen, T., Crow, D. M., et al. 2005. Tailoring the pharmacokinetics and positron emission tomography imaging properties of anti-carcinoembryonic antigen single-chain Fv-Fc antibody fragments. *Cancer Res.* 65:622–31.
129. Dall'Acqua, W. F., Woods, R. M., Ward, E. S., et al. 2002. Increasing the affinity of a human IgG1 for the neonatal Fc receptor: biological consequences. *J Immunol.* 169:5171–80.
130. Dall'Acqua, W. F., Kiener, P. A., Wu, H. 2006. Properties of human IgG1s engineered for enhanced binding to the neonatal Fc receptor (FcRn). *J Biol Chem.* 281:23514–24.
131. Hinton, P. R., Johlfs, M. G., Xiong, J. M., et al. 2004. Engineered human IgG antibodies with longer serum half-lives in primates. *J Biol Chem.* 279:6213–16.
132. Datta-Mannan, A., Witcher, D. R., Tang, Y., Watkins, J., Wroblewski, V. J. 2007. Monoclonal antibody clearance. Impact of modulating the interaction of IgG with the neonatal Fc receptor. *J Biol Chem.* 282:1709–17.
133. Chang, C. H., Sharkey, R. M., Rossi, E. A., et al. 2002. Molecular advances in pretargeting radioimunotherapy with bispecific antibodies. *Mol Cancer Ther.* 1:553–63.
134. Colcher, D., Pavlinkova, G., Beresford, G., Booth, B. J., Choudhury, A., Batra, S. K. 1998. Pharmacokinetics and biodistribution of genetically-engineered antibodies. *Q J Nucl Med.* 42:225–41.
135. Wu, A. M. 2004. Engineering multivalent antibody fragments for in vivo targeting. *Methods Mol Biol.* 248:209–25.
136. Wu, A. M., Yazaki, P. J. 2000. Designer genes: recombinant antibody fragments for biological imaging. *Q J Nucl Med.* 44:268–83.
137. Ward, E. S., Güssow, D., Griffiths, A. D., Jones, P. T., Winter, G. 1989. Binding activities of a repertoire of single immunoglobulin variable domains secreted from *Escherichia coli. Nature.* 341:544–46.
138. Holliger, P., Hudson, P. J. 2005. Engineered antibody fragments and the rise of single domains. *Nat Biotechnol.* 23:1126–36.
139. Bird, R. E., Hardman, K. D., Jacobson, J. W., et al. 1988. Single-chain antigen-binding proteins. *Science.* 242:423–26.
140. Brinkmann, U., Reiter, Y., Jung, S. H., Lee, B., Pastan, I. 1993. A recombinant immunotoxin containing a disulfide-stabilized Fv fragment. *Proc Natl Acad Sci U S A.* 90:7538–42.
141. Hollige, P., Prospero, T., Winter, G. 1993. "Diabodies": Small bivalent and bispecific antibody fragments. *Proc Natl Acad Sci U S A.* 90:6444–48.

142. Bera, T. K, Onda, M., Brinkmann, U., Pastan, I. 1998. A bivalent disulfide-stabilized Fv with improved antigen binding to erbB2. *J Mol Biol.* 281:475–83.
143. Chapman, A. P. 2002. PEGylated antibodies and antibody fragments for improved therapy: a review. *Adv Drug Deliv Rev.* 54:531–45.
144. Veronese, F. M., Mero, A. 2008. The impact of PEGylation on biological therapies. *BioDrugs.* 22 :315–29.
145. Bourne, T., Fossati, G., Nesbitt, A. 2008. A PEGylated Fab′ fragment against tumor necrosis factor for the treatment of Crohn disease: Exploring a new mechanism of action. *BioDrugs.* 22 :331–37.
146. Bargou, R., Leo, E., Zugmaier, G., Klinger, M., et al. 2008. Tumor regression in cancer patients by very low doses of a T cell-engaging antibody. *Science.* 321:974–77.
147. Amann, M., Brischwein, K., Lutterbuese, P., et al. 2008. Therapeutic window of MuS110, a single-chain antibody construct bispecific for murine EpCAM and murine CD3. *Cancer Res.* 68:143–51.
148. Kipriyanov, S. M., Cochlovius, B., Schäfer, H. J., et al. 2002. Synergistic antitumor effect of bispecific CD19 x CD3 and CD19 x CD16 diabodies in a preclinical model of non-Hodgkin's lymphoma. *J Immunol.* 169:137–44.
149. Wang, X. B., Zhao, B. F., Zhao, Q., et al. 2004. A new recombinant single chain trispecific antibody recruits T lymphocytes to kill CEA (carcinoma embryonic antigen) positive tumor cells in vitro efficiently. *J Biochem.* 135:555–65.

Section II

Biologics for Autoimmunity and Inflammation

2 Biologic Therapies for Multiple Sclerosis

Paulo Fontoura and Hideki Garren

CONTENTS

KEY WORDS: multiple sclerosis, pathophysiology, treatments, clinical trials

ABSTRACT

Multiple sclerosis (MS) remains a disease in which effective treatments are desperately needed. The currently approved treatments suffer from relatively modest efficacy and little if any impact on disability progression. Several promising new strategies have emerged in recent years, from newer nonspecific immunomodulators, to superselective agents with well-defined molecular targets. For this latter approach, biologic agents such as cells, proteins, monoclonal antibodies, and vaccines offer a combination of biologic potency with selectiveness of action, in the hope of conferring greater safety and efficacy. T cells continue to be a main target for many of the biologic drugs currently in development; either by blocking their migration, altering their phenotype, preventing antigen presentation, or simply killing them. A final area in which biologics can make a significant difference is in regeneration and remyelination of MS lesions, problems for which there are no effective current solutions. In this review we provide a background of MS pathogenesis with particular reference to potential targets of biologic agents, and then discuss available agents for MS treatment as well as biologic agents currently in mid- to late-stage clinical development.

MULTIPLE SCLEROSIS: CLINICAL MANIFESTATIONS AND PATHOLOGICAL SUBSTRATE

A simple definition of multiple sclerosis (MS) is not easily attainable. Any attempts to provide one come up against our ever-changing view of this disease, and relative lack of clear evidence for causation, evolution, and underlying pathophysiological mechanisms. As a working concept, we can define MS as a chronic inflammatory disease of the central nervous system (CNS) that predominantly affects the white matter causing discrete demyelinating and sclerotic lesions. These pathological hallmarks of the disease, called "plaques" since the nineteenth century, are randomly dispersed throughout the entire CNS, although some regions appear to be more predisposed to develop them: spinal cord, brainstem, cerebellum, optic nerve, and the periventricular white matter. Plaques are composed of a mixture of immune system related molecules and cells, including lymphocytes and macrophages, and evolve from acute inflammatory lesions with abundant periaxonal myelin destruction to chronic and sclerotic ones containing astroglial scar tissue.[1,2]

In acute as well as in chronic lesions, there is evident axonal loss besides myelin destruction, an aspect that has received more attention recently.[3] Axonal loss may result from direct immune-mediated attacks, as well as from less well understood processes, including chronic low-grade inflammation, loss of trophic support, glial scarring, ion channel redistribution, and glutamate excitotoxicity.[4–7] Finally, besides these classic MS plaques, there is increasing evidence for the importance of gray matter inflammatory lesions, which are widespread and may be significant contributors at later stages of the disease.[8,9]

The regenerative ability for MS lesions is probably one of the main factors responsible for accruing neurological disability. Although most acute inflammatory lesions evolve into chronic dense astroglial scars, there is direct evidence both for the presence of abundant oligodendrocyte precursors and for remyelination in a significant percentage of plaques.[10,11] However, it may be that several factors, including the absence of required transcription factors like Olig-1,[12] the activation of Jagged-Notch-Hes pathway signaling,[13] plus difficulties in paranodal myelin domain reorganization,[14] contribute to the failure of this process. Furthermore, the regenerative capacity of CNS axons may be severely impeded by the presence of myelin-derived axonal regrowth inhibitors, such as MAG, OMgp, and Nogo-A.[15] In conjunction, all these factors lead to the accumulation of irreversibly demyelinated sclerotic lesions containing few viable axons.

MS has pleomorphic clinical manifestations, given the random distribution of inflammatory plaques. In the initial stages of the disease, the majority of patients have clinical courses characterized by periods of rapid neurological worsening, called relapses. Although these can affect all CNS regions, some manifestations are relatively typical, especially in a young patient, such as an optic neuritis, partial transverse myelitis, or a brainstem syndrome (e.g., internuclear ophthalmoplegia, ataxia). The presence of acute plaques has some correlation with these clinical manifestations, and these inflammatory lesions can be easily detected by appropriate magnetic resonance imaging (MRI) techniques. Relapses are usually self-limited; clinical improvement usually follows, especially if sped up by steroid treatment. In some cases, there may be residual neurological deficits after a relapse, although it is currently uncertain if these contribute to long-term disability. This stage of the disease is termed relapsing–remitting MS (RRMS). With the passage of time, nearly all patients evolve into a progressive clinical course, characterized by insidious neurological worsening in the absence of relapses, manifesting itself as a chronic spinal cord syndrome (spastic paraparesis, sphincteric symptoms) and dementia.[16–18] The pathological substrate for this secondary phase of the disease (secondary progressive, SPMS) appears to be related to CNS atrophy, widespread axonal loss, and gray matter microglial-mediated inflammation.[9,19,20] Besides these more common clinical courses, in some patients the disease appears to progress from the outset, whereas others have few acute relapses superimposed on a progressive course. To clarify the matter, a simple classification in widespread use separates patients into the more common RRMS and SPMS types, and the rarer primary progressive (PPMS) and progressive relapsing (PRMS) types.[21] There appears to be a dichotomy between acute inflammation in RRMS and chronic progression and neurodegeneration in SPMS,[22] and recent evidence has supported a quasi-autonomous status for either of these two manifestations of the disease.[23] It is unclear, at present, what is the pathophysiological common ground for inflammation and degeneration.

MULTIPLE SCLEROSIS: AUTOIMMUNE PATHOPHYSIOLOGY

MS is habitually classified as an autoimmune disease occurring in genetically susceptible individuals, which after contact with an environmental stimulus such as an infection (current culprits include mainly the herpesvirus family) generate a dysregulated cross-reactive response against the CNS. Unlike what happens in acute postinfectious encephalomyelitis, phenomena such as molecular mimicry between viral and CNS antigens, epitope diversification of the immune response, and the existence of regulatory lymphocyte populations have been invoked to explain the chronic relapsing nature of the disease.[24] This view has evolved mainly from studies in the animal model experimental autoimmune encephalomyelitis (EAE), up to now the biggest source of pathophysiological insight into MS, and the testing ground for new therapies.[25]

The currently accepted standard model for MS pathophysiology proposes that acute MS lesions are caused by a perivascular inflammatory reaction based on an autoimmune attack against several proteic and lipid antigens present in the myelin sheath[26,27] (Figure 2.1). This immune-mediated attack is coordinated by proinflammatory T-helper lymphocytes possessing a Th1 or, more recently discovered, Th17 phenotype, both of which are specific for myelin antigens.[28] These cells can directly and indirectly lead to parenchymal destruction, through cytokine production and activation of macrophages and microglia. Antimyelin terminally differentiated T lymphocytes are probably generated in lymphoid organs outside the CNS after first contact with antigen; after migration though the blood–brain barrier (BBB), these cells reencounter their cognate antigen and execute their inflammatory genetic program. Transendothelial migration is a closely regulated process, involving a set of well-defined molecular signals, including cell-adhesion molecules and chemokines, as well as the expression of matrix metalloproteinases (MMP-2 and MMP-9) required for penetration into the parenchyma.[29,30] Within this vast group of molecules, some appear to be more relevant than others; the pair $\alpha4\beta1$ integrin–VCAM-1 appears to be the most relevant for lymphocyte adhesion, and blocking this interaction abrogates CNS infiltration.[31]

After coming in contact with antigen, T-cell phenotypic differentiation is dependent on several factors, including antigen type and dose, costimulation by antigen-presenting cells (APC), cytokines, and the presence of "danger signals" associated with infection, which can be sensed by receptors such as the Toll-like family (TLR).[32] Up until recently, the dominant Th1-Th2 paradigm posited that helper T cells had these two mutually exclusive differentiation pathways, and that for autoimmune diseases such as EAE and MS, Th1 was the predominant phenotype, with Th2 protective or therapeutic.[33] Th1 differentiation was driven mainly by exposure to interleukin-12 (IL-12), whereas Th2 by IL-4. This view was recently perfected by the discovery of Th17 cells, dependent on transforming growth factor-beta (TGFβ), IL-6, and IL-23, and suppressed by IL-27[28,34] (Figure 2.2). Th17 cells could have important roles in progression of disease, as opposed to Th1 cells, which might be responsible for acute lesion generation. Other T-cell populations, such as $CD8^+$ cytotoxic T cells (actually more abundant in MS lesions), could have an important role, especially given their destructive potential for oligodendrocytes, astrocytes, and neurons.[35]

There are several populations of regulatory T cells with potential relevance in controlling the autoimmune response in MS. These include T $CD4^+CD25^+FoxP3^+$ cells (Treg), as well as Th3 (TGFβ producing), Tr1 (IL-10 producing), $CD8^+$Qa-1 restricted T cells, and $CD1d^+$ NK-T cells.[36,37] Despite their presence in demyelinating lesions in EAE, these cells are not sufficient to control inflammation, and may actually promote Th17 differentiation;[38] there are also indications that this population may be deficient in patients with MS.[39] At present, we have very little information on how to judge the *in vivo* role of these cells in MS.

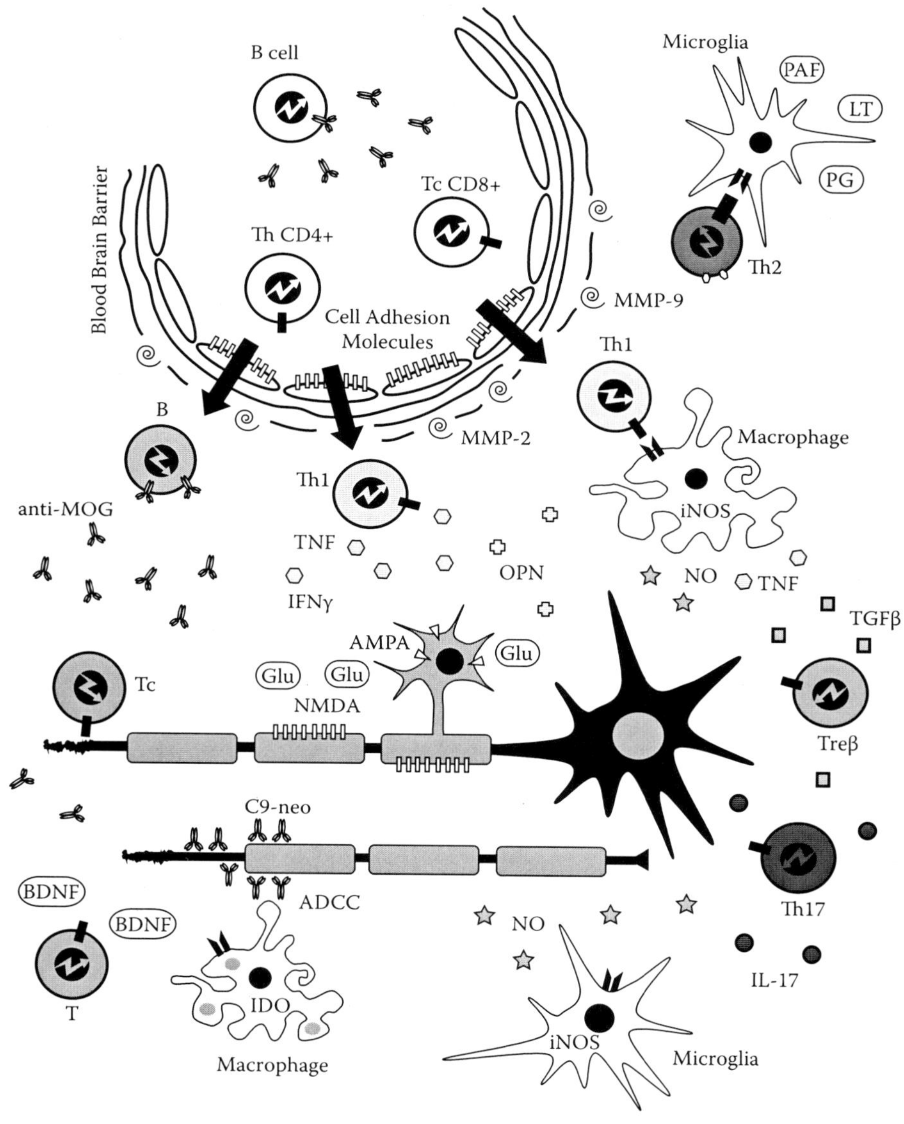

FIGURE 2.1 A potential model for autoimmune pathophysiology of MS.

Local antigen presentation in the CNS is fundamental for reactivation of self-reactive T-lymphocyte populations; however, there are strict control mechanisms for antigen presentation in the CNS, especially for major histocompatibility complex (MHC) class II expression, destined to promote an anti-inflammatory environment.[40] The perivascular compartment is the main gateway into CNS parenchyma, and locally resident microglia and dendritic cells have the capacity for acting as professional APC. After deeper T-cell penetration into the parenchyma, migrating

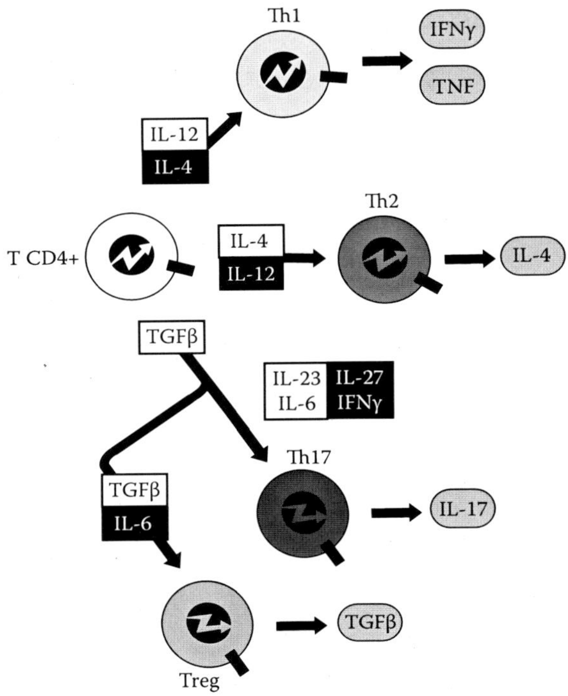

FIGURE 2.2 Differentiation of CD4+ T cells into different subsets.

microglial cells are responsible for maintaining the inflammatory response, not only by acting as APC, but also by generating reactive oxygen and nitrogen species, and secreting inflammatory cytokines.[40,41]

Antigen presentation results from the combination of a specific trimolecular complex of MHC, antigen, and the T-cell receptor (TCR). The MHC *locus* (6p21) confers the greatest known genetic risk for MS, especially the DR15 haplotype (DRB1*1501, DRB5*0101, DQA1*0102, DQB1*0602) for the European and North American Caucasian population.[24] This haplotype generates MHC class II molecules which have been shown to be locally involved in myelin antigen presentation in MS lesions.[42] Antigen recognition by T cells has traditionally been considered very specific, and therefore TCR repertoire restriction was expected to occur. However, the phenomenon of TCR degeneracy, used to explain molecular mimicry, has also made the conditions less stringent for formation of relevant trimolecular complexes with myelin antigens.[43] Nonetheless, restriction of the Vβ TCR region has been identified in a few situations in the EAE model[44,45] and in patients with MS.[46,47]

The myelin sheath has proved to be an antigenic hotspot, with several of its proteins and lipids implicated in the pathogenesis of the EAE model, and presumably MS. The main antigenic candidates include myelin basic protein (MBP), proteolipid protein (PLP), and myelin oligodendrocyte glycoprotein (MOG), although several other minor proteins have been implicated.[24] MBP, the first antigen to be isolated, is composed of five isoforms resulting from alternative splicing of the Golli-MBP gene, and has important structural roles in central and peripheral myelin. Peptides derived from its sequence can be used to induce EAE in several mammalian species, and in humans the main epitopes appear to be MBP83-99 (associated with HLA-DR2a and DR2b haplotypes) and MBP111-129. A humanized double transgenic EAE model for this epitope and the DRB1*0401 molecule results in a phenotype associated with brainstem and cerebellar lesions.[48] PLP is the most abundant myelin protein, and two isoforms can be identified: full-length PLP and DM20, which lacks the PLP139-151 immunodominant epitope. In humans, several epitopes have been identified, although none appears to be dominant. MOG is a CNS-specific immunoglobulin family protein, which localizes to the periphery of the myelin sheath, therefore making it more available for interaction with cells and antibodies.[49] Immunization with MOG epitopes can induce EAE in several species, including the marmoset model which accurately reproduces several of the clinical and histological aspects of MS.[50] Unlike other antigens, anti-MOG antibody transfer results in myelin vesiculation and aggravates EAE.[49] In patients with MS the main epitopes appear to be MOG1-20 and MOG35-55; there appears to be some correlation between the presence of anti-MOG B and T-cell reactivity and lesions in the optic nerve and spinal cord, as was shown in transgenic models as well as for the Asian opticospinal variant of MS.[51,52]

The B-cell response has received considerably less attention than the T-cell response, although the detection of oligoclonal immunoglobulins in cerebrospinal fluid (CSF) continues to be part of the diagnostic criteria; it is used as proof of immune activation and presumably of antigen-driven B-cell clonal expansion. It has been difficult to identify these antigens, which include several infectious agents, myelin components, and a diverse group of other molecules.[24] Recently, the presence of anti-MBP and anti-MOG antibodies in the CSF was briefly hailed as a biomarker for risk of conversion to clinically definite MS, although this has recently been disproved.[53,54] Nonetheless, anti-MOG antibodies and complement activation can be detected in a subset of MS lesions in association with myelin damage.[55] Given the lack of MHC restriction, the B-cell response against nonproteic antigens, such as lipids may prove to be important in MS. In fact, anti-sulfatide, glycolipid, and ganglioside antibodies have been detected in patients with MS, and recently, a specific reactivity pattern against several lipids (sulfatides, sphingomyelin, and phospholipids) was identified.[56] The presence of antilipid immunoglobulin M (IgM) oligoclonal antibodies has also been shown to predict a more aggressive clinical course in patients.[57] On the whole, all these studies point to the need to search beyond classical myelin antigens, especially when considering the B-cell response. Besides this role as antibody makers, it should not be forgotten that B cells can also act as professional APC, and that, therefore, therapeutic strategies aimed at them may work not only by reducing antibody production, but also by interfering with T-cell priming.

MULTIPLE SCLEROSIS TREATMENT STRATEGIES: OVERVIEW

Until recently, the field of MS therapies was littered with the failed remains of several promising agents, and generally MS neurologists adopted a nihilistic approach to the treatment of this disease. This was reflected in a less aggressive attitude toward diagnosis, which finally changed in the last two decades by the introduction of the first effective disease-modifying agents, interferon beta and glatiramer acetate (GA, also known as copolymer-1). Even these, however, were not developed as part of a rational design program, and clarification of the mechanisms of action for both continue to be the object of research. Furthermore, clinical efficacy is modest in the majority of patients, especially when measured by disease progression, as opposed to relapse rate reduction. This is clearly not a desirable situation, and there is need for improvement and better therapeutic alternatives.

Current disease-modifying therapy is based on the concept of MS as an autoimmune disease, and is mainly aimed at controlling inflammation. There are at present no approved therapies specifically targeted at stopping neurodegeneration, or at promoting regeneration and remyelination. Therapy is based on escalated use of immunomodulatory or immunosuppressive agents. For entry-level patients, options include the various formulations of beta interferon or GA; in cases of failure, alternatives include several combination strategies (with drugs such as intravenous immunoglobulins, steroids, or azathioprine), or immunosuppression with mitoxantrone or cyclophosphamide.[58–61] Optimal use of these drugs should include detection and control of potential side effects (especially for more aggressive drugs), as well as adherence monitoring and possibly neutralizing antibody detection for the interferons. For this drug, unofficial consensus is that patients can be divided into responders, partial responders, or nonresponders to these drugs, although a clear definition of this status is lacking. Also, remaining questions include the best route of administration, magnitude of the dose–response effect, when to start and stop therapy, and how to monitor clinical and MRI efficacy. Despite that, we are reaching the operating ceiling for all these agents, and further improvements to them will marginally increase efficacy.

Given our current standard model for MS pathogenesis, several intervention strategies can be devised based on target identification.[62] Even if most of these have shown efficacy on preclinical development, unfortunately most of them mainly serve as reminders of the limitations of animal models.[63] However, several promising new strategies have emerged in recent years, from newer nonspecific immunomodulators, to superselective agents with well-defined molecular targets.[64,65] For this latter approach, biologic agents such as cells, proteins, monoclonal antibodies, and vaccines offer a combination of biologic potency with selectiveness of action, in the hope of conferring greater safety and efficacy.

Among the upcoming nonspecific immunomodulators, promising agents in clinical development include PPAR agonists, statins, cladribine, laquinimod, teriflunomide, BG-12, and FTY720 (fingolimod). Results from ongoing or planned phase II/III trials for the latter five drugs are eagerly awaited. Cladribine, laquinimod, and teriflunomide are DNA synthesis interfering drugs with immunomodulatory effects at low dose, BG-12 is a fumaric acid derivative with interesting cytoprotective potential, and FTY720 is a sphingosine-1-receptor modulator that induces lymphocyte sequestration in secondary lymphoid organs. Besides having a favorable side-effect profile, one of the main attractions for these drugs is having an oral route of administration, with obvious advantages over currently available parenteral agents.[66]

In addition to interferon beta, other cytokines have also been tested as therapeutic agents, based on the classical Th1-Th2 paradigm. Initial trials of Th2 cytokines such as TGFβ, IL-4, and IL-10 were unsuccessful and have been discontinued.[63] Orally administered interferon tau is being tested in phase II trials currently, and may become an alternative to parenteral beta interferon. Blocking Th1 cytokines has also been tried, most famously with the anti-TNF (tumor necrosis factor) drugs infliximab (anti-TNF monoclonal antibody) and lenercept (soluble TNF receptor–immunoglobulin fusion protein), which led to paradoxical worsening of disease.[63] The fact that TNF, and several other cytokines, probably have complex and interplaying roles, at times promoting or suppressing inflammation, makes it difficult to predict efficacy based on theoretical models. Also, the discovery of newer Th phenotypes, such as Th17, may point to heretofore unknown targets.

Other simple molecular targets to aim for may be the complex signaling involved in immune cell trafficking across the BBB. Among these, chemokines and their receptors, and cell adhesion molecules have come under scrutiny, and a monoclonal antibody against α4-integrin (natalizumab, described in detail below) has been approved for treatment of RRMS. Relevant chemokine targets in early preclinical and clinical development include the receptors CCR1, CCR2, CCR5, and CXCR3.[67] It remains to be seen if any of these chemokines plays an essential role in CNS migration, akin to α4-integrin, or if the system behaves more like the complex cytokine network, which might mean that blocking a single target would not be sufficient.

T cells continue to be the main target for most of the drugs currently in development; either by blocking their migration, altering their phenotype, preventing antigen presentation, or simply killing them. Conceptually, therapeutic approaches can be divided into nonantigen specific and antigen specific; a good example of the first class is any nonselective immunomodulator or suppressor. Antigen-based approaches are currently represented solely by GA, whose molecular structure is loosely based on MBP. The random nature of its primary structure and the ability to bind promiscuously to several MHC molecules, as well as inducing degenerate T-cell recognition, have led to the proposition that GA may work as a "universal" altered peptide ligand.[68] Development of antigen-specific approaches to MS treatment has evolved from trials of oral tolerization with bovine myelin, to synthetic copolymers, altered peptide ligands, and DNA vaccination.[69] Despite several setbacks, from lack of efficacy to excessive biologic effects, antigen-specific therapies continue to retain the appeal of selectively affecting solely the deleterious aspects of the immune response in autoimmunity.

A final area in which biologics can make a significant difference is in regeneration and remyelination of MS lesions, problems for which there are no effective current solutions. Stem cell technology offers hope for tissue regeneration in MS; transplantation of neural precursor cells (neurospheres) has been shown to promote recovery in animal models, by differentiating into oligodendroglial precursors and exerting beneficial immunomodulatory effects.[70,71] Besides this approach, promotion of axonal regrowth and remyelination has taken a new direction with the identification of some of the main molecular signals involved in their obstruction in the CNS. Among the axonal regrowth inhibitors, Nogo-A has assumed great importance, and it has been shown in several animal models that blocking Nogo signaling by using specific anti-Nogo antibodies promotes axonal regrowth after CNS injury.[72] On the other hand, one of the molecules present in the Nogo receptor complex, LINGO-1, was recently identified as a key signal in stopping oligodendrocyte precursor differentiation and remyelination,[73] and anti-LINGO antibodies resulted in significant clinical improvement of the EAE model, accompanied by preservation of axons and promotion of remyelination.[74] Given the presence of abundant oligodendrocyte precursors, it may therefore be possible to induce remyelination in MS lesions by a similar strategy.

We proceed to discuss currently available biologic agents for MS treatment and their modes of action, as well as considering agents in mid- to late-stage development.

CURRENTLY APPROVED TREATMENT FOR MS

BETA-INTERFERON

At present the oldest and most widely used therapy for MS is with the beta-interferons. Because of the modest efficacy and relatively benign side-effect profile of the beta-interferons, they are generally considered as first-line therapy for RRMS and account for nearly $3 billion in worldwide sales. There are currently three different beta-interferons available with similar efficacy and side-effect profiles (Table 2.1). They are interferon beta-1b (Betaseron), intramuscular interferon beta-1a (Avonex), and subcutaneous interferon-beta 1a (Rebif). The main differentiating factor among these three drugs is the dose and frequency of administration. The side-effect profile of these drugs is nearly identical, with flulike side effects the most common, and the efficacy is comparable with some subtle differences which are highlighted below. Although the side effects are relatively benign, they are often intolerable, leading frequently to discontinuation or poor compliance. A new agent with comparable or improved efficacy or more tolerable side effects should be able to compete effectively against the beta-interferons.

TABLE 2.1
Comparison of Currently Approved Treatments for MS

	Interferon-Beta 1b	Interferon-Beta 1a	Interferon-Beta 1a	Glatiramer Acetate	Natalizumab
Brand name	Betaseron	Avonex	Rebif	Copaxone	Tysabri
Route	SC	IM	SC	SC	IV
Frequency of dose	Every other day	Once per week	Three times per week	Daily	Once per month
Side effects	Flulike symptoms	Flulike symptoms	Flulike symptoms	Injection-site reactions	PML
Relapse rate reduction	31%	18%	29%–32%	29%	68%
Reduction in disease progression	29%	37%	23%–31%	12%	42%

As the molecular structure of the three beta-interferons are nearly identical, it is felt that they share a common mechanism of action. Exactly what is that mechanism of action is unknown. Various studies *in vitro* or *ex vivo* have led to hypotheses including a suppression of autoreactive T-cell function, alteration of inflammatory cytokine production, and blockade of T-cell migration across the BBB.

Betaseron

Betaseron is the most frequently administered beta-interferon, with a subcutaneous dose administered every other day. Betaseron was the first disease-modifying therapy for RRMS to be approved in the United States, in 1993. The approval was based on a study published in 1993 of 372 patients in one of three arms: placebo, 1.6 million units, or 8 million units. In this study, compared to placebo, the high dose reduced relapses by 31 percent compared to placebo.[75] Betaseron was subsequently approved for use in RRMS in Europe in 1996 and in Japan in 2000. In 1999, Betaseron was also approved in Europe for SPMS with or without relapses and in 2003 was approved in the United States for the treatment of SPMS with relapses. Currently, Betaseron remains as the only beta-interferon approved for such use.

Betaseron has recently been tested in clinically isolated syndrome patients, who are at risk of developing MS. This trial, called the BENEFIT trial (Betaferon/Betaseron in Newly Emerging Multiple Sclerosis for Initial Treatment), demonstrated that Betaseron reduced the risk of conversion to RRMS by 50 percent compared with placebo.[76] These results led to the approval of Betaseron in both the United States and Europe for early-stage MS.

Current trials of Betaseron include a direct comparison to Avonex in the ABOVE trial (Interferon Beta-la versus Interferon Beta-1B Observation of Efficacy) and a comparison of two different doses of Betaseron, 250 versus 500 μg, to Copaxone in the BEYOND trial (Betaferon Efficacy Yielding Outcomes of a New Dose).

Avonex

Avonex is the least frequently administered beta-interferon at one dose per week, but it is also the only one administered intramuscularly, making it difficult for some patients to tolerate. Nevertheless, in 2007 it had the largest share of the MS market, capturing nearly one-third of all patients with MS. Avonex was approved in the United States in 1996, in Europe in 1997, and in Japan in 2006. The initial approval was based on a pivotal trial that demonstrated an 18 percent reduction in relapse rates versus placebo.[77] Avonex was also approved for early-stage MS in Europe in 2002 and in the United States 1 year later. This approval was based on a study that demonstrated a 49 percent reduction in conversion to MS compared to placebo, along with a 44 percent reduction in relapse rates compared to placebo.[78] One of the beneficial differentiating factors of Avonex compared to the other beta-interferons is that it has the lowest frequency of neutralizing antibodies, thus implying that it might have longer-term efficacy than the other beta-interferons.

Avonex has also been investigated in trials for SPMS and PPMS, but without substantial success, and thus is not approved for either of these indications. Currently, Avonex is being investigated in several trials in combination with other agents such as methylprednisolone, methotrexate, and topiramate. However, a combination trial of Avonex and natalizumab (Tysabri) had to be halted when two patients developed the serious and potentially fatal brain disease progressive multifocal leukoencephalopathy (PML). Thus, the use of Avonex in combination would likely be approached with caution.

Rebif

Rebif is the most recently approved beta-interferon and is administered subcutaneously at a frequency between that of Betaseron and Avonex, that is, three times per week. Rebif was approved in the United States in 2002 and in Europe in 1999. This was based on a study in which relapse rates were reduced 29 percent compared to placebo with the 22 μg dose and 32 percent with the 44 μg dose.[79] Rebif was directly compared to and shown superior in efficacy to Avonex in the EVIDENCE (Evidence for Interferon Dose-Effect: European-North American Comparative Efficacy) trial in 2001.[80] This trial demonstrated that more patients were relapse-free on Rebif compared to Avonex (74.9 versus 63.3 percent). Rebif was also approved in 2006 in Europe for use in early-stage MS based on a study which showed a reduction in conversion to MS of 61 percent compared to placebo, using the lower 22 μg dose of Rebif.[81]

Rebif has also been examined in patients with SPMS, but not yet approved. The SPECTRIMS trial (Secondary Progressive Efficacy Trial of Interferon-β-1a in MS) demonstrated that although there was no significant effect on disability scores, there was a significantly lower number of relapses compared to placebo.[82]

Current trials are examining the relative efficacy of Rebif to that of glatiramer acetate (Copaxone) in a 2-year trial.

Glatiramer Acetate (Copaxone)

Copaxone is a synthetic mixture of polypeptides consisting of only four amino acids, l-alanine, l-lysine, l-glutamic acid, and l-tyrosine. These four amino acids were selected because they are the most abundant amino acids in myelin basic protein (MBP). MBP is thought to be one of the dominant autoantigens present on CNS myelin, and thus an important target of the autoimmune attack in MS. Because of this resemblance in structure of Copaxone to MBP, it was initially thought that Copaxone might block the MHC molecule responsible for presenting MBP to the T-cell receptor (TCR), and thereby inhibiting T-cell activation by disrupting the MHC–TCR interaction.[83] However, recent evidence suggests that instead of an antagonist, Copaxone may act as an agonist by causing the induction of anti-inflammatory Th2 cells.[84]

Copaxone was approved in the United States in 1997 and in Europe in 2001 for RRMS. This was based on a 1995 study which showed that Copaxone reduced relapse rates by 29 percent compared to placebo, although the results were not significant with a p value of 0.055.[85] Copaxone is the most frequently administered of MS therapies, with subcutaneous dosing required daily. However, because of a low side-effect profile and high tolerability of the drug, its use has grown and enjoys over $1 billion in sales.

An oral formulation of Copaxone was tested in two phase II trials but failed to demonstrate efficacy. Copaxone was also examined in patients with SPMS and PPMS in the PROMISE trial (Copaxone in Primary Progressive Multiple Sclerosis), but again there was a lack of efficacy in delaying disease progression. Like the beta-interferons, Copaxone is also being tested in patients with early-stage MS, the results of which are pending. Finally, based on a phase II trial showing that higher doses of Copaxone (40 mg compared to the approved 20 mg) improved the efficacy profile without worsening the side-effect profile, a phase III trial of this higher dose is now being conducted.

NATALIZUMAB (TYSABRI)

Tysabri is the first monoclonal antibody approved for MS. It targets alpha 4-integrin, which is a molecule present on T cells that facilitates the entry of T cells into the brain across the BBB. Thus, the proposed mechanism is the inhibition of migration of inflammatory T cells across the BBB which reduces the number of inflammatory lesions within the CNS.[86]

Tysabri was originally approved in the United States in November 2004 for RRMS. However, in February 2005, Tysabri was voluntarily withdrawn from the market because three patients who had received Tysabri in clinical trials developed PML, a rare CNS disease caused by the JC virus.[87–89] There are no known cures for PML and the disease typically causes fatality. In fact, two of the three patients who developed PML while on Tysabri have died.

After an extensive review of the safety data, and from intense pressure by patient advocacy groups, Tysabri was relaunched in the United States and Europe in July 2006 with a black box warning about the risk of PML and under a strict risk management program.

Of all of the approved treatments available for MS, Tysabri has the best efficacy by far. In clinical trials Tysabri was shown to reduce relapse rates by 68 percent and reduce disease progression by 42 percent compared to placebo.[86] Further, on end points measured by MRI, Tysabri reduced T2 lesion volumes by 83 percent and Gd-enhancing lesions by 92 percent compared to placebo.

Because of the potential risk of PML and the inconvenience imposed by the risk management program as well as the intravenous dosing, the uptake of Tysabri has been slow. Most neurologists have reserved its use to those patients who have severe disease or have failed other therapy.

THERAPIES IN LATE-STAGE CLINICAL TRIALS FOR MS

The current therapies in late-stage clinical trials, that is, phase II or later, can broadly be divided into the large number of oral drugs being tested and the injectables. A brief review of each category of therapies is provided.

ORAL THERAPIES

FTY-720 (Fingolimod)

FTY-720 is a sphingosine-1-phosphase (S1P) receptor modulator that, similar to Tysabri, alters lymphocyte trafficking.[90] However, FTY-720 alters trafficking by preventing lymphocytes from exiting lymph nodes and thus keeps lymphocytes out of the CNS, whereas Tysabri prevents lymphocytes from entering the CNS by blocking access through the BBB. By virtue of possessing this similar mechanism of action to Tysabri, there exists the potential of similar side effects including PML, although none has yet to be reported in clinical trials. One potential advantage of FTY-720 over Tysabri is that FTY-720 has a short half-life and thus its adverse effects may be rapidly reversible. Currently, FTY-720 is in a phase III trial in RRMS.

BG-12

BG-12 is a fumarate derivative that inhibits macrophage infiltration in the CNS, possibly as a chemokine receptor antagonist. It has also been shown to increase the expression of the anti-inflammatory cytokine IL-10 and to decrease the expression of proinflammatory cytokines TNFα and IL-6.[91] In phase II trials in patients with RRMS, BG-12 exhibited a dose-dependent effect with the higher 720-mg dose the most effective. There were also relatively few side effects reported, most commonly elevation of liver enzymes, gastrointestinal side effects, and nasopharyngitis. The drug is currently being tested in phase III trials in patients with RRMS.

Laquinimod

Laquinimod is an immunomodulator that is structurally similar to roquinimex (Linomide) and has been shown to inhibit macrophage entry into the CNS as well as to shift T cells from Th1 to Th2.[92] Because of the similarity with Linomide there has been concern about the potential for cardiac side effects. In a phase II trial, the higher dose of laquinimod tested (0.3 mg) was shown to be effective at reducing active lesions on MRI. Laquinimod was well tolerated with relatively few side effects, including elevation of liver enzymes and infections.[93] There were no increases in myocardial infraction, which was the cause of discontinuation of Linomide during trials in MS. In a subsequent phase IIb trial, an even higher dose of laquinimod, 0.6 mg, was tested and showed significant improvement in active lesions compared to placebo, with a similar safety profile as in previous trials. The drug is considered to have modest efficacy, in part because there was no effect on relapses or disability, but concern still remains about potential cardiac toxicity.

Teriflunomide

Teriflunomide is an oral immunosuppressive that is the active metabolite of leflunomide (Arava), which is currently used for the treatment of rheumatoid arthritis. In a phase II trial in which two doses of terifluonomide were tested, both doses reduced the number of active lesions on MRI, and the higher dose (14 mg) was also effective in reducing relapses and disability.[94] The drug was well tolerated but concern remains because the parent compound, leflunomide, has been associated with serious side effects including agranulocytosis, thrombocytopenia, hepatotoxicity, and opportunistic infections. Currently, teriflunomide is being tested in phase III trials for MS.

Cladribine

Cladribine was originally indicated for the treatment of hairy-cell leukemia. It is a chemotherapeutic agent that inhibits the proliferation of activated T cells and B cells. Oral cladribine was tested in a phase I/II trial and subcutaneous cladribine was tested in a phase II trial, both with positive results.[95,96] However, the drug has shown relatively low efficacy in RRMS and better efficacy in SPMS. The drug was generally well tolerated but it has a lifetime dose limit because of suppression of platelet production. Oral cladribine is currently being tested in phase III trials in RRMS.

INJECTABLES

Rituximab

Rituximab is a mouse/human chimeric monoclonal antibody against the B-cell marker CD20 that has been approved for the treatment of B-cell non-Hodgkin's lymphoma and for rheumatoid arthritis. A phase II trial for RRMS was recently completed with rituximab and the results as presented at several conferences were encouraging with significant reductions in Gd-enhancing lesions compared to placebo. Although in this trial the drug was well tolerated with relatively few side effects, rituximab does carry a black box warning for potentially fatal infusion reactions, as well as other serious effects such as cardiac arrhythmias and renal failure. Therefore, although there is considerable excitement and interest for rituximab particularly because it is also being tested in PPMS, there is significant concern about its safety especially if used on a chronic basis.

Daclizumab

Daclizumab is a humanized monoclonal antibody against the alpha subunit of the IL-2 receptor on activated T cells. It has been approved and used for the control of kidney transplant rejection. The drug is currently being tested in MS and other autoimmune diseases such as uveitis and type 1 diabetes. Results from a phase II trial of patients with RRMS in which Daclizumab or placebo was added as an adjunct onto already existing beta-interferon therapy was presented at the 23rd European Conference for the Treatment and Research in Multiple Sclerosis (ECTRIMS) in October 2007. The results demonstrated that the higher 2 mg/kg dose showed significant improvement in lesion activity compared to beta-interferon alone and the drug was well tolerated. Daclizumab also showed promising results in an earlier small trial which involved patients with SPMS.[97] Of all the monoclonal antibodies being developed for MS, daclizumab appears to have the best safety profile. However, because its efficacy is generally considered to be modest and because monotherapy trials in RRMS have yet to be completed, daclizumab may have difficulty competing with the other available therapies.

Alemtuzumab

Alemtuzumab, more commonly known by its trade name Campath, is a monoclonal antibody against CD52, which is a glycoprotein of unknown function present on all mature lymphocytes and monocytes. It is currently used for the treatment of chronic lymphocytic leukemia. Alemtuzumab was tested in a phase II trial of RRMS in a direct comparison to Rebif. Interim results demonstrated superior efficacy to Rebif in terms of relapse risk reduction, reduction in risk of disability progression, as well as MRI parameters. However, several patients developed idiopathic thrombocytopenic purpura and one case proved to be fatal, and thus the trial was suspended. There was also an increase in incidence of Graves' disease with alemtuzumab.[98] Further affecting this drug negatively were results from a phase II trial in patients with SPMS where the results did not demonstrate any reduction in disability progression compared to placebo. Thus, mainly because of concern about its safety profile, alemtuzumab will have substantial difficulty competing with other therapies.

MBP-8298

MBP-8298 is a synthetic peptide that consists of amino acids number 82 to 98 of the MBP molecule. Thus the peptide mimics what many consider the dominant epitope of MBP and acts to downregulate T cells and antibodies specific for that epitope. In a phase II trial of patients with PPMS and SPMS, 500 mg of MBP-8298 given intravenously every 6 months was able to reduce disease progression in patients who were HLA-DR2 or HLA-DR4 positive.[99] The drug was also shown to reduce the level of anti-MBP antibodies in CSF of patients treated with MBP-8298. The drug was well tolerated with injection-site irritation as the most common side effect. The drug currently is being tested in a phase III trial of SPMS, but is also being tested in a phase II trial in patients with RRMS. Although there are skeptics because of the need for HLA testing and the small trials to date, there is enthusiasm for this drug because it is antigen specific and because of the large unmet medical need in the SPMS population.

NeuroVax

NeuroVax is a T-cell receptor vaccine that consists of three peptides in incomplete Freund's adjuvant that mimics the T-cell receptors of anti-MBP T cells. The proposed mechanism of action is that this vaccine will activate regulatory T cells that will then downregulate the pathogenic MBP-specific T cells. The drug has shown promising results in phase II trials in patients with RRMS, and also has been reported to increase the frequency of regulatory T cells. Currently, the drug is being tested in larger phase II trials in patients with RRMS as well as in SPMS, but one of the criticisms of this approach is that the T-cell receptor repertoire of humans is quite diverse and thus an approach that focuses on a small number of T-cell receptors may have limited broad applicability.

Tovaxin

Tovaxin is a T-cell vaccine that contains attenuated T cells specific to three of the major myelin antigens: MBP, PLP, and MOG. This method requires that the patient's own autologous T cells be collected from the patient's peripheral blood, expanded *ex vivo*, irradiated to inactivate these cells, and then used to immunize those same patients. Tovaxin has shown promising results in a small phase I/II clinical trial and is currently being tested in phase IIb trials in patients with RRMS. This approach is appealing in that it is antigen specific and thus should have a favorable safety profile, but it is rather cumbersome in that it requires processing of the patient's own T cells.

BHT-3009

BHT-3009 is the only DNA vaccine in development for MS.[100] BHT-3009 is a plasmid DNA that encodes for full-length MBP and is administered intramuscularly every 4 weeks. The proposed mechanism of action is the tolerization of MBP-specific T cells either by direct anergy or by the production of regulatory T cells. The latter mechanism is proposed because the results from clinical trials demonstrate tolerance not only to MBP, but also to other myelin-specific antigens such as PLP and MOG. A 30 patient phase I trial demonstrated that BHT-3009 was safe, well tolerated, and produced evidence of tolerance both on peripheral blood myelin-specific T cells and on autoantibodies against myelin antigens in the CSF.[101] A larger 289 patient phase II trial was completed and the results were presented at the 23rd ECTRIMS meeting in October 2007. The results demonstrated a 50 to 60 percent reduction in brain lesion activity as measured by Gd^+ lesions on MRI with the lower 0.5 mg dose tested. Further, this dose caused a significant reduction in myelin-specific autoantibody levels within the CSF of essentially all treated patients. Curiously, the higher 1.5 mg dose did not demonstrate efficacy on MRI nor did it reduce autoantibody levels in the CSF. Despite these results with the higher dose, further clinical testing is planned because of the advantages of convenience of dose (once monthly intramuscular injections), safety, and antigen-specific tolerance that BHT-3009 affords.

CONCLUSIONS

MS remains a disease in which effective treatments are desperately needed. The currently approved treatments suffer from relatively modest efficacy and little if any impact on disability progression. Many of the agents in clinical testing have the potential for improved efficacy, but also carry with them the propensity for more detrimental side effects. This is because several of these agents are also used in cancer chemotherapy where more side effects are tolerated or are agents that target the immune system broadly. Several innovative therapies are in mid-stage development that target the disease-causing immune cells in an antigen-specific manner, and thus presumably should be extremely well tolerated. Early-stage preclinical and clinical studies are being conducted with agents that promote myelin repair or inhibit neurodegeneration. It will be several years before we know whether these novel agents have clinical efficacy in pivotal trials. Because of the heterogeneity and chronicity of MS, it is likely that effective treatment will involve a combination of approaches designed to attenuate various aspects of the disease pathophysiology. The next decade in MS clinical research should prove to be exciting and productive as the results of these promising agents become available.

REFERENCES

1. D.A. Hafler, J.M. Slavik, D.E. Anderson, K.C. O'Connor, P. De Jager and C. Baecher-Allan, *Immunol Rev* 204 (2005) 208–31.
2. E.M. Frohman, M.K. Racke and C.S. Raine, *N Engl J Med* 354 (2006) 942–55.
3. B.D. Trapp, J. Peterson, R.M. Ransohoff, R. Rudick, S. Mork and L. Bo, *N Engl J Med* 338 (1998) 278–85.

4. J.W. Peterson and B.D. Trapp, *Neurol Clin* 23 (2005) 107–29, vi–vii.
5. E.M. Frohman, M. Filippi, O. Stuve, S.G. Waxman, J. Corboy, J.T. Phillips, C. Lucchinetti, J. Wilken, N. Karandikar, B. Hemmer, N. Monson, J. De Keyser, H. Hartung, L. Steinman, J.R. Oksenberg, B.A. Cree, S. Hauser and M.K. Racke, *Arch Neurol* 62 (2005) 1345–56.
6. J. Imitola, T. Chitnis and S.J. Khoury, *Arch Neurol* 63 (2006) 25–33.
7. H. Lassmann, *J Neurol Neurosurg Psychiatry* 74 (2003) 695–97.
8. M. Vercellino, F. Plano, B. Votta, R. Mutani, M.T. Giordana and P. Cavalla, *J Neuropathol Exp Neurol* 64 (2005) 1101–7.
9. I. Pirko, C.F. Lucchinetti, S. Sriram and R. Bakshi, *Neurology* 68 (2007) 634–42.
10. P. Patrikios, C. Stadelmann, A. Kutzelnigg, H. Rauschka, M. Schmidbauer, H. Laursen, P.S. Sorensen, W. Bruck, C. Lucchinetti and H. Lassmann, *Brain* 129 (2006) 3165–72.
11. A. Chang, W.W. Tourtellotte, R. Rudick and B.D. Trapp, *N Engl J Med* 346 (2002) 165–73.
12. H.A. Arnett, S.P. Fancy, J.A. Alberta, C. Zhao, S.R. Plant, S. Kaing, C.S. Raine, D.H. Rowitch, R.J. Franklin and C.D. Stiles, *Science* 306 (2004) 2111–15.
13. G.R. John, S.L. Shankar, B. Shafit-Zagardo, A. Massimi, S.C. Lee, C.S. Raine and C.F. Brosnan, *Nat Med* 8 (2002) 1115–21.
14. M. Simons and K. Trajkovic, *J Cell Sci* 119 (2006) 4381–89.
15. M.T. Filbin, *Nat Rev* 4 (2003) 703–13.
16. B.M. Keegan and J.H. Noseworthy, *Annu Rev Med* 53 (2002) 285–302.
17. J.H. Noseworthy, C. Lucchinetti, M. Rodriguez and B.G. Weinshenker, *N Engl J Med* 343 (2000) 938–52.
18. A. Compston and A. Coles, *Lancet* 359 (2002) 1221–31.
19. D.H. Miller, F. Barkhof, J.A. Frank, G.J. Parker and A.J. Thompson, *Brain* 125 (2002) 1676–95.
20. R.A. Bermel and R. Bakshi, *Lancet Neurol* 5 (2006) 158–70.
21. F.D. Lublin and S.C. Reingold, *Neurology* 46 (1996) 907–11.
22. L. Steinman, *Nat Immunol* 2 (2001) 762–64.
23. C. Confavreux and S. Vukusic, *Brain* 129 (2006) 606–16.
24. M. Sospedra and R. Martin, *Annu Rev Immunol* 23 (2005) 683–747.
25. L. Steinman and S.S. Zamvil, *Trends Immunol* 26 (2005) 565–71.
26. L. Steinman, *Cell* 85 (1996) 299–302.
27. B. Hemmer, J.J. Archelos and H.P. Hartung, *Nature Rev* 3 (2002) 291–301.
28. L. Steinman, *Nat Med* 13 (2007) 139–45.
29. R.M. Ransohoff, P. Kivisakk and G. Kidd, *Nature Revi* 3 (2003) 569–81.
30. B. Engelhardt and R.M. Ransohoff, *Trends Immunol* 26 (2005) 485–95.
31. T.A. Yednock, C. Cannon, L.C. Fritz, F. Sanchez-Madrid, L. Steinman and N. Karin, *Nature* 356 (1992) 63–66.
32. T. Kielian, *J Neurosci Res* 83 (2006) 711–30.
33. S.S. Zamvil and L. Steinman, *Annu Rev Immunol* 8 (1990) 579–621.
34. C.T. Weaver, R.D. Hatton, P.R. Mangan and L.E. Harrington, *Annu Rev Immunol* 25 (2007) 821–52.
35. H. Neumann, I.M. Medana, J. Bauer and H. Lassmann, *Trends Neurosci* 25 (2002) 313–19.
36. H. Jiang and L. Chess, *J Clin Invest* 114 (2004) 1198–208.
37. E.M. Shevach, *Annu Rev Immunol* 18 (2000) 423–49.
38. T. Korn, J. Reddy, W. Gao, E. Bettelli, A. Awasthi, T.R. Petersen, B.T. Backstrom, R.A. Sobel, K.W. Wucherpfennig, T.B. Strom, M. Oukka and V.K. Kuchroo, *Nat Med* 13 (2007) 423–31.
39. V. Viglietta, C. Baecher-Allan, H.L. Weiner and D.A. Hafler, *J Exp Med* 199 (2004) 971–79.

40. B. Becher, I. Bechmann and M. Greter, *J Mol Med* (Berlin, Germany) 84 (2006) 532–43.
41. F.L. Heppner, M. Greter, D. Marino, J. Falsig, G. Raivich, N. Hovelmeyer, A. Waisman, T. Rulicke, M. Prinz, J. Priller, B. Becher and A. Aguzzi, *Nat Med* 11 (2005) 146–52.
42. M. Krogsgaard, K.W. Wucherpfennig, B. Cannella, B.E. Hansen, A. Svejgaard, J. Pyrdol, H. Ditzel, C. Raine, J. Engberg and L. Fugger, *J Exp Med* 191 (2000) 1395–412.
43. K.W. Wucherpfennig, *Mol Immunol* 40 (2004) 1009–17.
44. S.S. Zamvil, D.J. Mitchell, N.E. Lee, A.C. Moore, M.K. Waldor, K. Sakai, J.B. Rothbard, H.O. McDevitt, L. Steinman and H. Acha-Orbea, *J Exp Med* 167 (1988) 1586–96.
45. H. Acha-Orbea, D.J. Mitchell, L. Timmermann, D.C. Wraith, G.S. Tausch, M.K. Waldor, S.S. Zamvil, H.O. McDevitt and L. Steinman, *Cell* 54 (1988) 263–73.
46. K.W. Wucherpfennig, K. Ota, N. Endo, J.G. Seidman, A. Rosenzweig, H.L. Weiner and D.A. Hafler, *Science* 248 (1990) 1016–19.
47. J.R. Oksenberg, M.A. Panzara, A.B. Begovich, D. Mitchell, H.A. Erlich, R.S. Murray, R. Shimonkevitz, M. Sherritt, J. Rothbard, C.C. Bernard et al., *Nature* 362 (1993) 68–70.
48. J.A. Quandt, M. Baig, K. Yao, K. Kawamura, J. Huh, S.K. Ludwin, H.J. Bian, M. Bryant, L. Quigley, Z.A. Nagy, H.F. McFarland, P.A. Muraro, R. Martin and K. Ito, *J Exp Med* 200 (2004) 223–34.
49. C.C. Bernard, T.G. Johns, A. Slavin, M. Ichikawa, C. Ewing, J. Liu and J. Bettadapura, *J Mol Med* 75 (1997) 77–88.
50. C.P. Genain and S.L. Hauser, *J Mol Med* (Berlin, Germany) 75 (1997) 187–97.
51. G. Krishnamoorthy, H. Lassmann, H. Wekerle and A. Holz, *J Clin Invest* 116 (2006) 2385–92.
52. M. Minohara, H. Ochi, S. Matsushita, A. Irie, Y. Nishimura and J. Kira, *Tissue Antigens* 57 (2001) 447–56.
53. T. Berger, P. Rubner, F. Schautzer, R. Egg, H. Ulmer, I. Mayringer, E. Dilitz, F. Deisenhammer and M. Reindl, *N Engl J Med* 349 (2003) 139–45.
54. J. Kuhle, C. Pohl, M. Mehling, G. Edan, M.S. Freedman, H.P. Hartung, C.H. Polman, D.H. Miller, X. Montalban, F. Barkhof, L. Bauer, S. Dahms, R. Lindberg, L. Kappos and R. Sandbrink, *N Engl J Med* 356 (2007) 371–78.
55. C.P. Genain, B. Cannella, S.L. Hauser and C.S. Raine, *Nat Med* 5 (1999) 170–75.
56. J.L. Kanter, S. Narayana, P.P. Ho, I. Catz, K.G. Warren, R.A. Sobel, L. Steinman and W.H. Robinson, *Nat Med* 12 (2006) 138–43.
57. L.M. Villar, M.C. Sadaba, E. Roldan, J. Masjuan, P. Gonzalez-Porque, N. Villarrubia, M. Espino, J.A. Garcia-Trujillo, A. Bootello and J.C. Alvarez-Cermeno, *J Clin Invest* 115 (2005) 187–94.
58. K. Bashir and J.N. Whitaker, *Arch Neurol* 59 (2002) 726–31.
59. E.M. Frohman, O. Stuve, E. Havrdova, J. Corboy, A. Achiron, R. Zivadinov, P.S. Sorensen, J.T. Phillips, B. Weinshenker, K. Hawker, H.P. Hartung, L. Steinman, S. Zamvil, B.A. Cree, S. Hauser, H. Weiner, M.K. Racke and M. Filippi, *Arch Neurol* 62 (2005) 1519–30.
60. R.A. Rudick, *Arch Neurol* 56 (1999) 1079–84.
61. S.A. Gauthier, G.J. Buckle and H.L. Weiner, *Neurol Clin* 23 (2005) 247–72, viii–ix.
62. M. Chofflon, *BioDrugs* 19 (2005) 299–308.
63. H. Wiendl and R. Hohlfeld, *BioDrugs* 16 (2002) 183–200.
64. B.C. Kieseier, B. Hemmer and H.P. Hartung, *Curr Opin Neurol* 18 (2005) 211–20.
65. P. Fontoura, L. Steinman and A. Miller, *Curr Opin Neurol* 19 (2006) 260–66.
66. B.A. Cohen and P. Rieckmann, *Int J Clin Pract* 61 (2007) 1922–30.
67. A. Szczucinski and J. Losy, *Acta Neurol Scand* 115 (2007) 137–46.
68. D.A. Hafler, *J Clin Invest* 109 (2002) 581–84.
69. P. Fontoura, H. Garren and L. Steinman, *Int Rev Immunol* 24 (2005) 415–46.

70. S. Pluchino, A. Quattrini, E. Brambilla, A. Gritti, G. Salani, G. Dina, R. Galli, U. Del Carro, S. Amadio, A. Bergami, R. Furlan, G. Comi, A.L. Vescovi and G. Martino, *Nature* 422 (2003) 688–94.
71. S. Pluchino, L. Zanotti, B. Rossi, E. Brambilla, L. Ottoboni, G. Salani, M. Martinello, A. Cattalini, A. Bergami, R. Furlan, G. Comi, G. Constantin and G. Martino, *Nature* 436 (2005) 266–71.
72. A.D. Buchli and M.E. Schwab, *Ann Med* 37 (2005) 556–67.
73. S. Mi, R.H. Miller, X. Lee, M.L. Scott, S. Shulag-Morskaya, Z. Shao, J. Chang, G. Thill, M. Levesque, M. Zhang, C. Hession, D. Sah, B. Trapp, Z. He, V. Jung, J.M. McCoy and R.B. Pepinsky, *Nat Neurosci* 8 (2005) 745–51.
74. S. Mi, B. Hu, K. Hahm, Y. Luo, E.S. Kam Hui, Q. Yuan, W.M. Wong, L. Wang, H. Su, T.H. Chu, J. Guo, W. Zhang, K.F. So, B. Pepinsky, Z. Shao, C. Graff, E. Garber, V. Jung, E.X. Wu and W. Wu, *Nat Med* 13 (2007) 1228–33.
75. The IFNB Multiple Sclerosis Study Group, *Neurology* 43 (1993) 655–61.
76. L. Kappos, C.H. Polman, M.S. Freedman, G. Edan, H.P. Hartung, D.H. Miller, X. Montalban, F. Barkhof, L. Bauer, P. Jakobs, C. Pohl and R. Sandbrink, *Neurology* 67 (2006) 1242–49.
77. L.D. Jacobs, D.L. Cookfair, R.A. Rudick, R.M. Herndon, J.R. Richert, A.M. Salazar, J.S. Fischer, D.E. Goodkin, C.V. Granger, J.H. Simon, J.J. Alam, D.M. Bartoszak, D.N. Bourdette, J. Braiman, C.M. Brownscheidle, M.E. Coats, S.L. Cohan, D.S. Dougherty, R.P. Kinkel, M.K. Mass, F.E. Munschauer, 3rd, R.L. Priore, P.M. Pullicino, B.J. Scherokman, R.H. Whitham et al., *Ann Neurol* 39 (1996) 285–94.
78. R.W. Beck, D.L. Chandler, S.R. Cole, J.H. Simon, L.D. Jacobs, R.P. Kinkel, J.B. Selhorst, J.W. Rose, J.A. Cooper, G. Rice, T.J. Murray and A.W. Sandrock, *Ann Neurol* 51 (2002) 481–90.
79. PRISMS Study Group, *Lancet* 352 (1998) 1498–504.
80. H. Panitch, D.S. Goodin, G. Francis, P. Chang, P.K. Coyle, P. O'Connor, E. Monaghan, D. Li and B. Weinshenker, *Neurology* 59 (2002) 1496–506.
81. G. Comi, M. Filippi, F. Barkhof, L. Durelli, G. Edan, O. Fernandez, H. Hartung, P. Seeldrayers, P.S. Sorensen, M. Rovaris, V. Martinelli and O.R. Hommes, *Lancet* 357 (2001) 1576–82.
82. D.K. Li, G.J. Zhao and D.W. Paty, *Neurology* 56 (2001) 1505–13.
83. R. Aharoni, D. Teitelbaum, R. Arnon and M. Sela, *Proc Natl Acad Sci USA* 96 (1999) 634–39.
84. P.W. Duda, M.C. Schmied, S.L. Cook, J.I. Krieger and D.A. Hafler, *J Clin Invest* 105 (2000) 967–76.
85. K.P. Johnson, B.R. Brooks, J.A. Cohen, C.C. Ford, J. Goldstein, R.P. Lisak, L.W. Myers, H.S. Panitch, J.W. Rose and R.B. Schiffer, *Neurology* 45 (1995) 1268–76.
86. C.H. Polman, P.W. O'Connor, E. Havrdova, M. Hutchinson, L. Kappos, D.H. Miller, J.T. Phillips, F.D. Lublin, G. Giovannoni, A. Wajgt, M. Toal, F. Lynn, M.A. Panzara and A.W. Sandrock, *N Engl J Med* 354 (2006) 899–910.
87. B.K. Kleinschmidt-DeMasters and K.L. Tyler, *N Engl J Med* 353 (2005) 369–74.
88. A. Langer-Gould, S.W. Atlas, A.J. Green, A.W. Bollen and D. Pelletier, *N Engl J Med* 353 (2005) 375–81.
89. G. Van Assche, M. Van Ranst, R. Sciot, B. Dubois, S. Vermeire, M. Noman, J. Verbeeck, K. Geboes, W. Robberecht and P. Rutgeerts, *N Engl J Med* 353 (2005) 362–68.
90. L. Kappos, J. Antel, G. Comi, X. Montalban, P. O'Connor, C.H. Polman, T. Haas, A.A. Korn, G. Karlsson and E.W. Radue, *N Engl J Med* 355 (2006) 1124–40.
91. S. Schilling, S. Goelz, R. Linker, F. Luehder and R. Gold, *Clin Exp Immunol* 145 (2006) 101–7.
92. J.S. Yang, L.Y. Xu, B.G. Xiao, G. Hedlund and H. Link, *J Neuroimmunol* 156 (2004) 3–9.

93. C. Polman, F. Barkhof, M. Sandberg-Wollheim, A. Linde, O. Nordle and T. Nederman, *Neurology* 64 (2005) 987–91.
94. P.W. O'Connor, D. Li, M.S. Freedman, A. Bar-Or, G.P. Rice, C. Confavreux, D.W. Paty, J.A. Stewart and R. Scheyer, *Neurology* 66 (2006) 894–900.
95. J.S. Romine, J.C. Sipe, J.A. Koziol, J. Zyroff and E. Beutler, *Proc Assoc Am Physicians* 111 (1999) 35–44.
96. G.P. Rice, M. Filippi and G. Comi, *Neurology* 54 (2000) 1145–55.
97. J.W. Rose, H.E. Watt, A.T. White and N.G. Carlson, *Ann Neurol* 56 (2004) 864–67.
98. A.J. Coles, M. Wing, S. Smith, F. Coraddu, S. Greer, C. Taylor, A. Weetman, G. Hale, V.K. Chatterjee, H. Waldmann and A. Compston, *Lancet* 354 (1999) 1691–95.
99. K.G. Warren, I. Catz, L.Z. Ferenczi and M.J. Krantz, *Eur J Neurol* 13 (2006) 887–95.
100. H. Garren and L. Steinman. In Fathman, C.G., ed., *Biologic and Gene Therapy of Autoimmune Disease*, Karger, Basel 2000, pp. 203–16.
101. A. Bar-Or, T. Vollmer, J. Antel, D.L. Arnold, C.A. Bodner, D. Campagnolo, J. Gianettoni, F. Jalili, N. Kachuck, Y. Lapierre, M. Niino, J. Oger, M. Price, S. Rhodes, W.H. Robinson, F.D. Shi, P.J. Utz, F. Valone, L. Weiner, L. Steinman and H. Garren, *Arch Neurol* 64 (2007) 1407–15.

3 Biologic Therapies in Rheumatoid Arthritis

Mark J. Houseman and John D. Isaacs

CONTENTS

KEY WORDS: rheumatoid arthritis, anti-TNF, biologic drugs

ABSTRACT

Rheumatoid arthritis (RA) is a chronic inflammatory condition primarily affecting synovial joints. The panacea of medical treatment remains clinical remission with inhibition of joint destruction. Over the last few years, a better understanding into the molecular pathogenesis of RA has led to the development of biologic therapies targeted at specific components of the inflammatory and immune cascades. These include monoclonal antibodies or fusion proteins that neutralize the function of specific cytokines (such as tumor necrosis factor alpha, or TNFα, and interleukin-1), deplete particular cell types, or inhibit cell activation. Biologic therapy of RA has met with significant success. The quality of the response is often better than with traditional drugs, particularly in terms of preventing joint destruction, although not all patients respond for reasons that remain to be elucidated. Furthermore, the specific nature of the treatments has provided further insights into disease pathogenesis as well as the critical roles of the mediators and cells that are targeted. For example, TNFα blockade revealed the nonredundant role of TNFα in host response against *Mycobacterium* infection. Despite these issues, the advent of biologic drugs has brought significant new hope to patients suffering with RA.

INTRODUCTION

Rheumatoid arthritis (RA) is a chronic inflammatory joint disease that affects approximately 1 percent of the population in the industrialized world.[1] It primarily targets the synovium, with secondary damage to the cartilage and bone, in the small diarthrodial joints of the hands and feet. This typically occurs in a symmetrical pattern. The clinical presentation reflects the underlying inflammatory processes with symptoms of joint pain and stiffness and signs of joint swelling, heat, and loss of function. Active, chronic disease culminates in severe disability and increased mortality. Consequently, medical therapy has been aimed at symptom control and prevention of joint destruction. However, the complexity and heterogeneity of RA in terms of genetic susceptibility, long-term outcome, and response to treatment have presented researchers striving to fully understand the condition with many challenges. The recent development of biologic therapies has offered two major benefits. First, it has enhanced prognosis and quality of life, particularly in those individuals with RA refractory to traditional treatment regimens. Second, it has provided an opportunity to observe the effects of *in vivo* targeted modulation of the inflammatory cascade. This has offered important new insights into the pathogenesis of RA.[2] This chapter briefly outlines RA pathogenesis and relates this to established biologic therapies, which are discussed by their intended immunological target. Finally, we will take an opportunity to highlight emerging biologic therapies and their potential role in this rapidly progressing and exciting field of medicine.

PATHOGENESIS OF RHEUMATOID ARTHRITIS

The highly vascularized synovial membrane lines the nonweight-bearing surface of the joint. It produces synovial fluid that facilitates gliding of the articular surfaces and supplies nutrients to the adjacent avascular cartilage. In RA, the normally hypocellular membrane is transformed by the influx of $CD4^+$ T cells, B cells, and macrophages. In addition, there is marked hyperplasia and hypertrophy of type A (macrophage-like) and type B (fibroblast-like) synoviocytes. This hypercellular structure or pannus, driven by proinflammatory cytokines (including tumor necrosis factor-α (TNFα), interleukin-1 β (IL-1β), IL-6, IL-15, and IL-17), becomes locally invasive and with time leads to destruction of articular cartilage and bone (Figure 3.1). This is mediated by the action of degradative enzymes (matrix metalloproteinases) and activation of bone-resorbing osteoclasts. A unifying hypothesis for the pathogenesis of RA remains elusive. However, observations from animal models, *in vitro* analysis, and the clinical setting suggest antigen-driven T cells and autoreactive B cells producing proinflammatory cytokines and autoantibodies play a central

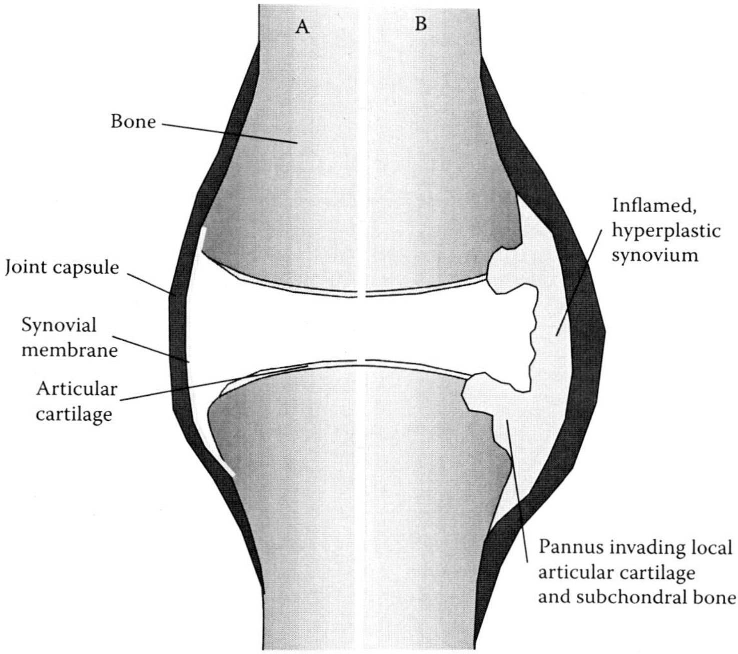

FIGURE 3.1 Illustration of a normal joint and RA joint. The normal joint (A) has a thin and delicate synovial membrane lining the joint capsule and other nonweight-bearing surfaces of the joint. In RA (B), the synovial membrane is infiltrated with $CD4^+$ T-cells, B cells, and macrophages. It eventually forms into "pannus," and becomes locally invasive leading to permanent joint damage.

role. A more detailed overview into the pathogenesis of RA can be found in other reviews of rheumatoid arthritis.[3,4]

Monitoring Disease Activity and Outcome

The clinical variability in the manifestations of RA presents potential difficulty when measuring clinical response. This has led to the development and validation of composite measures, which incorporate multiple aspects associated with "disease" state.[5] The American College of Rheumatology (ACR) response criteria measure percentage improvement from baseline (e.g., ACR50 approximates to 50 percent improvement), but does not provide information on current disease activity (Table 3.1). In contrast, the European League against Rheumatism (EULAR) disease activity score (DAS), including 28 joint counts (DAS28), provides an indication of current disease activity. In trials, an ACR20 is used to indicate a minimal measure of clinical response, greater than that expected from the "placebo" effect. More useful measures of efficacy are ACR50 or EULAR good responses, which correlate with substantial levels of improvement. Additional outcome measures include progression of joint damage utilizing serial radiographic imaging. Joint erosions in the small

TABLE 3.1
Response Criteria and Disease Activity Scores Used in the Study of Rheumatoid Arthritis

Criteria	Assessment Method
ACR response criteria	ACR20, 50, 70, 90 indicate improvement of 20%, 50%, 70%, or 90% from baseline, respectively. For example, an ACR20 response requires ≥20% improvement in tender joint count; ≥ 20% improvement in swollen joint count, plus ≥20% improvement in three of following: • Patient pain assessment on VAS • Patient global assessment (PGA) VAS • Physician global assessment VAS • Patient self-assessed disability, such as Health Assessment Questionnaire • Acute-phase reactant (ESR or CRP)
EULAR response criteria	EULAR response requires an initial assessment of DAS28: $DAS28 = 0.56 \times \sqrt{(TJC)} + 0.29 \times \sqrt{(SJC)} + 0.70 \times \ln(ESR) + 0.014 \times (PGA)$ • A good EULAR response requires ≥1.2 units improvement from baseline in DAS28 with a final absolute score < 3.2 • A moderate EULAR response requires ≥ 0.6 units improvement from baseline in DAS28 with a final absolute score < 5.1 • DAS < 2.6 is classified as remission; DAS < 3.2 is classified as low disease activity; DAS ≤ 5.1 is classified as moderate disease activity

Abbreviations: TJC, 28 tender joint count; SJC, 28 swollen joint count; PGA, patient global assessment; VAS, visual analogue scale; ESR, erythrocyte sedimentation rate; CRP, C-reactive protein.

bones of the hands and feet can often be detected within 6 months of disease onset and generally progress in the absence of effective therapy. In addition, measurement of functional impairment, and quality of life, using validated questionnaires also provides useful information on disease activity and treatment efficacy.

TUMOR NECROSIS FACTOR-α BLOCKADE

The proinflammatory and arthritogenic effects of TNFα are well recognized. Early *in vitro* studies of disaggregated mononuclear cells demonstrated a marked reduction of downstream cytokines following the addition of neutralizing quantities of anti-TNFα antibody.[6] In addition, observations from a number of animal studies have supported the important role of TNFα in RA. First, administration of exogenous TNFα in the established mouse collagen type-II induced arthritis model led to exacerbation of disease. More significantly, a clinical anti-inflammatory effect was noted following TNFα blockade with an anti-TNFα monoclonal antibody.[7] This extended to an apparent protective effect on joint damage, as indicated by histological examination. These findings were supported using a different mouse model carrying modified human TNFα transgenes and demonstrating dysregulated expression of TNFα.[8] The Tgl97 mice developed a polyarthritis at 3 to 4 weeks of age. Impairment in leg movement progressed to complete loss of movement of the hind legs at around 9 to 10 weeks of age. Treatment of these arthritic mice with a monoclonal antibody against human TNFα arrested development of this disease phenotype.

TNFα appears to function both as an autocrine stimulator and paracrine inducer of other cytokines.[9] TNFα also increases expression of endothelial adhesion molecules, which are essential for recruitment of additional effector immune cells, from the bloodstream, to areas of inflammation. It is produced mainly by monocytes and macrophages, but also B cells, T cells, and fibroblasts. Posttranslation, TNFα is directed to the plasma cell membrane. It has biologically active function both as a membrane-bound molecule and in soluble form following enzymatic cleavage of the anchoring domain. The cytokine binds two cell surface receptors (TNFαR), namely, p55 and p75, expressed on a number of cell types. Activation of p55 or p75 is thought to produce different downstream effects given the marked variation in structure of the cytoplasmic domains. Interestingly, the transmembrane region of both receptors is also susceptible to proteolytic cleavage, which culminates in the release of soluble receptor into the extracellular fluid and serum where it is free to bind soluble and transmembrane TNFα. Although predominantly neutralizing, the soluble receptors have also been shown to stabilize the bioactivity of TNFα.[10] These experimental data provided compelling evidence for the involvement of TNFα in RA synovitis.

Efficacy of Anti-TNFα Drugs

Infliximab (Remicade, Centocor)

Infliximab is a chimeric human/murine (human constant region, mouse variable region) immunoglobulin G_1 (IgG_1) kappa isotype monoclonal antibody (mAb) against TNFα, which does not recognize TNFβ (lymphotoxin). It has been shown to bind both the soluble and transmembrane forms of the cytokine.[9] Infliximab has a half-life of approximately 10 days and biologic effect lasting up to 2 months. The first open-label study recruited 20 patients with severe active RA, who had not responded to many of the traditional disease-modifying antirheumatic drugs (DMARDs).[11] Patients were enrolled onto the trial following at least 4 weeks during which they discontinued their current treatment regimen. Infliximab (20 mg/kg) was administered by two or four intravenous (iv) infusions over 2 weeks. This schedule had been extrapolated from previous animal studies. The clinical response of patients completing the treatment was marked. Improvement in a number of clinical indices of disease activity was significant at 1 and 6 weeks, compared to baseline. Biochemical markers of inflammation reduced correspondingly. Response was noted within a week, peaked after 6 weeks, and appeared sustained for a period up to 22 weeks. The presence of mouse peptide sequences within the infliximab molecule posed a potential issue for development of human antichimeric antibodies (HACAs). However, only 1 of 20 individuals developed an antiglobulin response suggesting infliximab was not especially immunogenic.

To confirm these promising observations, a multicenter, randomized double-blind, placebo-controlled trial of a single infusion of 1 or 10 mg/kg of infliximab in 73 patients with active rheumatoid arthritis was undertaken.[12] Assessment at 4 weeks indicated 79 percent of individuals receiving 10 mg/kg and 44 percent receiving

1 mg/kg, compared to 8 percent given the placebo, attained the predetermined response criteria. Notably, the incidence of infections was greater in patients treated with infliximab compared with placebo (12 vs. 4 percent). Also, not all individuals demonstrated clinical benefit and all of those who responded ultimately relapsed. To investigate possible effects of repeat infliximab treatment on the latter patient group, a smaller follow-up study was performed using individuals from the original open-label cohort.[13] Eight patients who previously received 20 mg/kg doses were permitted a reduced 10 mg/kg treatment up to four times during a 1-year period. Following each administration of infliximab there was a marked clinical improvement; although in some patients a reduction in the duration of benefit was noted. HACAs specific to the murine segment of infliximab were eventually detected in four of eight subjects. However, their role in limiting the efficacy of subsequent infliximab treatment was not clear. More significantly, four patients failed to complete the therapy for safety reasons. These included procedure-related vasovagal syncope, posttreatment urticaria, chronic sinusitis, and development of dsDNA autoantibodies (with no clinical evidence of systemic lupus erythematosus, SLE). These preliminary studies highlighted a role for infliximab in management of RA refractory to traditional DMARD treatment and possible longer-term therapeutic strategies. However, they also raised questions regarding safety, immunogenicity, and efficacy on multiple dosing.

A subsequent placebo-controlled dose-ranging trial of infliximab was undertaken to further investigate some of the above issues.[14] Patients were recruited to the study with active RA exhibiting an incomplete response or flare of disease activity while receiving low-dose methotrexate (MTX). In addition, stable doses of nonsteroidal anti-inflammatory drugs and/or corticosteroids were permitted. Each individual ($n = 101$) was randomized to one of seven groups: infliximab alone (1, 3, or 10 mg/kg dosing groups), infliximab (same dose regimen) + low-dose (7.5 mg weekly) MTX, or low dose MTX + placebo infusion. Patients received five infliximab (or placebo) infusions over a 14-week period and were assessed routinely until 26 weeks. The groups administered infliximab at 3 and 10 mg/kg doses + supplemental MTX achieved a similar ACR50 response (~40 percent patients) at 26 weeks. Significant differences in clinical improvement were present at the majority of time points compared to the MTX + placebo group. Also, the duration of response was consistently longer when infliximab was combined with MTX versus infliximab monotherapy. However, this did not reach statistical significance ($p = 0.26$ at 3 mg/kg and $p = 0.11$ at 10 mg/kg). More interesting findings were seen in the group receiving low-dose infliximab (1 mg/kg) alone. Approximately 60 percent achieved an ACR20 response at 2 weeks. However, the improvement reduced dramatically, and at 6 weeks was comparable with the low-dose MTX + placebo group. This trend continued until 26 weeks despite two further infliximab infusions. In contrast, ~60 percent of the group receiving infliximab at 1 mg/kg + MTX achieved an ACR20 response that was maintained until 16 weeks. This frequency declined to 20 percent at week 26. In the groups receiving infliximab, the rate of HACA development was inversely proportional to the dosage. This occurred in 53, 21, and 7 percent of the patients who were receiving repeated treatment with infliximab at 1, 3, and 10 mg/kg, respectively. Concomitant therapy with low-dose MTX greatly diminished the frequency of HACA, with incidence rates of 15, 7, and 0 percent at the above doses of infliximab.

This suggested a mechanism of immunologic tolerance with higher doses of infliximab. The study also highlighted an apparent synergistic effect in clinical response between infliximab and low-dose MTX, possibly because the immunosuppressive effect of MTX reduced HACA responses.

The efficacy of infliximab + MTX combined therapy was further demonstrated in the pivotal Anti–Tumour necrosis factor Trial in Rheumatoid Arthritis with Concomitant Therapy (ATTRACT) study.[15] This involved a cohort of patients (n = 428) with RA inadequately controlled with MTX alone. The study recruited patients on stable therapeutic doses of MTX (>12.5 mg weekly). All patients continued MTX, with individuals randomized to receive placebo or infliximab (either 3 or 10 mg/kg) at 0, 2, and 6 weeks. The groups received infusions at 4 or 8 week intervals over a 30-week period. An ACR20 response was achieved in 50 to 60 percent of patients receiving infliximab + MTX compared with 20 percent on MTX alone (Figure 3.2). In addition, the eight weekly 3 mg/kg + MTX group appeared equally efficacious as those receiving 10 mg/kg at 4 or 8 week intervals. The response appeared sustained

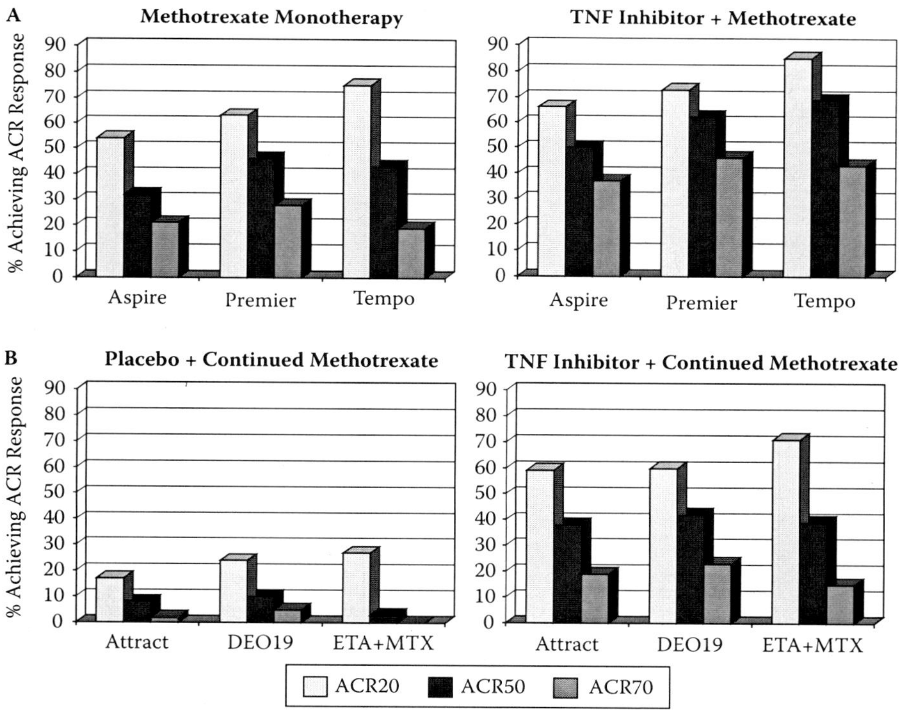

FIGURE 3.2 Frequency and level of ACR response observed with TNFα blockade. (A) Newly instituted MTX therapy (plus placebo infusion or injection; left panel) and MTX plus TNF-inhibitor therapy (right panel) from clinical trials of infliximab (ASPIRE),[16] adalimumab (PREMIER),[19] and etanercept (TEMPO).[25] (B) Patients with active disease despite MTX who continued this drug and received either placebo (left panel) or TNF inhibitor (right panel): infliximab (ATTRACT),[15] adalimumab (de019),[18] or etanercept (ETA).[24] (Adapted from Smolen, J. et al. *Lancet* 2007;370:1861–74. With permission from Elsevier.)

over the treatment period. A relatively high proportion of individuals (36 percent) were unable to complete the study in the placebo group, largely due to lack of efficacy. The authors reported that 8 percent of patients receiving infliximab + MTX developed antibodies to infliximab during the trial. However, despite the presence of antibodies, similar proportions of patients with and without antibodies to infliximab achieved an ACR20 improvement.

A role for infliximab + MTX as first-line treatment of early RA has also been demonstrated. A study involving patients (n = 1049) with onset of RA symptoms in the previous 3 years and naive to MTX treatment was performed.[16] The patients received either regular infliximab (3 or 6 mg/kg) or placebo infusions with concurrent doses of MTX (target of 20 mg weekly). At week 54, 62 and 66 percent of patients receiving infliximab + MTX (3 and 6 mg/kg, respectively) achieved ACR20 responses compared to 54 percent of patients receiving placebo + MTX. In addition, greater reductions in DAS28 scores and higher remission rates were demonstrated in groups receiving combined infliximab. Significantly, both this study and the ATTRACT study demonstrated that individuals receiving infliximab + MTX showed a significant reduction in radiographic progression of disease compared to patients receiving MTX + placebo infusion.

Adalimumab (Humira, Abbott Laboratories)

Adalimumab is a fully human mAb against TNFα. This was genetically engineered through phage display biotechnology and is indistinguishable in structure and function from natural human IgG_1. It was developed under the premise that humanized or fully human mAbs should be less immunogenic than their chimeric counterparts. Adalimumab has been shown to effectively inhibit a broad range of biologic activities of TNFα and impair binding to its receptors.[17] Three early phase I therapeutic trials evaluated efficacy in a total of 198 patients with active long-standing RA. The mean duration of RA symptoms extended from 10 to 12 years and the mean number of previous DMARDs received by patients in the trials was approximately 3.5. Between 56 and 80 percent of patients treated with adalimumab, either by single or multiple infusions, demonstrated an ACR20 clinical response. In contrast, only 0 to 16 percent of placebo-treated patients achieved a similar outcome.[17]

The efficacy of combined adalimumab + MTX treatment was demonstrated in a large multicenter trial (n = 619) of patients with long-standing active RA inadequately controlled on stable MTX (mean 17 mg weekly).[18] Subjects were assigned to receive subcutaneous adalimumab (40 mg every two weeks or 20 mg weekly) + MTX or MTX + placebo. At 52 weeks, the frequency of individuals achieving ACR20 response was greater in those groups administered adalimumab (40 mg—59 percent and 20 mg—55 percent) compared with placebo (24 percent) (Figure 3.2). Therapeutic efficacy was apparent 24 hours after administration of adalimumab and response reached maximum effect within 2 weeks. Also, the rate of radiographic progression of disease was diminished in individuals receiving adalimumab. Clinical and radiographic efficacy was significantly greater compared to the placebo group. The rate of adverse events (both serious and nonserious) was comparable in the adalimumab and placebo groups, although the proportion of patients reporting serious infections was higher in patients receiving adalimumab (3.8 percent) than in those receiving placebo (0.5 percent) and was highest in the patients receiving 40 mg every other week. Discontinuations occurred in 22 percent of adalimumab-treated patients and in 30 percent of placebo-treated patients. The observed rate of formation of anti-adalimumab antibodies was 0.8 percent (3 of 397), which is approximately 10-fold less than observed in some trials with infliximab. However, 12 percent of adalimumab-treated patients (48 of 397) and 9.1 percent of placebo-treated patients (17 of 186) converted from antinuclear antibody (ANA) negative to ANA positive, whereas 5.5 percent of adalimumab-treated patients (22 of 397) and 5.9 percent of placebo-treated patients (11 of 186) converted from ANA positive to ANA negative. Among patients who had an increased baseline ANA titer, anti-dsDNA antibodies were detected in 12 percent of the adalimumab-treated patients (14 of 120) and undetected in the placebo-treated patients. No individuals developed symptoms of lupus-like illness.

A large phase III clinical trial has demonstrated the benefit of combined adalimumab + MTX therapy in patients with early active RA (<3 years onset) naive to MTX.[19] Subjects were randomized to receive adalimumab (40 mg every two weeks) + MTX (≥7.5 mg weekly), adalimumab, or MTX alone. The combination treatment arm was superior to monotherapy in all outcomes measured at 1 year. For instance, 73, 54, and 63 percent achieved ACR20 in the combined, adalimumab, and MTX groups, respectively. This trend continued to ACR90 for 2 years. The MTX dosage reached approximately 16.5 mg weekly over the course of the study. Furthermore, at 1 year of treatment, clinical remission (defined as DAS28 ≤ 2.6) was achieved in 43 percent of patients receiving combination therapy, compared with 23 percent of patients receiving adalimumab monotherapy and 21 percent of patients receiving MTX monotherapy. After 2 years of treatment, 49 percent of patients receiving combination therapy exhibited a major clinical response, compared with 25 and 27 percent of patients, respectively, in the adalimumab and MTX monotherapy groups. There was no significant difference in efficacy observed between the monotherapy trial arms. In addition, there was significantly less radiographic progression among patients receiving adalimumab than MTX alone. No statistically significant differences were observed for serious adverse events between the treatment groups. One patient in the adalimumab monotherapy group developed a lupus-like reaction with positive ANA. This study showed that combined therapy was important for both symptom control and reduction in disease progression determined by radiographic analysis in patients with early RA. Furthermore, MTX and adalimumab monotherapy proved equally efficacious in terms of clinical response. However, the latter also reduced joint destruction over a 2-year period.

Etanercept (Enbrel, Wyeth)

Etanercept is a genetically engineered fusion protein consisting of two recombinant human p75 TNF receptor extracellular domains fused with a human IgG_1-Fc moiety. This dimeric structure has been shown to be approximately 1,000 times more efficient than monomeric soluble p75 TNF receptor at neutralizing TNFα.[20] Etanercept also shows binding affinity for lymphotoxin (LT), a cytokine with high levels of homology to TNF in both structure and function. Following early human dose finding and safety studies, a trial of etanercept in patients with refractory RA demonstrated an ACR20 response of up to 75 percent at 4 months compared to 14 percent in the placebo group.[21] The treatment was given twice weekly via subcutaneous injection and was well tolerated.

Subsequent trial data demonstrated efficacy of etanercept in patients with long-standing active RA despite treatment with MTX. For instance, a 6-month study (n = 234) demonstrated 59 percent of patients receiving etanercept (25 mg twice weekly) achieved ACR20 compared to 11 percent administered placebo.[22] A subsequent larger randomized trial (n = 632) investigated efficacy of etanercept (10 or 25 mg twice weekly) versus MTX in patients with early RA naïve to MTX.[23] The frequency of patients receiving 25 mg of etanercept who had ACR20, ACR50, and ACR70 responses was significantly greater than individuals receiving MTX within the first 6 months. However, the numbers demonstrating a response after this period was similar in both groups until 1 year—ACR20 of 72 percent vs. 65 percent (P = 0.16) for etanercept and MTX, respectively. More importantly, etanercept therapy delayed the progression of joint destruction as assessed by radiographic analysis from baseline. This suggested early intervention with etanercept, as with infliximab and adalimumab, could prevent joint destruction. A number of individuals (<3 percent) were positive intermittently for nonneutralizing anti-etanercept antibodies. Their presence was not demonstrated to alter clinical response or predispose patients to adverse effects.

The synergistic effect of combined etanercept + MTX was shown in a small trial of patients with active RA on MTX (≥10 mg) for at least 6 months prior to enrollment.[24] Patients were randomized to receive either etanercept (25 mg twice weekly) + MTX or placebo + MTX. At 24 weeks, 71 percent of patients receiving combined therapy (etanercept + MTX) attained an ACR20 clinical response. In contrast, only 27 percent of patients on monotherapy (MTX + placebo) achieved the same outcome (Figure 3.2). The TEMPO (Trial of Etanercept and MTX with radiographic Patient Outcomes) study also investigated combined versus monotherapy over 1 year.[25] The trial population (n = 686) included individuals with RA ranging from 6 months to 20 years duration (mean 6.6 years). Also, patients were not excluded if they had previous exposure to MTX, provided this was stopped 6 months prior to study entry. Patients were randomized to three treatment arms: etanercept (25 mg twice weekly), MTX (7.5 mg escalated to 20 mg weekly if symptomatic), or etanercept + MTX. At 1 year, 85 percent in the combination group achieved ACR20 responses compared to 75 and 76 percent in the MTX and etanercept monotherapy treatment arms, respectively. This finding was statistically significant for combined versus monotherapy. The same trend was also seen with the disease activity score and health assessment questionnaire (a measurement of function). Patients receiving combination treatment demonstrated fewer changes associated with RA on radiographic analysis compared with groups receiving monotherapy. Previous exposure to MTX in the total study cohort did not have a significant effect on the trial outcomes. Overall, etanercept and MTX combination therapy provided greater clinical response and a reduction in radiological joint erosion than either treatment alone, which is consistent with findings using infliximab and adalimumab. Recently, 3-year data have been published from the TEMPO study, which demonstrate sustained benefit of combination therapy in terms of symptoms and joint damage.[26]

Treatment Response

Anti-TNFα biologic therapies have undoubtedly improved the quality of life and disease outcome for patients with RA. Trial data and clinical observations indicate that benefit is conferred on patients both with a recent diagnosis of RA and, more importantly, those with disease refractory to traditional DMARD treatment. Previously, many included in the latter group would have progressed to severe and disabling RA.

However, as encouraging as the introduction of the anti-TNFα drugs has been, they do not appear to be equally effective in all patients with clinically similar disease. These individuals have been stratified into distinct groups: primary nonresponders who fail to show any response, and secondary nonresponders who demonstrate initial improvement but relapse despite continued treatment. A proportion of these individuals demonstrate benefit (>ACR20 response) when switched to an alternative anti-TNFα drug. This was shown in a study of individuals treated first with infliximab + MTX who, following poor response, were switched to etanercept + MTX. The cohort included 34 primary nonresponders and 38 secondary nonresponders. After 12 weeks on etanercept + MTX, 42, 30, and 15 percent achieved ACR20, ACR50, and ACR70 responses, respectively; the percentages for the secondary nonresponse group were 34, 21, and 14 percent, respectively.[27] Synovial TNFα expression is present in the vast majority of patients with RA.[28] Consequently, these observations suggest either inadequate TNFα blockade or alternative proinflammatory cytokines driving the disease process in certain individuals. The former might be related to variability in pharmacokinetic properties of individual biologic drugs or patient-specific pharmacogenetic factors. For instance, there is some evidence to suggest polymorphisms in *TNFα* and *FcγRIIIa* might influence clinical response.[29,30] Also, there are undoubtedly pharmacodynamic differences *in vivo* between the anti-TNFα drugs. This is exemplified in other inflammatory conditions; for example, infliximab produces a dramatic clinical response in patients with Crohn's disease[31,32] but the same effect is not seen with etanercept. These observations indicate modes of action that cannot simply be equated with neutralization of soluble TNFα and likely reflect distinct effects via membrane-bound TNFα, for example, related to apoptosis of proinflammatory cell types.[33,34] It has also now been demonstrated, in both RA and Crohn's disease, that antiglobulins to infliximab are associated with infusion reactions as well as shortening the half-life and reducing overall efficacy.[35,36]

Adverse Effects of Anti-TNF Biologic Drugs

Predicting side effects of any new treatment is a difficult process. Clinical trials scrutinize subjects closely pre- and postadministration of the drug, but these studies involve relatively small sample sizes. Therefore, rare adverse events, and those which occur only with chronic treatment, may not be identified. It has been shown that patients with severe RA have an increased background risk of infection requiring hospital treatment, compared to individuals without RA.[37] However, in the case of the anti-TNFα drugs the most significant consequence of therapy has been the increased incidence of tuberculosis (TB) and other intracellular infections.[38] Extrapulmonary and disseminated forms of TB have occurred and histology is characterized by the lack of granuloma formation. Observations from murine studies have suggested the importance of transmembrane TNFα, rather than the soluble form, for the control of *Mycobacterium* infections in the acute phase.[39] However, both forms of TNFα are required to maintain long-term granuloma integrity, providing an explanation for the clinical observations.[40] The variation in incidence of TB between the anti-TNF drugs remains less well defined. At present, prospective evaluation of biologic registries suggests moderately higher rates with antibodies compared to soluble receptor TNFα antagonists.[41,42] Supportive findings from *in vitro* whole blood cultures have shown that infliximab and adalimumab have a greater inhibitory effect on TB-responsive $CD4^+$ T cells and reduce interferon gamma production to a greater extent than etanercept.[43]

The most robust way to determine the incidence of rare adverse effects is through drug registries. A number of anti-TNF registries have been established in Europe. The British Society for Rheumatology Biologics Registry (BSRBR) has so far demonstrated no overall increase in the rate of serious infections in patients receiving anti-TNF drugs, although the rate of soft tissue and skin infections is four times higher than in a control group receiving DMARDs.[41] Other complications of TNFα blockade have also been identified. There appears to be exacerbation of preexisting demyelinating central nervous system (CNS) lesions in patients with multiple sclerosis, and onset of *de novo* demyelination in a small proportion of patients.[44,45] Similarly, congestive cardiac failure is worsened by TNFα blockade.[46–48] Both effects are paradoxical because the disease state is characterized by elevated TNFα levels, blockade of which was predicted to provide benefit. Last, patients with severe RA are at increased risk of malignant lymphomas and possibly leukaemia and multiple myeloma.[49] At present, it is unclear whether TNFα blockade will increase or, via a protracted reduction in disease activity, reduce this risk.

INTERLEUKIN-1 RECEPTOR BLOCKADE—ANAKINRA (AMGEN)

Studies have indicated a pivotal role for interleukin-1 (IL-1) in the pathogenesis of RA. Interleukin-1α (IL-α) and IL-β molecules both activate downstream effector cell function through binding of IL-1 receptor (IL-1R). Endogenous IL-1 receptor antagonist (IL-1Ra) serves to regulate this proinflammatory pathway through competitive binding for IL-1R. It has been shown that >95 percent receptor occupancy is necessary to block IL-1 signaling.[50] The availability of recombinant human IL-1Ra, which differs from the native nonglycosylated human IL-1Ra only by the addition of an N-terminal methionine, has enabled evaluation of IL-1 antagonism in patients with RA.

The initial trial of anakinra indicated only modest clinical improvement in patients with active RA.[51] Patients with symptoms for 6 months to 8 years were permitted to continue nonsteroidal anti-inflammatory drugs and corticosteroids. However, DMARDs were stopped in the preceding 6 weeks to enrolment in the study. Patients were randomized to receive placebo or anakinra (30, 75, or 150 mg) daily via subcutaneous injection. At 24 weeks, 43 percent receiving 150 mg daily anakinra and 27 percent administered placebo demonstrated an ACR20 clinical response. This finding was significant ($P = 0.014$). Notably, there was a significant reduction in radiographic progression of disease from baseline even in the group receiving 30 mg anakinra daily compared to the placebo group. The latter effects were consistent with findings from *in vitro* studies on explant cartilage treated with human recombinant IL-1Ra.[52] Common adverse events included injection site reactions (25 percent placebo vs. 81 percent anakinra) and infection requiring antibiotic therapy (12 percent placebo vs. 15 to 17 percent anakinra). Increased side-effect frequency was observed in patients receiving the higher dose of anakinra. Interestingly, only 0.6 percent of patients developed anti IL-1Ra antibody after participation in the study.

The increased efficacy of combined anakinra + MTX treatment compared to placebo was demonstrated in a dosing trial incorporating patients with moderate to severe RA.[53] Subjects were randomized to five groups and received anakinra (ranging from 0.04 to 2 mg/kg daily) + MTX (15 to 25 mg weekly). At 24 weeks, 42 percent of individuals in the 2 mg/kg anakinra + MTX group achieved ACR20 response compared to 23 percent in the placebo + MTX group ($P = 0.02$) (Figure 3.3). Similar side-effect profiles were observed to the previous study. Overall, IL-1 represents an important cytokine involved in joint destruction associated with RA. Anakinra is mildly to moderately effective and well tolerated in patients with active RA when used as monotherapy or in combination with methotrexate.

TARGETING B CELLS—RITUXIMAB (MABTHERA, ROCHE)

Rituximab (RTX) is a genetically engineered chimeric (murine V-region, human C-region) anti-CD20 mAb. CD20 is a B-cell surface antigen expressed exclusively on pre-B and mature B cells. Significantly, it is not found on stem cells and is absent

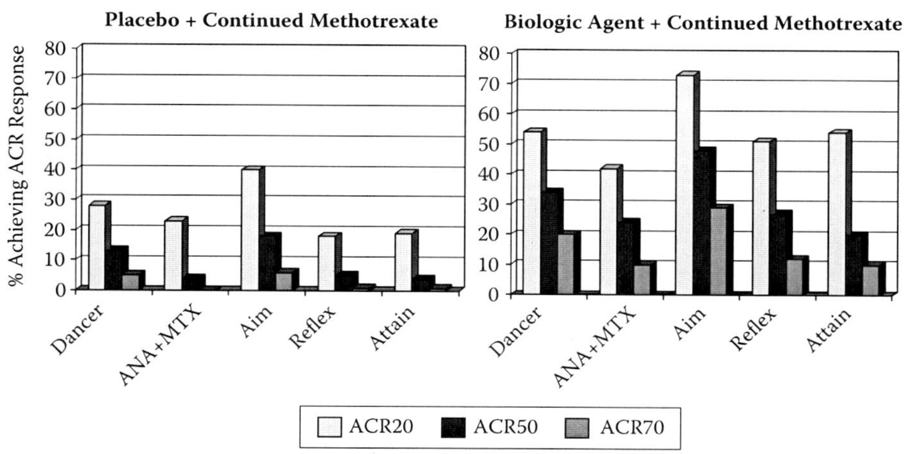

FIGURE 3.3 ***A color version of this figure follows page 176.*** Frequency and level of ACR response observed with non-TNF antagonist biologic therapies. Patients with active RA despite MTX who continued with this drug and received placebo (left panel) or biologic agent (right panel): rituximab (DANCER and REFLEX),[59,60] anakinra (ANA),[53] and abatacept (AIM and ATTAIN).[66,68] (Adapted from Smolen J, et al. *Lancet* 2007;370:1861–74. With permission from Elsevier.)

on antibody-producing plasma cells. As such, RTX targets and depletes the CD20 expressing B-cell population.[54] Currently, data suggest that cytotoxicity is a consequence of antibody-dependent cell-mediated cytotoxicity and complement-dependent lysis, with a further contribution from apoptosis.[55] Rituximab was originally licensed for the treatment of relapsed or refractory, low-grade, or follicular $CD20^+$ B cell non-Hodgkin's lymphoma and for diffuse, large B cell $CD20^+$ non-Hodgkin's lymphoma in combination with approved chemotherapy. The precise role of B cells in the pathogenesis of RA is not fully understood, although antigen presentation, (auto)antibody production, and cytokine secretion are each likely to play a part.[56]

In a small open study, five patients with refractory RA were treated with RTX, cyclophosphamide, and prednisolone, and followed for between 12 and 17 months.[57] At 26 weeks, all patients achieved at least ACR50 responses and required no additional further therapy. At 1 year, three of five patients maintained an ACR70 response and rheumatoid factor (RF) levels normalized. Interestingly, the remaining two patients relapsed with the return of B lymphocytes in the presence of elevated RF levels. However, both demonstrated ACR70 response after re-treatment. These promising findings provided the basis for a 24-week, double-blind, randomized controlled trial in 161 patients with active RA despite treatment with MTX.[58] Rituximab was administered as two 1,000 mg infusions on days 1 and 15 as monotherapy, or combined with weekly MTX or two doses of cyclophosphamide consecutively with RTX. The control group was given a stable dose of MTX alone. Prior to each RTX infusion, subjects received an infusion of methylprednisolone and were also given oral prednisolone (60 mg daily on days 2 to 7 and 30 mg daily on days 8 to 14). The trial design was based around lymphoma protocols and glucocorticoids were administered as prophylaxis against infusion reactions. ACR20 response rate was significantly ($p = <0.05$) greater in those groups receiving RTX (65, 76, and 73 percent in RTX alone, plus cyclophosphamide or MTX, respectively) compared with MTX (38 percent) monotherapy. Also, both ACR50 (43 vs. 13 percent) and ACR70 (23 vs. 5 percent) responses were significantly higher for the combined RTX + MTX group compared with MTX alone ($P = <0.01$). The highest incidence of serious adverse events occurred in the group receiving RTX + cyclophosphamide (15 percent), with similar frequencies in the remaining treatment arms (5 to 8 percent). Significantly, one patient from the group receiving RTX monotherapy died from bronchopulmonary pneumonia with no causative organism identified. Human antichimeric antibodies developed in approximately 4 percent of patients with no specific clinical consequence. The observations from this and the pilot study clearly identified B cells as key contributors in the pathogenesis of RA.

A subsequent trial investigated the influence of RTX dosage and also the role of glucocorticoids used in first dose reaction prophylaxis. The phase IIb Dose-ranging Assessment: iNternational Clinical Evaluation of Rituximab in RA (DANCER) study randomized patients ($n = 465$) on stable MTX to one of nine groups.[59] Briefly, individuals were assigned to receive placebo infusions, 500 mg of RTX or 1,000 mg of RTX on days 1 and 15. Furthermore, in each of these three groups: one third of patients received no prophylaxis against first-dose reactions, one third received 100 mg iv methylprednisolone prior to each RTX/placebo infusion, and one third received both iv and oral corticosteroid prophylaxis. Patients with mean disease duration of

11 years were enrolled in the study having previously received 2.5 DMARDs (mean value for RTX arms). Approximately 30 percent of subjects had also previously been treated with TNFα blockade and demonstrated lack or loss of response. At the 24 week primary end point, ACR response rates for placebo, 500 mg, and 1 g RTX groups were 28, 55, 54 percent (ACR20), 13, 33, 34 percent (ACR50), and 5, 13, 20 percent (ACR70) (Figure 3.3). First dose reactions were clearly reduced (although not prevented) with corticosteroid premedication but there was no advantage to receiving oral as well as iv prophylaxis. Furthermore, glucocorticoid premedication did not contribute toward overall efficacy. There were more serious adverse events reported in the RTX groups, but these were distributed evenly across body systems and did not follow a particular pattern. There was no difference between the groups in serious infection rate.

The Randomised Evaluation oF Long-term Efficacy of rituXimab in RA (REFLEX) phase III trial incorporated patients with active RA and an inadequate response to one or more anti-TNF agents.[60] Individuals received either two infusions of RTX (1,000 mg) or placebo on day 1 and 15 and remained on MTX (10 to 25 mg weekly). Oral and iv glucocorticoids were administered as prophylaxis against infusion reactions. At 24 weeks, the proportion of patients with an ACR20 response was significantly higher for individuals receiving RTX + MTX (51 percent) than placebo + MTX (18 percent). In addition, the frequency of patients achieving ACR50 and ACR70 responses at week 24 was also significantly higher in the former group (27 percent vs. 5 percent and 12 percent vs. 1 percent, for ACR50 and ACR70, respectively) (Figure 3.3). Also, more patients receiving RTX achieved good or moderate EULAR responses compared with placebo (65 percent vs. 22 percent, $P = <0.01$). Notably, despite a decrease in the rheumatoid factor (RF) level (mean 55 percent) of the group receiving RTX from baseline, linear regression analysis suggested the effect of treatment on the ACR response was not dependent on RF status. Also, selective B-cell depletion was not found to produce a decrease in mean levels of circulating immunoglobulin in this study. Although approximately 4 percent of individuals were found to have positive HACA titers, this did not have any apparent effect on ACR response, side effects, or peripheral B-cell counts. The frequency of adverse events appeared similar in both treatment arms with no reports of TB or opportunistic infections. In addition, an extension to 56 weeks in this study indicated a statistically significant slowing of radiographic progression in the RTX group as measured by the Sharp-Genant score.[61]

As the DANCER and REFLEX studies have demonstrated, RTX treatment appears to be efficacious in patients with established RA refractory to numerous DMARDs and anti-TNF biologic therapy, providing patients and clinicians with a distinct treatment approach. The outcome of repeat RTX administration in patients who have relapsed has also been investigated.[62] Re-treatment required symptomatic relapse because no reliable biomarkers of impending disease flare, such as CD20 count of RF levels, have been identified. The open-label extension study observed the clinical response after 24 weeks from additional treatment (1,000 mg dose RTX 2 weeks apart). In total, 570 received two courses, 191 received three courses, and 40 received four courses. ACR20 responses were comparable at week 24 after course 1 and at week 24 after course 2 (65 percent vs. 72 percent). EULAR moderate/good responses were also comparable in course 2 relative to course 1 (88 percent vs. 79 percent). Notably, the proportion of patients with circulating IgM and IgG levels below the lower limit of normal increased with subsequent courses. However, infection rates in these patients were comparable with those in which immunoglobulin levels remained within the normal range. Also, the patients who developed HACA (9.2 percent) did not exhibit decreased efficacy of response or present additional safety concerns.

COSTIMULATION BLOCKADE—ABATACEPT (ORENCIA, BRISTOL-MYERS SQUIBB)

T cells play a central role in immune responses. They are broadly divided into $CD4^+$ and $CD8^+$ subsets; $CD4^+$ T cells are involved in coordinating the downstream immune response and $CD8^+$ T cells have a cytotoxic function. T-cell activation requires at least two signals to be delivered via surface receptors. The first is an antigen-specific signal received when the unique, clonotypic, T-cell receptor recognizes an appropriate MHC–peptide complex on an antigen-presenting cell (APC). The second, or costimulatory, signal needs to be delivered to the T cell by CD80 (B7.1) and CD86 (B7.2) on the APC binding to T-cell CD28.[63] An important concept in T-cell biology is that partial T-cell activation (e.g., signal 1 without signal 2) may temporarily or permanently prevent subsequent T-cell activation, even if all necessary signals are delivered at a later time point.[64] Following activation, T cells upregulate a cell surface molecule cytotoxic T-lymphocyte associated antigen-4 (CTLA-4). This is a negative regulator of T-cells that has a higher affinity than CD28 for CD80 and CD86 (Figure 3.4). This led to the concept and development of abatacept, a fusion protein containing the extracellular domain of CTLA-4 and the human IgG_1 Fc. The Fc domain of abatacept has been modified to prevent complement activation, thereby potentially reducing infusion reactions. This molecule has the potential to disrupt CD28:CD80/86-mediated costimulation and prevent, even possibly reverse, T-cell activation.

In a phase IIb trial, patients with long-standing RA on a stable dose of MTX and persisting disease activity were randomized to receive infusions of either abatacept (2 or 10 mg/kg) or placebo, on days 1, 15, and 29 and then every 28 days.[65] The primary end point at 6 months indicated an ACR20 response of 42 and 60 percent in the abatacept + MTX treatment groups (2 and 10 mg/kg, respectively) compared to 35 percent in the placebo + MTX group. The rates of improvement for 10 mg/kg abatacept + MTX were significantly superior to the placebo response ($P = <0.01$). There was no difference between the three groups in adverse response rates.

Two phase III clinical trials of abatacept have been reported. In the Abatacept in Inadequate responders to Methotrexate (AIM) trial, 652 patients with active RA on a stable dose of MTX were randomized to receive 10 mg/kg abatacept or placebo according to the regimen in the phase IIb study.[66] After 1 year, response rates in placebo and abatacept groups were 40 and 73 percent (ACR20), 18 and 48 percent (ACR50), and 6 and 29 percent (ACR70) (Figure 3.3). Importantly, abatacept-treated patients showed reduction of progression of radiological joint damage compared to placebo-treated patients, measured using the Sharp-Genant score. Overall adverse

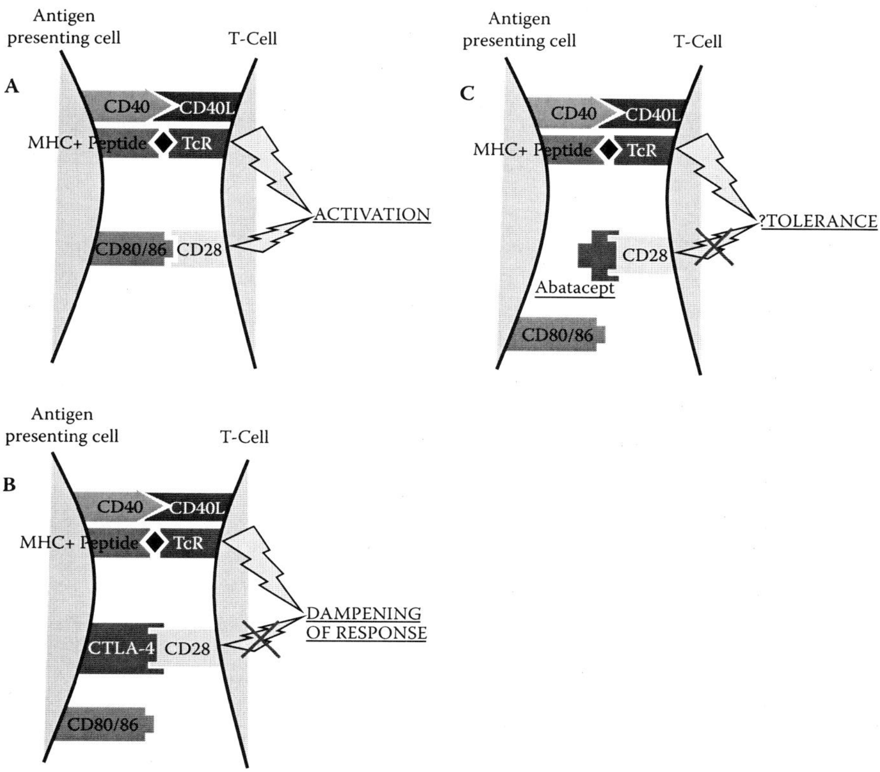

FIGURE 3.4 Critical surface interactions leading to T-cell activation by an APC. (A) In addition to interaction of the TcR and MHC-peptide complex, a second, or costimulatory signal, is required to produce full T-cell activation. The main second signal is generated through interaction of CD80 and CD86 on the APC with CD28 on the T cell. CD40 and CD40 ligand (CD40L) provide additional signals. (B) T-cell activation upregulates CTLA-4, which has greater affinity for binding CD28 than CD80/86. CTLA-4 serves to downregulate the immune response. (C) Abatacept, a CTLA-4 fusion protein, is hypothesized to block production of the second signal following the TcR/MHC-peptide complex interaction. This inhibits T-cell activation and might lead to immune tolerance or anergy of autoreactive T cells.

event rates were similar in the two groups although abatacept-treated patients had a higher incidence of acute infusion reactions (8.8 vs. 4.1 percent) and prespecified serious infections (2.5 vs. 0.9 percent) compared to placebo recipients. The radiological benefits of abatacept have been shown to extend to 2 years.[67]

The phase III Abatacept Trial in Treatment of Anti-TNF INadequate responders (ATTAIN) study randomized patients with a current or previous inadequate response to a TNFα antagonist to receive placebo or 10 mg/kg abatacept, again using the same dosing regime.[68] Patients were washed out from TNFα antagonists and continued to receive stable doses of a variety of DMARDs, including anakinra. At 6 months, response rates in the placebo and abatacept groups were 19 and 54 percent (ACR20), 4 and 20 percent (ACR50), and 1 and 10 percent (ACR70) (Figure 3.3). Acute infusion reactions and infections were nonsignificantly higher in abatacept recipients but serious adverse events and serious infections were similar in both groups.[69]

Abatacept is now licensed for the treatment of RA, the first biologic drug that targets the T cell–APC interaction. Intriguingly and, unlike most other biologics, the response rate and quality of response appear to increase with time on abatacept. For example, patients achieving an ACR20 response at 3 months may achieve ACR50 at 6 months and there is a steady increase in response rate, at least during the first year of therapy. Furthermore, data from long-term extension studies suggest a sustained rate of response at 2 and 3 years by intention-to-treat analysis, and an increased rate of response according to the less robust "as-observed" analysis. Time will tell whether abatacept is truly a tolerance-inducing drug. In this case, if a patient taking abatacept achieves remission, it should be possible to interrupt therapy without loss of therapeutic benefit.

Thus far, the side-effect profile of abatacept appears acceptable when used in combination with MTX or other conventional DMARDs. Infusion reactions are rare and there does not appear to be a significantly increased risk of infection or malignancy. Its use in combination with TNFα blockade, however, was associated with a doubling of serious adverse events without a corresponding improvement in efficacy.[70] The latter finding was investigated further in a randomized, placebo-controlled, double-blind trial of patients on etanercept (25 mg twice weekly) with active RA.[71] Subjects (n = 121) received either supplemental low-dose abatacept (2 mg/kg) or placebo for 1 year. No significant difference in ACR response was noted between the two groups at the primary endpoint. However, a greater frequency of individuals receiving abatacept + etanercept compared with placebo + etanercept experienced SAE (16.5 versus 2.8 percent, respectively). Based on these observations, the authors recommend that abatacept should not be used together with etanercept for treatment of RA. The mechanism by which this therapeutic combination increases adverse side effects is not yet clear.

EMERGING THERAPIES

A number of biologic therapies are in phase II and III development for RA. Tocilizumab (TCZ) is a humanized mAb that binds to the IL-6 receptor. The downstream effects of IL-6 receptor signaling include B-cell stimulation, T-cell growth, dendritic cell maturation, acute phase response induction, angiogenesis, and bone resorption.[72] IL-6 levels are elevated in the blood and synovial fluid of patients with RA, and correlate with symptoms and radiographic damage.[73,74] Therefore, the blockade of IL-6 or its receptor should have beneficial effects on RA. The phase II CHARISMA study compared three dose levels of TCZ (2, 4, and 8 mg/kg by monthly infusion), with or without MTX, with MTX alone in patients with active RA and a previously unsatisfactory response to MTX.[75] At 16 weeks, with the exception of patients receiving 2 mg/kg TCZ monotherapy, all active treatment groups demonstrated a superior reduction in DAS28 score compared to patients continuing MTX monotherapy. By using ACR20 responses, all TCZ/MTX combinations were superior to MTX monotherapy but 8 mg/kg + MTX was the only group to demonstrate superiority at ACR70. A study from Japan (SAMURAI) has shown protection from joint damage in patients receiving TCZ.[76]

Yet to be licensed anti-TNF agents include golimumab, a humanized mAb,[77] and certolizumab pegol, an antibody Fab fragment conjugated to polyethylene glygol.[78] The latter agent is unique in having no Fc domain and, therefore, acting as a pure neutralizing agent devoid of effector function. In clinical trials it has demonstrated efficacy similar to the currently licensed anti-TNF drugs.

Last, a number of drugs in development are designed to target B cells or their growth factors. Ocrelizumab is a humanized anti-B-cell mAb currently in phase III trials[79] and ofatumumab is a fully human mAb derived from human immunoglobulin transgenic mice.[80] B-cell growth factors include B Lymphocyte Stimulator (BLyS; also known as B-cell Activating Factor of the TNF Family, or BAFF) and A PRoliferation Inducing Ligand (APRIL). BLyS binds to all three growth factor receptors on B cells: BAFF-R, B-cell maturation antigen (BCMA), and Transmembrane Activator and CalcIum modulator and cyclophilin ligand activator (TACI). APRIL binds only BCMA and TACI. Compounds in early stage development include an anti-BLyS mAb (belimumab), and soluble forms of BAFF-R (B3-Fc) and TACI (TACI-Ig). The former soluble receptor binds and neutralizes BLyS whereas TACI-Ig binds both BLyS and APRIL, with potentially more potent effects.[81]

CONCLUSIONS

Biologic therapies have had a major impact on rheumatological disease. They have brought new hope to tens of thousands of RA sufferers who have experienced suboptimal responses to conventional therapies. The ability of these new compounds to prevent joint damage is just as important as their anti-inflammatory effects, promising reduced disability as well as a greatly reduced need for joint surgery. We continue to learn about and understand the therapeutic ratios of these drugs, whose potency at blocking key pathways could also reveal unexpected adverse effects. A major bonus to having targeted therapies such as these is a newfound ability to dissect pathological processes in the chronic inflammatory environment that is RA. For example, the advent of B-cell targeting confirmed and extended prevailing thoughts as to the role of B cells in RA. A less satisfying aspect of these drugs is their current cost, and the fact that not all patients experience beneficial outcomes of therapy. In the not-too-distant future we can expect to reap the rewards of large-scale pharmacogenetic studies which, on the one hand, will allow us to prescribe these drugs more rationally while, on the other, reveal further, fascinating insights into disease heterogeneity.

REFERENCES

1. Gabriel SE. The epidemiology of rheumatoid arthritis. *Rheum Dis Clin North Am* 2001;27:269–81.
2. Choy EH, Panayi GS. Cytokine pathways and joint inflammation in rheumatoid arthritis. *N Engl J Med* 2001;344:907–16.
3. Smolen JS, Aletaha D, Koeller M, Weisman MH, Emery P. New therapies for treatment of rheumatoid arthritis. *Lancet* 2007;370:1861–74.

4. McInnes IB, Schett G. Cytokines in the pathogenesis of rheumatoid arthritis. *Nature Rev* 2007;7:429–42.
5. Aletaha D, Smolen JS. The definition and measurement of disease modification in inflammatory rheumatic diseases. *Rheum Dis Clin North Am* 2006;32:9–44, vii.
6. Brennan FM, Chantry D, Jackson A, Maini R, Feldmann M. Inhibitory effect of TNF alpha antibodies on synovial cell interleukin-1 production in rheumatoid arthritis. *Lancet* 1989;2:244–47.
7. Williams RO, Feldmann M, Maini RN. Anti-tumor necrosis factor ameliorates joint disease in murine collagen-induced arthritis. *Proc Natl Acad Sci USA* 1992;89:9784–88.
8. Keffer J, Probert L, Cazlaris H, et al. Transgenic mice expressing human tumour necrosis factor: A predictive genetic model of arthritis. *EMBO J* 1991;10:4025–31.
9. Tracey D, Klareskog L, Sasso EH, Salfeld JG, Tak PP. Tumor necrosis factor antagonist mechanisms of action: A comprehensive review. *Pharmacol Therap* 2008;117:244–79.
10. Aderka D, Engelmann H, Maor Y, Brakebusch C, Wallach D. Stabilization of the bioactivity of tumor necrosis factor by its soluble receptors. *J Exp Med* 1992;175:323–29.
11. Elliott MJ, Maini RN, Feldmann M, et al. Treatment of rheumatoid arthritis with chimeric monoclonal antibodies to tumor necrosis factor alpha. *Arthritis Rheum* 1993;36:1681–90.
12. Elliott MJ, Maini RN, Feldmann M, et al. Randomised double-blind comparison of chimeric monoclonal antibody to tumour necrosis factor alpha (ca2) versus placebo in rheumatoid arthritis. *Lancet* 1994;344:1105–10.
13. Elliott MJ, Maini RN, Feldmann M, et al. Repeated therapy with monoclonal antibody to tumour necrosis factor alpha (ca2) in patients with rheumatoid arthritis. *Lancet* 1994;344:1125–27.
14. Maini RN, Breedveld FC, Kalden JR, et al. Therapeutic efficacy of multiple intravenous infusions of anti-tumor necrosis factor alpha monoclonal antibody combined with low-dose weekly methotrexate in rheumatoid arthritis. *Arthritis Rheum* 1998;41:1552–63.
15. Lipsky PE, van der Heijde DM, St Clair EW, et al. Infliximab and methotrexate in the treatment of rheumatoid arthritis. Anti-tumor necrosis factor trial in rheumatoid arthritis with concomitant therapy study group. *N Engl J Med* 2000;343:1594–602.
16. St Clair EW, van der Heijde DM, Smolen JS, et al. Combination of infliximab and methotrexate therapy for early rheumatoid arthritis: A randomized, controlled trial. *Arthritis Rheum* 2004;50:3432–43.
17. Kempeni J. Update on d2e7: A fully human anti-tumour necrosis factor alpha monoclonal antibody. *Ann Rheum Dis* 2000;59 Suppl 1:i44–45.
18. Keystone EC, Kavanaugh AF, Sharp JT, et al. Radiographic, clinical, and functional outcomes of treatment with adalimumab (a human anti-tumor necrosis factor monoclonal antibody) in patients with active rheumatoid arthritis receiving concomitant methotrexate therapy: A randomized, placebo-controlled, 52-week trial. *Arthritis Rheum* 2004;50:1400–11.
19. Breedveld F, Weisman M, Kavanaugh A, et al. The premier study: A multicenter, randomised double-blind clinical trial of combination therapy with adalimumab plus methotrexate versus methotrexate alone or adalimumab alone in patients with early, aggressive rheumatoid arthritis who had not had previous methotrexate treatment. *Arth Rheum* 2006;54:26–37.
20. Mohler KM, Torrance DS, Smith CA, et al. Soluble tumor necrosis factor (TNF) receptors are effective therapeutic agents in lethal endotoxemia and function simultaneously as both TNF carriers and TNF antagonists. *J Immunol* 1993;151:1548–61.
21. Moreland LW, Baumgartner SW, Schiff MH, et al. Treatment of rheumatoid arthritis with a recombinant human tumor necrosis factor receptor (p75)-fc fusion protein. *N Engl J Med* 1997;337:141–47.

22. Moreland LW, Schiff MH, Baumgartner SW, et al. Etanercept therapy in rheumatoid arthritis. A randomized, controlled trial. *Ann Intern Med* 1999;130:478–86.
23. Bathon JM, Martin RW, Fleischmann RM, et al. A comparison of etanercept and methotrexate in patients with early rheumatoid arthritis. *N Engl J Med* 2000;343:1586–93.
24. Weinblatt ME, Kremer JM, Bankhurst AD, et al. A trial of etanercept, a recombinant tumor necrosis factor receptor:Fc fusion protein, in patients with rheumatoid arthritis receiving methotrexate. *N Engl J Med* 1999;340:253–59.
25. Klareskog L, van der Heijde D, de Jager JP, et al. Therapeutic effect of the combination of etanercept and methotrexate compared with each treatment alone in patients with rheumatoid arthritis: Double-blind randomised controlled trial. *Lancet* 2004;363:675–81.
26. van der Heijde D, Klareskog L, Landewe R, et al. Disease remission and sustained halting of radiographic progression with combination etanercept and methotrexate in patients with rheumatoid arthritis. *Arthritis Rheum* 2007;56:3928–39.
27. Buch MH, Bingham SJ, Bejarano V, et al. Therapy of patients with rheumatoid arthritis: Outcome of infliximab failures switched to etanercept. *Arthritis Rheum* 2007;57:448–53.
28. Ulfgren AK, Andersson U, Engstrom M, Klareskog L, Maini RN, Taylor PC. Systemic anti-tumor necrosis factor alpha therapy in rheumatoid arthritis down-regulates synovial tumor necrosis factor alpha synthesis. *Arthritis Rheum* 2000;43:2391–96.
29. Mugnier B, Balandraud N, Darque A, Roudier C, Roudier J, Reviron D. Polymorphism at position -308 of the tumor necrosis factor alpha gene influences outcome of infliximab therapy in rheumatoid arthritis. *Arthritis Rheum* 2003;48:1849–52.
30. Tutuncu Z, Kavanaugh A, Zvaifler N, Corr M, Deutsch R, Boyle D. Fcgamma receptor type iiia polymorphisms influence treatment outcomes in patients with inflammatory arthritis treated with tumor necrosis factor alpha-blocking agents. *Arthritis Rheum* 2005;52:2693–96.
31. Targan S, Hanauer S, van Deventer S, et al. A short-term study of chimeric monoclonal antibody ca2 to tumor necrosis factor for Crohn's disease. *N Engl J Med* 1997;337:1029–36.
32. Present D, Rutgeerts P, Targan S, et al. Infliximab for the treatment of fistulas in patients with Crohn's disease. *N Engl J Med* 1999;340:1398–405.
33. Scallon B, Moore M, Trinh H, Knight D, Ghrayeb J. Chimeric anti-TNF-alpha monoclonal antibody ca2 binds recombinant transmembrane TNF-alpha and activates immune effector functions. *Cytokine* 1995;7:251–59.
34. Shen C, Assche GV, Colpaert S, et al. Adalimumab induces apoptosis of human monocytes: A comparative study with infliximab and etanercept. *Aliment Pharmacol Ther* 2005;21:251–58.
35. Baert F, Noman M, Vermeire S, et al. Influence of immunogenicity on the long-term efficacy of infliximab in Crohn's disease. *N Engl J Med* 2003;348:601–8.
36. Wolbink GJ, Vis M, Lems W, et al. Development of antiinfliximab antibodies and relationship to clinical response in patients with rheumatoid arthritis. *Arthritis Rheum* 2006;54:711–15.
37. Doran MF, Crowson CS, Pond GR, O'Fallon WM, Gabriel SE. Frequency of infection in patients with rheumatoid arthritis compared with controls: A population-based study. *Arthritis Rheum* 2002;46:2287–93.
38. Keane J, Gershon S, Wise RP, et al. Tuberculosis associated with infliximab, a tumor necrosis factor alpha-neutralizing agent. *N Engl J Med* 2001;345:1098–104.
39. Olleros ML, Guler R, Vesin D, et al. Contribution of transmembrane tumor necrosis factor to host defense against *Mycobacterium bovis* bacillus Calmette-Guerin and *Mycobacterium tuberculosis* infections. *Am J Pathol* 2005;166:1109–20.

40. Saunders BM, Tran S, Ruuls S, Sedgwick JD, Briscoe H, Britton WJ. Transmembrane TNF is sufficient to initiate cell migration and granuloma formation and provide acute, but not long-term, control of *Mycobacterium tuberculosis* infection. *J Immunol* 2005;174:4852–59.
41. Dixon WG, Watson K, Lunt M, Hyrich KL, Silman AJ, Symmons DP. Rates of serious infection, including site-specific and bacterial intracellular infection, in rheumatoid arthritis patients receiving anti-tumor necrosis factor therapy: Results from the British Society for Rheumatology Biologics Register. *Arthritis Rheum* 2006;54:2368–76.
42. Dinarello CA. Differences between anti-tumor necrosis factor-alpha monoclonal antibodies and soluble TNF receptors in host defense impairment. *J Rheumatol* 2005;74:40–47.
43. Saliu OY, Sofer C, Stein DS, Schwander SK, Wallis RS. Tumor-necrosis-factor blockers: Differential effects on mycobacterial immunity. *J Infect Dis* 2006;194:486–92.
44. Mohan N, Edwards ET, Cupps TR, et al. Demyelination occurring during anti-tumor necrosis factor alpha therapy for inflammatory arthritides. *Arthritis Rheum* 2001;44:2862–69.
45. Robinson W, Genovese M, Moreland L. Demyelinating and neurologic events reported in association with tumor necrosis factor alpha antagonism: By what mechanisms could tumor necrosis factor alpha antagonists improve rheumatoid arthritis but exacerbate multiple sclerosis? *Arthritis Rheum* 2001;44:1977–83.
46. Chung E, Packer M, Lo K, Fasanmade A, Willerson J. Anti-TNF therapy against congestive heart failure investigators. Randomized, double-blind, placebo-controlled, pilot trial of infliximab, a chimeric monoclonal antibody to tumor necrosis factor-alpha, in patients with moderate-to-severe heart failure: Results of the anti-TNF therapy against congestive heart failure (ATTACH) trial. *Circulation* 2003;107:3133–40.
47. Mann D, McMurray J, Packer M, et al. Targeted anticytokine therapy in patients with chronic heart failure: Results of the randomized etanercept worldwide evaluation (renewal). *Circulation* 2004;109:1594–602.
48. Kwon H, Coté T, Cuffe M, Kramer J, Braun M. Case reports of heart failure after therapy with a tumor necrosis factor antagonist. *Ann Intern Med* 2003;138:807–11.
49. Askling J, Fored CM, Baecklund E, et al. Haematopoietic malignancies in rheumatoid arthritis: Lymphoma risk and characteristics after exposure to tumour necrosis factor antagonists. *Ann Rheum Dis* 2005;64:1414–20.
50. Arend WP, Malyak M, Guthridge CJ, Gabay C. Interleukin-1 receptor antagonist: Role in biology. *Annu Rev Immunol* 1998;16:27–55.
51. Bresnihan B, Alvaro-Garcia JM, Cobby M, et al. Basic science—Treatment of rheumatoid arthritis with recombinant human interleukin-1 receptor antagonist. *Arthritis Rheum* 1998;41:2196.
52. Smith RJ, Chin JE, Sam LM, Justen JM. Biologic effects of an interleukin-1 receptor antagonist protein on interleukin-1-stimulated cartilage erosion and chondrocyte responsiveness. *Arthritis Rheum* 1991;34:78–83.
53. Cohen S, Hurd E, Cush J, et al. Treatment of rheumatoid arthritis with anakinra, a recombinant human interleukin-1 receptor antagonist, in combination with methotrexate: Results of a twenty-four-week, multicenter, randomized, double-blind, placebo-controlled trial. *Arthritis Rheum* 2002;46:614–24.
54. Silverman GJ. Therapeutic B cell depletion and regeneration in rheumatoid arthritis: Emerging patterns and paradigms. *Arthritis Rheum* 2006;54:2356–67.
55. Maloney DG, Smith B, Rose A. Rituximab: Mechanism of action and resistance. *Semin Oncol* 2002;29:2–9.
56. Panayi GS. B cells: A fundamental role in the pathogenesis of rheumatoid arthritis? *Rheumatology* (Oxford) 2005;44 Suppl 2:ii3–ii7.

57. Edwards JC, Cambridge G. Sustained improvement in rheumatoid arthritis following a protocol designed to deplete B lymphocytes. *Rheumatology* (Oxford) 2001;40:205–11.
58. Edwards JC, Szczepanski L, Szechinski J, et al. Efficacy of B-cell-targeted therapy with rituximab in patients with rheumatoid arthritis. *N Engl J Med* 2004;350:2572–81.
59. Emery P, Fleischmann R, Filipowicz-Sosnowska A, et al. The efficacy and safety of rituximab in patients with active rheumatoid arthritis despite methotrexate treatment: Results of a phase IIb randomized, double-blind, placebo-controlled, dose-ranging trial. *Arthritis Rheum* 2006;54:1390–400.
60. Cohen SB, Emery P, Greenwald MW, et al. Rituximab for rheumatoid arthritis refractory to anti-tumor necrosis factor therapy: Results of a multicenter, randomized, double-blind, placebo-controlled, phase III trial evaluating primary efficacy and safety at twenty-four weeks. *Arthritis Rheum* 2006;54:2793–806.
61. Cohen S, Dougados M, Genovese M, et al. Consistent inhibition of structural damage progression by rituximab in medically important subgroups of patients with an inadequate response to TNF inhibitors: Week 56 reflex results. *Ann Rheum Dis* 2007;66:428.
62. Keystone E, Fleischmann R, Emery P, et al. Safety and efficacy of additional courses of rituximab in patients with active rheumatoid arthritis: An open-label extension analysis. *Arthritis Rheum* 2007;56:3896–908.
63. Bretscher PA. A two-step, two-signal model for the primary activation of precursor helper T cells. *Proc Natl Acad Sci USA* 1999;96:185–90.
64. Mueller DL, Jenkins MK, Schwartz RH. Clonal expansion versus functional clonal inactivation: A costimulatory signalling pathway determines the outcome of T cell antigen receptor occupancy. *Annu Rev Immunol* 1989;7:445–80.
65. Kremer JM, Westhovens R, Leon M, et al. Treatment of rheumatoid arthritis by selective inhibition of T-cell activation with fusion protein CTLA4Ig. *N Engl J Med* 2003;349:1907–15.
66. Kremer JM, Genant HK, Moreland LW, et al. Effects of abatacept in patients with methotrexate-resistant active rheumatoid arthritis: A randomized trial. *Ann Intern Med* 2006;144:865–76.
67. Genant HK, Peterfy CG, Westhovens R, et al. Abatacept inhibits structural damage progression in rheumatoid arthritis: Results from the long-term extension of the AIM trial. *Ann Rheum Dis* 2008;67:1084–9.
68. Genovese MC, Becker JC, Schiff M, et al. Abatacept for rheumatoid arthritis refractory to tumor necrosis factor alpha inhibition. *N Engl J Med* 2005;353:1114–23.
69. Genovese MC, Schiff M, Luggen M, et al. Efficacy and safety of the selective co-stimulation modulator abatacept following 2 years of treatment in patients with rheumatoid arthritis and an inadequate response to anti-TNF therapy. *Ann Rheum Dis* 2008;67:547–54.
70. Weinblatt M, Combe B, Covucci A, Aranda R, Becker JC, Keystone E. Safety of the selective costimulation modulator abatacept in rheumatoid arthritis patients receiving background biologic and nonbiologic disease-modifying antirheumatic drugs: A one-year randomized, placebo-controlled study. *Arthritis Rheum* 2006;54:2807–16.
71. Weinblatt M, Schiff M, Goldman A, et al. Selective costimulation modulation using abatacept in patients with active rheumatoid arthritis while receiving etanercept: A randomised clinical trial. *Ann Rheum Dis* 2007;66:228–34.
72. Smolen JS, Steiner G. Therapeutic strategies for rheumatoid arthritis. *Nat Rev Drug Discov* 2003;2:473–88.
73. Houssiau FA, Devogelaer JP, Van Damme J, de Deuxchaisnes CN, Van Snick J. Interleukin-6 in synovial fluid and serum of patients with rheumatoid arthritis and other inflammatory arthritides. *Arthritis Rheum* 1988;31:784–88.

74. Madhok R, Crilly A, Watson J, Capell HA. Serum interleukin 6 levels in rheumatoid arthritis: Correlations with clinical and laboratory indices of disease activity. *Ann Rheum Dis* 1993;52:232–34.
75. Maini RN, Taylor PC, Szechinski J, et al. Double-blind randomized controlled clinical trial of the interleukin-6 receptor antagonist, tocilizumab, in European patients with rheumatoid arthritis who had an incomplete response to methotrexate. *Arthritis Rheum* 2006;54:2817–29.
76. Nishimoto N, Hashimoto J, Miyasaka N, et al. Study of active controlled monotherapy used for rheumatoid arthritis, an IL-6 inhibitor (samurai): Evidence of clinical and radiographic benefit from an x ray reader-blinded randomised controlled trial of tocilizumab. *Ann Rheum Dis* 2007;66:1162–67.
77. Zhou H, Jang H, Fleischmann RM, et al. Pharmacokinetics and safety of golimumab, a fully human anti-TNF-alpha monoclonal antibody, in subjects with rheumatoid arthritis. *J Clin Pharmacol* 2007;47:383–96.
78. Choy EH, Hazleman B, Smith M, et al. Efficacy of a novel pegylated humanized anti-TNF fragment (cdp870) in patients with rheumatoid arthritis: A phase II double-blinded, randomized, dose-escalating trial. *Rheumatology* (Oxford) 2002;41:1133–37.
79. Genovese M, Kaine J, Lowenstein M, et al. Safety, pharmacodynamics, and clinical activity of ocrelizumab, a humanized anti-CD20 antibody in patients with moderate-severe rheumatoid arthritis receiving methotrexate: A phase I/II dose-ranging controlled trial. *Ann Rheum Dis* 2007;66:430.
80. Østergaard M, Baslund B, Rigby W, et al. Ofatumumab (humax-CD20), a fully human CD20 monoclonal antibody, in the treatment of rheumatoid arthritis. Early results from an ongoing, double-blind, randomized, placebo controlled clinical trial. *Ann Rheum Dis* 2007;66:124.
81. Tak PP, Thurlings RM, Rossier C, et al. Atacicept in patients with rheumatoid arthritis: Results of a multicenter, phase Ib, double-blind, placebo-controlled, dose-escalating, single- and repeated-dose study. *Arthritis Rheum* 2008;58:61–72.

4 Biologic Therapies in Systemic Lupus Erythematosus

Robert Eisenberg

CONTENTS

KEY WORDS: systemic lupus erythematosus, B cells, biologics

ABSTRACT

Systemic lupus erythematosus (SLE) is a complex autoimmune disease that has been difficult to treat. Because no new therapies have been approved for decades, the potential for new biologic agents that are active in this disease represents an exciting opportunity. Potential targets include B cells or T cells, growth factors, costimulatory molecules, cytokines, interferons, complement, and specific antigen reactivities. Several agents are in advanced stages of development, and positive results from pharmaceutical industry-sponsored randomized clinical trials could be forthcoming over the next several years. We will need to understand better the biology of these agents, and develop additional biomarkers to monitor their use. If a novel biologic proves to be sufficiently efficacious to receive Food and Drug Administration (FDA) approval for the treatment of SLE, it would represent a major advance for patients with SLE and their physicians.

SYSTEMIC LUPUS ERYTHEMATOSUS

Systemic lupus erythematosus (SLE) is a multisystem autoimmune disease that particularly affects young women in their third and fourth decades of life. It is characterized by a spectrum of autoantibodies that target many normal cellular components, particularly constituents of intranuclear complexes of nucleic acids and protein. Although certain autoantibodies, such as anti-DNA and anti-Sm, are sufficiently closely associated with SLE to have diagnostic utility, most of the several dozen specificities identified have an unclear role in the pathogenesis of the clinical manifestations of the diseases. Glomerulonephritis, one of the more common serious internal organ manifestations in SLE, involves the extensive deposition of immunoglobulins in the glomerular tufts, and this is at least in part related to anti-dsDNA autoantibodies.[1] However, other autoantibody species almost certainly play a role in renal involvement.[2] Furthermore, the pathogenesis of many of the other organ dysfunctions has not been convincingly demonstrated to be antibody mediated. Nevertheless, it is accepted that B cells are central to the overall pathogenesis, and thus they or molecules associated with them are reasonable targets for therapy. T cells are also almost certainly critically involved, although what role they might play beyond providing help for autoantibody production is still unclear. It has been frustratingly difficult to prove that autoreactive T cells exist in patients with SLE, or, if they do, that they are specific to them.

The investigation of SLE has been greatly aided by the existence of animal models, almost all in mice.[3] Several inbred mouse strains spontaneously develop the autoantibodies characteristic of SLE, including the diagnostic anti-dsDNA and anti-Sm specificities. The best-studied examples are the MRL/*lpr*, the NZM2410, and the (NZBxNZW)F1 hybrid. The disease is driven by the genetic makeup of these strains, most with multiple loci each contributing a modest part of the overall genetic risk. The *lpr* (a mutant form of the Fas death receptor) gene is an exception, in that it can make a large difference when added to an otherwise susceptible genetic background as in the MRL strain.[4] However, abnormalities in *Fas* do not appear to contribute to SLE in humans.[5] Many of the other, less potent, genetic loci are beginning to be identified by genetic mapping studies in mice and in humans.[6] They will likely turn out to be polymorphic variants, rather than dysfunctional mutations, and thus their potential as therapeutic targets might be unpredictable. In addition, the contributing genes in different SLE mouse models are likely to be largely nonoverlapping. In fact, the loci identified in a given model depend very much on the "normal" strain used for backcrossing, and paradoxically, some of the identified gene variants that are identified as risk factors actually come from the nonautoimmune breeding partner. In addition, it is striking how many single gene knockouts or overexpressing transgenes lead to at least some of the serological manifestations of SLE.[3,7] Like the *Fas* mutation, these are probably not direct counterparts of multigenic human SLE. The complete deficiencies of the early complement components of the classical pathway are exceptions to this generalization, but the mechanism of this effect is still mysterious. Although most people identified to be C1q, C4, or C2 deficient have SLE or a similar autoimmune syndrome, these defects account for only a tiny percentage of the human SLE population.

The immune system is delicately balanced between effective recognition of foreign antigens and self-tolerance, and a series of subtle changes in several parameters of immunoregulation could allow self-reactive clones to express pathological autoreactivity (Figure 4.1). Furthermore, in SLE, the immunological processes that lead to autoantibodies are apparently largely regulated by the normal rules of the immune response. For example, the autoantibodies to dsDNA or to Sm appear to be T-cell dependent and antigen driven, such that they show clonal expansion and somatic mutations that are affinity selected.[8,9] The basic abnormality is the failure of negative selection to cull these autospecificities at the appropriate checkpoints. In fact, the combination of partial failures at several sequential stages of tolerance induction would together increase the probability of an anti-dsDNA B cell becoming activated to secrete autoantibodies to the point that this specificity appears in the serum. In

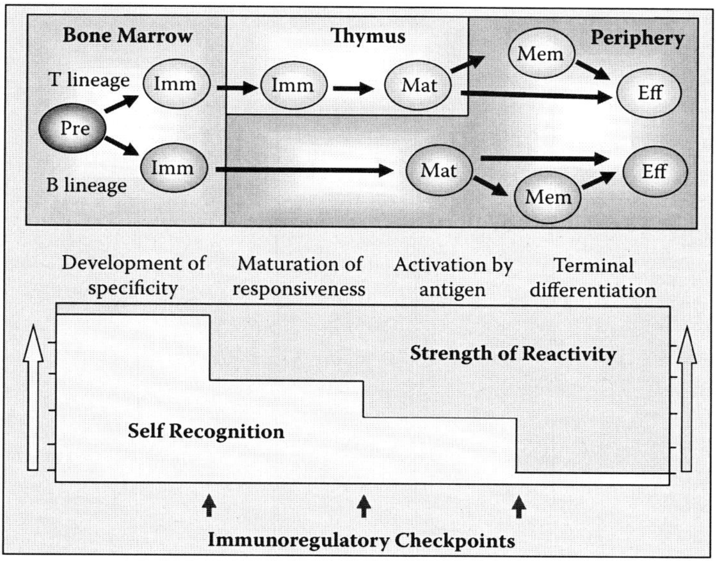

FIGURE 4.1 ***A color version of this figure follows page 176.*** Schematic representation of the ontogeny of the adaptive immune system and the avoidance of autoreactivity. Initially uncommitted precursor cells develop into specific B cells and T cells in the bone marrow and thymus (T cells only). They mature further to be able to respond with appropriate intracellular signaling to encounter with antigen. After such an encounter, they further differentiate to become effector cells, which carry out the protective functions of the immune system. As they pass through the various stages of ontogeny, they become increasingly reactive to extracellular stimuli, and they are culled of their autoreactive specificities through various checkpoints. Alterations in any of these steps can cause inefficiencies in the tolerization checkpoints, and eventually result in the activation of potentially pathogenic autoreactive effector cells. Pre, precursor cell; Imm, immature cell; Mat, mature cell; Eff, effector cell.

part, this process would be stochastic, so that even with a defined genetic predisposition, the resultant autoantibody specificities would be predictable only on a population level.[10] Environmental contributions play little role in the mouse models where it has been investigated,[11] and no epidemiological data in humans has yet been definitive, although widespread agents such as Epstein-Barr (EB) virus may be involved.[12] Drug-induced SLE clearly has an external cause, and several mouse models may be experimentally induced by chemicals or graft versus host reaction.[13]

The essential feature of SLE can be considered to be the loss of B-cell tolerance, that is, a failure of the adaptive immune system. It is reasonable then to search for defects in the regulation of this system, and to target its cells and molecules therapeutically. However, it is becoming increasingly clear that the innate immune system, which has many levels of cross talk with the adaptive system, also participates in self-antigen recognition, particularly for antigens that contain nucleic acids. The Toll-like receptors, for example, can bind DNA or RNA, including autologous molecules in certain circumstances, and can be shown to alter the course of SLE in several mouse models.[14] Thus, elements of innate immunity may well become therapeutic targets, along with those in the adaptive immune system.

THERAPIES

The therapies traditionally used in SLE have been thought to treat either the immune system (immunosuppressives) or the downstream tissue inflammation (anti-inflammatories) (Table 4.1). Drugs such as cyclophosphamide, cyclosporine, mycophenolic acid, or azathioprine would fall in the former category, whereas aspirin and other nonsteroidal anti-inflammatory drugs (NSAIDs) would be the latter type. Corticosteroids, a mainstay of therapy, probably have both immunosuppressive and anti-inflammatory actions, while hydroxychloroquine and other antimalarials may alter antigen processing. The current therapies suffer from a number of defects. First, in many cases, their mechanisms are poorly understood. For example, cyclophosphamide suppresses a number of hematopoietic cell lineages. But which one or ones are responsible for its potent therapeutic effect in SLE: B cells, T cells, antigen presenting cells? This is not a purely academic issue, because it is likely that much of

TABLE 4.1
Pharmacologic Agents Traditionally Used to Treat SLE

Type	Therapeutic Agent
Anti-inflammatories	NSAIDs, including aspirin
	Corticosteroids, including topical, oral, and parenteral
Antimalarials	Hydroxychloroquine, quinacrine
Cytotoxics	Cyclophosphamide, mycophenolic acid, azathioprine, methotrexate leflunomide
Calcineurin inhibitors	Cyclosporine, tacrolimus
Other	IVIg

the toxicity of cyclophosphamide is the result of its effects on tissues that are not part of the pathogenesis of SLE (e.g., the bladder epithelium). Furthermore, not knowing the key therapeutic target of our therapy makes it difficult to have biologic markers that help guide its use. We titrate cyclophosphamide dosage against bone marrow toxicity (chiefly neutrophil counts), but it is unlikely that peripheral blood polymorphonuclear leukocytes (PMNs) are the most important therapeutic target for this intervention.

Second, the level of toxicity of the present therapies is unacceptable. Long-term steroid use inevitably leads to serious side effects, including osteoporosis, obesity, and diabetes. Third, a significant subset of patients with severe SLE do not adequately respond to current therapies, and progress to renal failure, long-term disability, or death. Thus, the unmet therapeutic need exists in the SLE population, and effective targeted biologics would be very welcome.

Biologics in SLE

No new therapy has been approved for SLE by the Food and Drug Administration (FDA) in over 30 years. What's more, until the last decade, no attempts were made by the pharmaceutical industry to develop new therapies for SLE, or even to extend the indications of existing agents to SLE. Although some recent interest, in terms of prospective, randomized double-blind controlled clinical trials (RCTs), has focused on small molecule, nonbiologic therapy, such as MMF and DHEA,[15,16] the application of biologic therapeutics to the treatment of SLE is clearly the most exciting development for patients and practitioners since the advent of diagnostic serological tests starting about 50 years ago. We now have agents that have been approved for other indications being seriously tested for a label indication for SLE, and other agents that are actually being developed principally for this disease. The extraordinary, and not generally predicted, success of the anti-TNF (tumor necrosis factor) biologics in rheumatoid arthritis (see Chapter 3) was a key milestone that gave the SLE community and the pharmaceutical industry the impetus required to consider trying biologics in SLE. Table 4.2 summarizes many of the agents in play, which we discuss briefly below.

The targets for biologic therapy derive from our understandings of the pathogenesis of the disease, both in human and in murine models. Thus, cellular elements, particular B cells, are important. Soluble mediators such as cytokines, chemokines, interferons, and growth factors are also relevant. Some therapies are directed at particular autoreactive specificities, although as we shall see this approach is fraught with difficulties. Other therapies interfere with cell–cell interactions, either relating to adhesion, trafficking, or signaling.

Anti-B Cell

Rituximab had certain advantages as a potential B-cell targeting agent in SLE, the most important of which was its existing approval (as of 1997) for use in non-Hodgkin's B-cell lymphoma and the subsequent extensive clinical experience and safety record.[17] In addition, the early provocatively "successful" anecdotal experience of this drug in rheumatoid arthritis, followed by clearly positive RCTs leading to FDA

TABLE 4.2
Targeted Biologic Therapies in Development for SLE

Target Area	Specific Target	Agents
B cells	CD20	Rituximab, others
	CD22	Epratuzumab
	BLyS/April	Belimumab, others
Costimulation/cell adhesion	CD80/86	Abaticept (CTLA4-Ig)
	CD154 (CD40L)	Anti-CD40L
	CD11a	Efalizumab
Cytokines	IL-6	Tociluzumab (anti-IL-6r)
	IL-10	Anti-IL-10
	TNFα	Infliximab, others
Complement	C5	Eculizumab
Interferons	Type I interferons	Anti-IFNα, soluble INFR
Specific autoantibodies	Anti-DNA	LJP394, edratide

approval for this indication, further encouraged this development. More than a dozen uncontrolled series from around the world have found that patients with SLE treated with rituximab improved clinically.[18–23] Two large, multicenter RCTs in renal and nonrenal SLE, respectively, have recently been completed. Unfortunately, the non-renal trial failed to reach any of its primary or secondary endpoints for efficacy. The data analysis for the renal trial should be available some time in 2009.[24] Although rituximab continues to be used frequently off-label in difficult SLE cases, the results of the RCTs will have a major effect on its future in this disease.

Rituximab is a chimeric monoclonal antibody (mAb) that binds with CD20 at high affinity. CD20 is a B-cell-specific cell surface tetraspan protein that has unknown function, but is found on all B cells from the pre-B cell stage until they terminally differentiate into plasma cells.[25] Rituximab is highly effective in depleting B cells from the peripheral blood and the peripheral lymphoid organs, although the *in vivo* mechanisms of this effect are not entirely understood. They include antibody dependent cellular cytotoxicity (ADCC), complement-mediated lysis, induction of apoptosis, and sensitization to chemotherapy; but which mechanism is most important in patients, or even in a given patient, remains to be determined. In any case, the treatment of patients with a single course of rituximab (about 2 g total over 2 to 4 weeks) causes profound B-cell depletion from the peripheral blood, which can last up to 6 months and longer. B cells are also depleted from the spleen, lymph node, bone marrow, and other organs, such as the rheumatoid synovium, but much less data are available about these effects, and the peripheral blood depletion is probably not fully reflected in what happens elsewhere in the body.[26–29] Furthermore, patients with SLE appear to respond differently to rituximab than do other patient groups. They deplete less consistently, repopulate their B cells earlier, and much more frequently develop an immune response to rituximab, human antichimeric antibody, or HACA.[20,22] In most cases, the SLE autoantibody levels do not appear to decrease with rituximab therapy, although in the English series, which generally include cyclophosphamide administration along with rituximab, antibody levels did fall.[30] The recent recognition of two cases of progressive multifocal leukoencephalopathy (PML) among the SLE patients treated with repeated courses of rituximab off-label, along with a further case of PML in a patient with vasculitis so treated, has led to additional caution about redosing patients.[31] Because it is unlikely that most patients with SLE will achieve long-term remission with one or two courses of rituximab, redosing will be an important issue to face if the RCTs are successful.

Why rituximab works in SLE (if it does) is not clear, particularly since autoantibodies levels do not consistently fall, and some patients appear to respond in the first month of therapy. Presumably, the drug is interrupting some of the other B-cell mechanisms in this disease, such as antigen presentation or immunoregulation. The recent development of reagents to target CD20 in mice should help to clarify this important question.[32]

Other anti-CD20 agents are in development, including several fully human mAbs, and smaller, antibody-like molecular constructs.[33] Although these new agents have theoretical advantages, such as possible decreased immunogenicity (which may be particularly important in patients with SLE with their high propensity to make HACAs), their superiority remains to be demonstrated. Radioisotope conjugated anti-CD20, such as tositumomab (iodine-131 tagged) or ibritumomab (yttrium-90 tagged), have also been used in patients, at least in oncology.[34,35] They may prove to have limited use in SLE either for more aggressive therapy or for diagnostic imaging.[36,37]

CD22 is another target that could be useful in SLE.[38] Epratuzumab (anti-CD22 mAb) has shown efficacy against B-cell lymphomas, and single anecdotal series report encouraging uncontrolled results in SLE and in Sjögren's.[39–41] The expression of CD22 largely overlaps with that of CD20 during B-cell ontogeny. However, the binding of CD22 to its ligands can provide either a positive or negative signal to B cells, and some mAb against CD22 can be agonistic. In the published results, the B-cell depletion with epratuzumab has been modest in the peripheral blood, even when patients improved clinically. In addition, the antilymphoma action of epratuzumab is probably less than that of rituximab. Thus, the mechanism of action of anti-CD22 may not be strongly linked to B-cell depletion, but rather to B-cell signaling.[42] There have not been any definite safety concerns with this agent, and its further testing in phase III trials in SLE should proceed in the near future. Interestingly, the combination of rituximab and epratuzumab has shown a possible synergistic effect in the treatment of lymphoma, and such an approach may be applicable to SLE.[43]

Another way to target B cells is to block essential survival factors. Most of the recent interest has been in the BAFF pathway. BAFF (also known as THANK, TALL-1, BLyS, zTNF4, or TNF13B) is a member of the TNF family that binds to three receptors found on B cells: BCMA, TACI, and BAFF-R.[44] A second homologous ligand, APRIL, also binds to the first two of these receptors.[45] The relative roles of the two ligands are not yet fully elucidated, although the more extreme phenotype of the BAFF knockout mouse compared to the APRIL knockout mouse suggests that BAFF is much more important.[46] Anti-BLyS mAb, belimumab (Human Genome Sciences) is furthest along in development.[47] It has shown modest effectiveness in rheumatoid arthritis in an RCT,[48] but the SLE trials to date have failed to reach their primary clinical end points. Several other nonantibody biologic agents are in earlier phases of development, including soluble receptor fusion proteins. The potential therapeutic efficacy of anti-BLyS probably does not only depend on overall B-cell depletion, but may also profit from denying this important survival factor to antibody-forming cells.[49] In addition, since B-cell depletion with anti-CD20 leads to increased serum BLyS levels,[50] a combination of targeting of BLyS along with B-cell surface molecules may have synergistic effects.[51,52] Preclinical evidence in mice and nonhuman primates provides some support for this possibility.[32] Other studies in mice indicate some different effects from inhibiting BAFF alone or BAFF and APRIL, but the human implications for this are as yet unknown.[53]

Anti-Costimulation/Cell Adhesions

Because the production of autoantibodies in SLE requires T cell–B cell collaboration, it is reasonable to target this process with therapeutic agents. Two mAbs to CD154 (also known as CD40 ligand) were tested in RCTs about 5 years ago. CD154 is normally expressed on T cells and interacts with CD40 on B cells to promote B-cell maturation and immunoglobulin class switching during a T-dependent antibody response. CD154 is abnormally overexpressed in SLE. In mouse models, blocking CD154–CD40 interactions have been shown to be therapeutically effective.[54] Despite this encouraging background, both human trials failed to show efficacy. The Biogen trial was stopped prematurely because of the occurrence of several unanticipated serious thrombotic events in drug-treated patients.[55] This adverse reaction may be related to the expression of CD154 on platelets.[56] Anecdotally, several of the drug-treated patients from this trial studied in detail at the National Institutes of Health (NIH) showed clinical improvement. The IDEC trial, with a different anti-CD154 mAb, showed no evidence for improvement in the drug-treated group versus control, although the problem with arterial thromboses was not seen.[57] At the present time it is not clear if the possibility of utilizing this target will be revisited in the future.

CD28–CD80/86 is another costimulatory pathway. In this system, the main signaling occurs through the T-cell-expressed CD28 receptor by interaction with CD80 or CD86 on antigen-presenting cells. CTLA4 binds the same ligands at even higher affinity than C28. It is also expressed on T cells, although only after activation, except on regulatory T cells. In addition, it delivers a negative signal to the T cell. Abatacept is a fusion molecule between CTLA4 and the Fc portion of immunoglobulin G_1 (IgG_1). As such, it is a soluble molecule that binds CD80 and CD86 strongly and persists for a relatively long time in the body. It blocks the ability of these ligands to bind CD28 and deliver the necessary costimulatory signals for T-cell activation. Abatacept is thus immunosuppressive, and has been approved for treatment of rheumatoid arthritis. Like anti-CD154, a murine homolog has been found to be effective in murine models of SLE.[58] An RCT testing the efficacy of abatacept in SLE is currently under way.

CD11a is more of an interaction molecule than a costimulatory receptor. It forms the α chain of LFA1, which interacts with ICAM-1, and thus plays a role in T-cell trafficking. It has shown substantial efficacy in treating psoriasis.[59] Interestingly, although several articles claim efficacy in uncontrolled treatment of cutaneous SLE with anti-CD11a (efatizumab), including subacute cutaneous lupus erythematosus

(SCLE) and discoid lupus,[60,61] another article discusses a case in which SCLE was allegedly induced by efatizumab.[62] Clearly, this agent would require more study for application to SLE.

Anticytokines

A number of cytokines have been shown to be elevated in the serum of patients with SLE, and thus efforts have been made to target them therapeutically. At this time, tocilizumab, which is an mAb against the interleukin 6 (IL-6) receptor, is the furthest along in development. It is approved in Japan for treatment of Castleman's disease, and has shown efficacy against rheumatoid arthritis in RCTs.[63] So far in SLE, it has been tested only in a small open-labeled, uncontrolled trial. A small uncontrolled series of anti-IL-10 therapy also showed evidence of clinical improvement in patients with SLE.[64] However, IL-10 is known to have important anti-inflammatory effects, such that blocking its action may be problematic. IL-12, IL-15, IL-17, IL-18, and IL-23 are each potentially attractive targets,[65–67] and specific biologics to block their action are being developed, but no clinical data are available in human SLE. Genetic deletion of IL-12 leads to diminished manifestations of SLE in spontaneous mouse models.[68,69]

The use in SLE of the anti-TNF agents approved for the treatment of rheumatoid arthritis is generally avoided, since it is known that they can induce antinuclear antibodies, and even occasional symptomatic drug-induced SLE. Moreover, the experience with the murine SLE models indicates that TNF can have either an enhancing or suppressing effect on the syndrome. Nevertheless, some investigators have seen clinical improvement, particularly in arthritis and proteinuria, in patients with SLE treated with infliximab.[70] Although disease flares were not seen, most of these treated patients showed increased levels of anti-dsDNA and other autoantibodies.[71] This approach is being further evaluated at several institutions, and its future remains uncertain.

Anticomplement

Immune complexes formed between autoantibodies in SLE and autologous antigens fix complement, and sometimes deposit at the sites of ongoing inflammation, such as the renal glomerulus. In addition, complement activation products can be detected in the serum or bound to blood cells, and low serum complement is a relatively good indicator of active SLE or a predictor of poor response to therapy. Paradoxically, however, complete congenital deficiencies in the early components of the classical complement pathway are perhaps the strongest genetic risk factors for human SLE.[72] The mechanism of this association is poorly understood, even though it is rather nicely reflected in mouse models.[73] Thus, the interest in targeting complement therapeutically has been tempered, and efforts in this direction have been halting. Eculizumab is an anti-C5 mAb that completely blocks the terminal complement pathway (C5–C9).[74] It appears to be quite effective in paroxysmal nocturnal hemoglobinemia, which results from the lack of a complement regulatory protein (CD59) and consequent recurrent episodes of excessive complement-mediated lysis of red blood cells (RBCs) . Unfortunately, its use in SLE has not been investigated so far. A concern about such an agent is the known association of genetic deficiencies in the terminal complement components with systemic Gram-negative coccemias.[74,75]

Other complement-directed strategies are in earlier stages of development, although it is not clear if or when they will be tested for SLE.[76]

Antigen-Specific Therapies

The adaptive and innate immune systems and their proinflammatory effector mechanisms are the principal players in the progression of SLE. Because these same pathways are critically involved in normal physiological functions, particularly the recognition of microorganisms, targeting them obviously runs the risk of interfering with host defense. In fact, as the treatment of SLE has become more aggressive over the last decades, infectious complications have become more common and more serious. This is perhaps even more of a concern with the new biologic therapies, which are exquisitely specific and highly effective in blocking their intended targets. The experience with tocilizumab in the treatment of multiple sclerosis (MS) and Crohn's disease is highly instructive. This is discussed further in Chapter 2 and Chapter 5. Briefly, this mAb targets an integrin ($\alpha_4\beta_7$) important for T-cell trafficking (presumably, in MS, to the central nervous system, CNS). Although it proved to be impressively efficacious against MS, it also resulted in ~0.1 percent incidence of PML, a usually fatal CNS infection that is extremely rare except in the population with acquired immune deficiency syndrome (AIDS). This epidemiological association of PML with tocilizumab is much more convincing than for rituximab (see above), and the biology makes more sense, since the causative JC virus is believed to be controlled by T cells and not B cells.[77] Nevertheless, the general point is that a therapeutic agent that can strongly block a key player in the immune system, in order to downregulate a pathologic immune process, carries a risk of serious infectious adverse events, including exotic ones like PML.

Therefore, it is at least theoretically desirable to be able to target the immune dysregulation of SLE by interfering specifically with the reactivity to autoantigens. Although many such possibilities have been proposed, and some tested in mouse models, two have been developed to the point of RCTs in human SLE patients. LJP394 (abetimus) is a tetravalent antigen construct consisting of a polysaccharide backbone linked to oligonucleotide DNA epitopes. It has been considered to be an anti-DNA tolerance-inducing agent, although only limited preclinical evidence supports this assertion.[78] It does bind anti-DNA antibodies, and in murine and human SLE has been shown to lower the titers of anti-DNA antibodies in the serum.[79] In several RCTs, however, it has failed to meet its primary end points, although secondary analyses were considered to show a suggestion of clinical efficacy.[80,81] Several aspects of the biology of this agent are of concern. First, the reestablishment of tolerance in the face of an ongoing (auto)immune response is not equivalent to the induction of tolerance to a foreign antigen in naïve individuals. Almost all the demonstrations of tolerance "induction" to disease-related autoantigens in animal models have required pretreatment before the administration of a tolerance-breaking challenge. Second, not only do many patients with SLE not make anti-dsDNA antibodies, but it is also not clear how important that specificity is in each anti-dsDNA-positive individual.[1] Again in mouse models, a number of manipulations have clearly dissociated anti-dsDNA serological from clinical disease (mainly glomerulonephritis).[82,83] So it is not obvious that the reestablishment of tolerance to dsDNA in a patient with SLE, even if it could be achieved, would have great therapeutic benefit in most cases. The concept of LJP394 is novel, but its future is still not established.

Edratide is another specific agent that targets the DNA specificity. It consists of a peptide from the CDR1 region of the heavy chain of a particular anti-dsDNA mAb, originally isolated from a patient with SLE.[84] It has been used in mouse models to suppress disease,[85] but a recent human RCT did not reach its primary end point of decreasing disease activity. It suffers from the same two basic reservations as LJP394, discussed above. In addition, it is not generally accepted that the anti-dsDNA response in humans and mice is idiotypically restricted. Where the anti-dsDNA repertoire has been evaluated in multiple SLE mice, it has been found to be rather diverse, although an individual animal may show its own particular oligoclonal B-cell expansion.[86] Therefore, a priori it does not seem compelling to predict that targeting a given anti-dsDNA idiotype (that is, a CDR1 primary sequence) would be a useful therapeutic approach except in a rare individual that happened to express mainly the same idiotype. Like LJP394, edratide has some attractive aspects, but some theoretical reservations, and, now, a failed clinical trial.

Anti-Interferons

The interferons have attracted a fair amount of interest recently, in part because of the identification in patients with SLE of the so-called "interferon signature," which refers to increased levels of mRNA in a number of genes known to be upregulated by Type I interferons (chiefly α and β).[87] In general, this has been found in unseparated peripheral blood leukocytes (PBLs) in SLE. Some evidence suggests that it may correlate with disease activity.[88] It presumably reflects an increased synthesis of Type I interferons in these patients, although these compounds themselves have been more difficult to measure directly in the blood. In SLE mice, blocking Type I interferons genetically or by biologics can lead to amelioration of disease, but in certain murine models, just the reverse is seen: for example, MRL/*lpr* mice that are Type I interferon receptor deficient have *more* severe disease.[89–91] Type I interferons themselves are used as therapies in chronic hepatitis C and in MS, and the induction of autoimmunity is a known side effect.[92] Type II interferon (γ) has also been implicated in some murine models, but has not been investigated as thoroughly in humans.[93]

At present, anti-interferons are in early stages of development. They include mAb to α interferon, mAb to the interferon receptor, or soluble interferon receptor. There are many different Type I interferons, but only one receptor. It is not obvious at this time whether it might be possible to target one or a small number of interferons that could be particularly important in SLE, or whether a broader targeting of the receptor or all interferons that bind the receptor (with the soluble receptor) might be necessary to affect the disease. Of course, the broader approach risks blocking the antiviral effects of the Type I interferons.

BIOMARKERS

The increasing understanding of the complex pathogenesis of SLE has not only permitted the rational development of biologic therapies, as we have discussed, but also suggests better ways to measure the effects of such therapies and to determine the response in individual patients. Several aspects of this quest for such *biomarkers* are relevant. Before initiating therapy, it would be helpful if we could determine which patients would be most likely to respond to a given biologic, either on the basis of their genetic makeup, or on the phenomenology of their disease. After beginning treatment, we would want to know if the biologic had its intended physiologic effect. Did anti-CD20, for example, deplete B cells in the peripheral blood? In the peripheral lymph nodes? In the kidneys? And so on. These are not binary measurements, but rather the degree of depletion might well provide useful information regarding whether the patient has been properly treated, and if so, when the patients might need to be treated again. Furthermore, biomarkers could give us better insight into the response of the disease to the particular therapy. Perhaps certain changes in B cell subsets could herald a good, or a bad, response to B-cell-targeted therapies. Some biomarkers might provide better objective measures of disease activity itself. Urinary protein levels are often a good guide to renal disease activity, but we need more sophisticated, general indicators that we can use to monitor disease to determine when to treat and when our treatment works. Of course, these are not new concerns, but they become all the more pressing in view of the potential to use a variety of novel therapies such as the biologics and others, and the importance of understanding the biology of how we are affecting the disease in an individual patient in order to provide the patient with optimal therapy. Other biomarkers might monitor potential adverse reactions. Liver function tests are routinely used to screen for liver toxicity with many drugs, but how do we predict who will develop PML or other unusual but serious events. If we could determine who was losing control of this latent virus (JC) early on, perhaps we could stop or reverse our immunosuppressive therapy to permit full recovery.

OUTCOMES

One of the hurdles in developing new biologics for SLE is the difficulty in selecting appropriate outcome measures in RCTs. To prove efficacy, the outcomes must be clinical and show improvement in patient well-being. So-called surrogate outcomes can be used, if it has been previously demonstrated that they reliably predict eventual patient well-being, and if subsequent studies are undertaken to show that in patients treated with the medicine under investigation, the surrogate retains its reliable predictive value. In the case of renal involvement in SLE, a response to medication can be tested by surrogate outcomes such as decreased proteinuria, clearing of the urine sediment, and improvement or at least stabilization of the creatinine clearance. A response defined by some combination of these relatively objective measures can be reasonably assumed to predict the desired clinical outcome, which is prevention of renal failure. So one could indeed plan a treatment trial aimed solely at renal SLE. However, this would limit recruitment in an already uncommon disease, and the generality regarding other organ systems of results obtained would be open to question.

Unfortunately, the options for selecting outcomes for manifestations of SLE other than renal involvement are much less clear. With a few exceptions, such as thrombocytopenia, the quantification of involvement in most organ systems is not objective, or has not been validated for use in clinical trials. Treatment interventions have thus often used composite indices of clinical disease activity, such as the SLEDAI, SLAM, or BILAG. The details of these evaluation systems are beyond the scope of the present review, but in general they represent attempts to codify the assessment of the degree of disease activity in each of the organ systems affected by SLE, and assign numerical scores that then either can be summed for an overall disease activity score or at least can be used equivalently in judging whether a patient has improved, remained the same, or deteriorated.[94] In this way, SLE patients with a spectrum of disease manifestations can be included in a single protocol. Thus, a patient whose severe skin involvement clears completely might be considered to have a response equivalent to someone whose profoundly depressed platelet count normalizes. Such a broad approach does favor patient recruitment into trials, and provides for some general applicability of a demonstration of efficacy. However, it risks mixing SLE subsets that may be quite different in their immunopathological mechanisms, and therefore vary in their response to a given targeted biologic therapy.

APPROACH TO THE DEVELOPMENT OF NEW THERAPIES IN SLE

It is useful to synthesize an overview of the process of developing a new drug for SLE, in order to summarize the specific discussions above. We consider only biologic agents, which inherently have an exquisite specificity. The first step is to select an appropriate target. This choice must come from a broad consideration of the understandings of the disease mechanisms in SLE, based on immunological and genetic studies in murine models as well as in human disease. Basically, it should be a molecule that appears to play an important role in the aberrant autoreactivities of SLE. Then the agent that would interact with such a target would most often be an mAb or a soluble decoy receptor. This agent would need to show its expected biologic activity *in vitro* on mouse and human cells. It would then need to show activity in mouse SLE models, certainly in preventing disease development when given early, but also in reversing active disease later. At this point it would be very useful if the dosage could be determined based on an *in vivo* biologic activity, other than its effect on disease. Phase I studies in humans would need to establish toxicity, pharmacokinetics, and pharmacodynamics. The last consists of measuring the time course of the expected biologic effects *in vivo*. Given that SLE is such a difficult disease to work with, it would be very helpful if the agent had been extensively tested, or even approved for use, in other rheumatic diseases, or at least in other human conditions. With such an available therapy, it is likely that anecdotal reports would appear that would provide precedent for the potential extension of the agent's use to SLE.

The initial phase I/II testing begins to encounter some of the complexity of working with SLE. What severity of patients does one enter in such a trial? If their disease is quite active, then there is likelihood they will suffer a serious adverse event, and it will be impossible to know that the drug itself did not make them worse. If they have no or minimal disease activity, it is harder to justify exposing them to a novel agent for which they do not have need. The outcomes of these early trials are obviously skewed toward safety, and thus must measure all reasonable parameters that could uncover problems. However, even at this stage, one would like to get some suggestions of efficacy. Unfortunately, the definition of clinical improvement is not simple, as is discussed above. If treated patients all fail to improve clinically, then the therapy is unlikely to be useful, even in the absence of a placebo control. If patients improve, that is consistent with the drug being efficacious, but obviously *proves* nothing and thus demands a true RCT. However, the patterns of improvement may be helpful. The combined experience of several uncontrolled trials, each looked at from various perspectives, while it provided no definitive answers, might suggest sufficient useful hypotheses to facilitate the structuring of future RCTs, including eventually pivotal trials.

The complexities of the phase I and II trials also apply to the design of the critical RCTs, except that here the stakes are much higher. The primary end point has to be defined in advance in terms of measures and time; power calculations need to be made based on no firm information; and the financial outlay is orders of magnitude higher. In addition, the lack of an accepted standard of care for severe disease (beyond steroids) that has FDA approval limits trial design. Similarly, the ethical considerations of making sure no one is harmed, either by the drug itself or by its (possible) lack of efficacy in a sick patient, further dictate the parameters of inclusion and exclusion. It is highly important that as many biologic parameters be measured as reasonably possible, so that a large trial will provide insight into the mechanism of the agent tested and, as well, that of the disease itself. This would be particularly important if the trial were to fail its primary end point for apparent reasons of trial design (rather than true lack of efficacy). Then, the mechanistic knowledge gained from such a trial would inform the design of potential future trials with the same therapies, the possible use of concomitant therapies, and the development of new agents directed at the same or different targets. Overall, the complexities of SLE trial design, combined with the relative infrequency of this disorder (compared, say, to rheumatoid arthritis), means that it is unlikely that the pharmaceutical industry will tolerate several failures of large RCTs without a clear success. The next few years will be critical in this regard.

CONCLUSIONS

The happy event we hope for in the near future is the definitive demonstration that a biologic is efficacious in SLE. This indeed will be a major advance, as no drug has been approved for this condition for decades. It will validate not only the agent involved, but also the trial design and the outcomes chosen. Thereafter, it will be easier to decide how to test further therapies, and the successful agent will provide a validated "standard of care." It is likely that in a disease as diverse and unpredictable as SLE, we will need several new therapeutic options. We foresee that with new biomarkers we will be able to individualize therapy efficiently. Of course, if the first approved agent shows striking efficacy, it could reduce the unmet need for new treatments, and make it more difficult to recruit patients into trials for additional therapies. For a disease as difficult as SLE, we should be so lucky as to have to deal with that predicament!

REFERENCES

1. Isenberg DA, Manson JJ, Ehrenstein MR, Rahman A. Fifty years of anti-ds DNA antibodies: Are we approaching journey's end? *Rheumatology* (Oxford). 2007 Jul;46(7):1052–56.

2. Maddison PJ, Reichlin M. Deposition of antibodies to a soluble cytoplasmic antigen in the kidneys of patients with systemic lupus erythematosus. *Arthritis Rheum.* 1979 Aug;22(8):858–63.
3. Liu K, Mohan C. What do mouse models teach us about human SLE? *Clin Immunol.* 2006 May;119(2):123–30.
4. Cohen PL, Eisenberg RA. Lpr and gld: Single gene models of systemic autoimmunity and lymphoproliferative disease. *Annu Rev Immunol.* 1991;9:243–69.
5. Cohen PL. Autoimmunity and lymphoproliferation: two genes are worse than one. *Mol Interv.* 2002 Nov;2(7):427–30.
6. Fairhurst AM, Wandstrat AE, Wakeland EK. Systemic lupus erythematosus: Multiple immunological phenotypes in a complex genetic disease. *Adv Immunol.* 2006;92:1–69.
7. Mohan C. Murine lupus genetics: Lessons learned. *Curr Opin Rheumatol.* 2001;13(5):352–60.
8. Bloom DD, Davignon JL, Retter MW, Shlomchik MJ, Pisetsky DS, Cohen PL, et al. V region gene analysis of anti-Sm hybridomas from MRL/Mp-lpr/lpr mice. *J Immunol.* 1993;150(4):1591–610.
9. Bloom DD, Davignon JL, Cohen PL, Eisenberg RA, Clarke SH. Overlap of the anti-Sm and anti-DNA responses of MRL/Mp-lpr/lpr mice. *J Immunol.* 1993;150(4):1579–90.
10. Eisenberg RA, Craven SY, Warren RW, Cohen PL. Stochastic control of anti-Sm autoantibodies in MRL/Mp-lpr/lpr mice. *J Clin Invest.* 1987;80(3):691–97.
11. Maldonado MA, Kakkanaiah V, MacDonald GC, Chen F, Reap EA, Balish E, et al. The role of environmental antigens in the spontaneous development of autoimmunity in MRL-lpr mice. *J Immunol.* 1999;162(11):6322–30.
12. James JA, Harley JB, Scofield RH. Epstein-Barr virus and systemic lupus erythematosus. *Curr Opin Rheumatol.* 2006 Sep;18(5):462–67.
13. Eisenberg RA. The chronic graft-versus-host model of systemic autoimmunity. In: Nemazee D, ed. *B Cell Biology in Autoimmunity.* Basel: Karger; 2002. pp. 228–43.
14. Rahman AH, Eisenberg RA. The role of Toll-like receptors in systemic lupus erythematosus. *Springer Semin Immunopathol.* 2006 Oct;28(2):131–43.
15. Ginzler EM, Dooley MA, Aranow C, Kim MY, Buyon J, Merrill JT, et al. Mycophenolate mofetil or intravenous cyclophosphamide for lupus nephritis. *N Engl J Med.* 2005 Nov 24;353(21):2219–28.
16. van Vollenhoven RF, Park JL, Genovese MC, West JP, McGuire JL. A double-blind, placebo-controlled, clinical trial of dehydroepiandrosterone in severe systemic lupus erythematosus. *Lupus.* 1999;8(3):181–87.
17. Eisenberg R, Albert D. B-cell targeted therapies in rheumatoid arthritis and systemic lupus erythematosus. *Nat Clin Pract Rheumatol.* 2006 Jan;2(1):20–27.
18. Gunnarsson I, Sundelin B, Jonsdottir T, Jacobson SH, Henriksson EW, van Vollenhoven RF. Histopathologic and clinical outcome of rituximab treatment in patients with cyclophosphamide-resistant proliferative lupus nephritis. *Arthritis Rheum.* 2007 Mar 28;56(4):1263–72.
19. Leandro MJ, Cambridge G, Edwards JC, Ehrenstein MR, Isenberg DA. B-cell depletion in the treatment of patients with systemic lupus erythematosus: A longitudinal analysis of 24 patients. *Rheumatology* (Oxford). 2005 Dec;44(12):1542–45.
20. Looney RJ, Anolik JH, Campbell D, Felgar RE, Young F, Arend LJ, et al. B cell depletion as a novel treatment for systemic lupus erythematosus: A phase I/II dose-escalation trial of rituximab. *Arthritis Rheum.* 2004 Aug;50(8):2580–89.
21. Sfikakis PP, Boletis JN, Lionaki S, Vigklis V, Fragiadaki KG, Iniotaki A, et al. Remission of proliferative lupus nephritis following B cell depletion therapy is preceded by down-regulation of the T cell costimulatory molecule CD40 ligand: An open-label trial. *Arthritis Rheum.* 2005 Feb;52(2):501–13.

22. Sutter JA, Kwan-Morley J, Dunham J, Du Y-Z, Kamoun M, Albert D, et al. A longitudinal analysis of SLE patients treated with rituximab (anti-CD20): Factors associated with B lymphocyte recovery. *Clin Immunol.* 2008;126:282–90.
23. Albert D, Dunham J, Khan S, Stansberry J, Kolasinski S, Tsai D, et al. Variability in the biological response to anti-CD20 B-cell depletion in SLE. *Ann Rheum Dis.* 2008;67:1724–31.
24. ClinicalTrials.gov [website]. U.S. National Institutes of Health. Available from: http://wwwclinicaltrialsgovct. 2007.
25. Cragg MS, Walshe CA, Ivanov AO, Glennie MJ. The biology of CD20 and its potential as a target for mAb therapy. *Curr Dir Autoimmun.* 2005;8:140–74.
26. Kavanaugh A, Rosengren S, Lee SJ, Hammaker D, Firestein GS, Kalunian K, et al. Assessment of rituximab's immunomodulatory synovial effects (the ARISE trial). I: Clinical and synovial biomarker results. *Ann Rheum Dis.* 2007 Jul 20.
27. Vos K, Thurlings RM, Wijbrandts CA, van Schaardenburg D, Gerlag DM, Tak PP. Early effects of rituximab on the synovial cell infiltrate in patients with rheumatoid arthritis. *Arthritis Rheum.* 2007 Mar;56(3):772–78.
28. Leandro MJ, Cooper N, Cambridge G, Ehrenstein MR, Edwards JC. Bone marrow B-lineage cells in patients with rheumatoid arthritis following rituximab therapy. *Rheumatology* (Oxford). 2007 Jan;46(1):29–36.
29. Teng YK, Levarht EW, Hashemi M, Bajema IM, Toes RE, Huizinga TW, et al. Immunohistochemical analysis as a means to predict responsiveness to rituximab treatment. *Arthritis Rheum.* 2007 Dec;56(12):3909–18.
30. Cambridge G, Leandro MJ, Teodorescu M, Manson J, Rahman A, Isenberg DA, et al. B cell depletion therapy in systemic lupus erythematosus: Effect on autoantibody and antimicrobial antibody profiles. *Arthritis Rheum.* 2006 Nov;54(11):3612–22.
31. Calabrese LH, Molloy ES, Huang D, Ransohoff RM. Progressive multifocal leukoencephalopathy in rheumatic diseases: Evolving clinical and pathologic patterns of disease. *Arthritis Rheum.* 2007 Jul;56(7):2116–28.
32. Gong Q, Ou Q, Ye S, Lee WP, Cornelius J, Diehl L, et al. Importance of cellular microenvironment and circulatory dynamics in B cell immunotherapy. *J Immunol.* 2005 Jan 15;174(2):817–26.
33. Tahir H, Rohrer J, Bhatia A, Wegener WA, Isenberg DA. Humanized anti-CD20 monoclonal antibody in the treatment of severe resistant systemic lupus erythematosus in a patient with antibodies against rituximab. *Rheumatology* (Oxford). 2005 Apr;44(4):561–62.
34. Kaminski MS, Tuck M, Estes J, Kolstad A, Ross CW, Zasadny K, et al. ^{131}I-tositumomab therapy as initial treatment for follicular lymphoma. *N Engl J Med.* 2005 Feb 3;352(5):441–49.
35. Gordon LI, Molina A, Witzig T, Emmanouilides C, Raubtischek A, Darif M, et al. Durable responses after ibritumomab tiuxetan radioimmunotherapy for CD20$^+$ B-cell lymphoma: Long-term follow-up of a phase 1/2 study. *Blood.* 2004 Jun 15;103(12):4429–31.
36. Dietlein M, Pels H, Schulz H, Staak O, Borchmann P, Schomacker K, et al. Imaging of central nervous system lymphomas with iodine-123 labeled rituximab. *Eur J Haematol.* 2005 Apr;74(4):348–52.
37. Iwamoto FM, Schwartz J, Pandit-Taskar N, Peak S, Divgi CR, Zelenetz AD, et al. Study of radiolabeled indium-111 and yttrium-90 ibritumomab tiuxetan in primary central nervous system lymphoma. *Cancer.* 2007 Dec 1;110(11):2528–34.
38. Tedder TF, Poe JC, Haas KM. CD22: A multifunctional receptor that regulates B lymphocyte survival and signal transduction. *Adv Immunol.* 2005;88:1–50.
39. Leonard JP, Coleman M, Ketas JC, Chadburn A, Ely S, Furman RR, et al. Phase I/II trial of epratuzumab (humanized anti-CD22 antibody) in indolent non-Hodgkin's lymphoma. *J Clin Oncol.* 2003 Aug 15;21(16):3051–59.

40. Dörner T, Kaufmann J, Wegener WA, Teoh N, Goldenberg DM, Burmester GR. Initial clinical trial of epratuzumab (humanized anti-CD22 antibody) for immunotherapy of systemic lupus erythematosus. *Arthritis Res Ther.* 2006;8(3):R74.
41. Steinfeld SD, Tant L, Burmester GR, Teoh NK, Wegener WA, Goldenberg DM, et al. Epratuzumab (humanised anti-CD22 antibody) in primary Sjogren's syndrome: An open-label phase I/II study. *Arthritis Res Ther.* 2006 Jul 20;8(4):R129.
42. Eisenberg R. Targeting B cells in systemic lupus erythematosus: Not just deja vu all over again. *Arthritis Res Ther.* 2006;8(3):108.
43. Leonard JP, Coleman M, Ketas J, Ashe M, Fiore JM, Furman RR, et al. Combination antibody therapy with epratuzumab and rituximab in relapsed or refractory non-Hodgkin's lymphoma. *J Clin Oncol.* 2005 Aug 1;23(22):5044–51.
44. Mackay F, Silveira PA, Brink R. B cells and the BAFF/APRIL axis: Fast-forward on autoimmunity and signaling. *Curr Opin Immunol.* 2007 Jun;19(3):327–36.
45. Dillon SR, Gross JA, Ansell SM, Novak AJ. An APRIL to remember: Novel TNF ligands as therapeutic targets. *Nat Rev Drug Discov.* 2006 Mar;5(3):235–46.
46. Bossen C, Schneider P. BAFF, APRIL and their receptors: Structure, function and signaling. *Semin Immunol.* 2006 Oct;18(5):263–75.
47. Ding C, Jones G. Belimumab Human Genome Sciences/Cambridge Antibody Technology/GlaxoSmithKline. *Curr Opin Invest Drugs.* 2006 May;7(5):464–72.
48. Looney RJ. B cell-targeted therapy for rheumatoid arthritis: An update on the evidence. *Drugs.* 2006;66(5):625–39.
49. Darce JR, Arendt BK, Wu X, Jelinek DF. Regulated expression of BAFF-binding receptors during human B cell differentiation. *J Immunol.* 2007 Dec 1;179(11):7276–86.
50. Lavie F, Miceli-Richard C, Ittah M, Sellam J, Gottenberg JE, Mariette X. Increase of B cell-activating factor of the TNF family (BAFF) after rituximab treatment: Insights into a new regulating system of BAFF production. *Ann Rheum Dis.* 2007 May;66(5):700–3.
51. Lin WY, Gong Q, Seshasayee D, Lin Z, Ou Q, Ye S, et al. Anti-BR3 antibodies: A new class of B-cell immunotherapy combining cellular depletion and survival blockade. *Blood.* 2007 Dec 1;110(12):3959–67.
52. Eisenberg R. Combination biologics: 1 stone, 2 birds. *Blood.* 2007 Dec 1;110(12):3817.
53. Ramanujam M, Wang X, Huang W, Liu Z, Schiffer L, Tao H, et al. Similarities and differences between selective and nonselective BAFF blockade in murine SLE. *J Clin Invest.* 2006 Mar;116(3):724–34.
54. Davidson A, Diamond B, Wofsy D, Daikh D. Block and tackle: CTLA4Ig takes on lupus. *Lupus.* 2005;14(3):197–203.
55. Boumpas DT, Furie R, Manzi S, Illei GG, Wallace DJ, Balow JE, et al. A short course of BG9588 (anti-CD40 ligand antibody) improves serologic activity and decreases hematuria in patients with proliferative lupus glomerulonephritis. *Arthritis Rheum.* 2003 Mar;48(3):719–27.
56. Tanaka T, Kuroiwa T, Ikeuchi H, Ota F, Kaneko Y, Ueki K, et al. Human platelets stimulate mesangial cells to produce monocyte chemoattractant protein-1 via the CD40/CD40 ligand pathway and may amplify glomerular injury. *J Am Soc Nephrol.* 2002 Oct;13(10):2488–96.
57. Kalunian KC, Davis JC, Jr., Merrill JT, Totoritis MC, Wofsy D. Treatment of systemic lupus erythematosus by inhibition of T cell costimulation with anti-CD154: A randomized, double-blind, placebo-controlled trial. *Arthritis Rheum.* 2002 Dec;46(12):3251–58.
58. Daikh DI, Wofsy D. Cutting edge: Reversal of murine lupus nephritis with CTLA4Ig and cyclophosphamide. *J Immunol.* 2001 Mar 1;166(5):2913–16.

59. Lebwohl M, Tyring SK, Hamilton TK, Toth D, Glazer S, Tawfik NH, et al. A novel targeted T-cell modulator, efalizumab, for plaque psoriasis. *N Engl J Med.* 2003 Nov 20;349(21):2004–13.
60. Usmani N, Goodfield M. Efalizumab in the treatment of discoid lupus erythematosus. *Arch Dermatol.* 2007 Jul;143(7):873–77.
61. Clayton TH, Ogden S, Goodfield MD. Treatment of refractory subacute cutaneous lupus erythematosus with efalizumab. *J Am Acad Dermatol.* 2006 May;54(5):892–95.
62. Bentley DD, Graves JE, Smith DI, Heffernan MP. Efalizumab-induced subacute cutaneous lupus erythematosus. *J Am Acad Dermatol.* 2006 May;54(5 Suppl):S242–43.
63. Ohsugi Y. Recent advances in immunopathophysiology of interleukin-6: An innovative therapeutic drug, tocilizumab (recombinant humanized anti-human interleukin-6 receptor antibody), unveils the mysterious etiology of immune-mediated inflammatory diseases. *Biol Pharm Bull.* 2007 Nov;30(11):2001–6.
64. Llorente L, Richaud-Patin Y, Garcia-Padilla C, Claret E, Jakez-Ocampo J, Cardiel MH, et al. Clinical and biologic effects of anti-interleukin-10 monoclonal antibody administration in systemic lupus erythematosus. *Arthritis Rheum.* 2000 Aug;43(8):1790–800.
65. Aringer M, Stummvoll GH, Steiner G, Koller M, Steiner CW, Höfler E, et al. Serum interleukin-15 is elevated in systemic lupus erythematosus. *Rheumatology* (Oxford). 2001 Aug;40(8):876–81.
66. Huang X, Hua J, Shen N, Chen S. Dysregulated expression of interleukin-23 and interleukin-12 subunits in systemic lupus erythematosus patients. *Mod Rheumatol.* 2007;17(3):220–23.
67. Dinarello CA. Interleukin-18 and the pathogenesis of inflammatory diseases. *Semin Nephrol.* 2007 Jan;27(1):98–114.
68. Pollard KM, Hultman P, Kono DH. Using single-gene deletions to identify checkpoints in the progression of systemic autoimmunity. *Ann N Y Acad Sci.* 2003 Apr;987:236–39.
69. Kikawada E, Lenda DM, Kelley VR. IL-12 deficiency in MRL-Fas(lpr) mice delays nephritis and intrarenal IFN-gamma expression, and diminishes systemic pathology. *J Immunol.* 2003 Apr 1;170(7):3915–25.
70. Aringer M, Graninger WB, Steiner G, Smolen JS. Safety and efficacy of tumor necrosis factor alpha blockade in systemic lupus erythematosus: An open-label study. *Arthritis Rheum.* 2004 Oct;50(10):3161–69.
71. Aringer M, Steiner G, Graninger WB, Hofler E, Steiner CW, Smolen JS. Effects of short-term infliximab therapy on autoantibodies in systemic lupus erythematosus. *Arthritis Rheum.* 2007 Jan;56(1):274–79.
72. Truedsson L, Bengtsson AA, Sturfelt G. Complement deficiencies and systemic lupus erythematosus. *Autoimmunity.* 2007 Dec;40(8):560–66.
73. Manderson AP, Botto M, Walport MJ. The role of complement in the development of systemic lupus erythematosus. *Annu Rev Immunol.* 2004;22:431–56.
74. Rother RP, Rollins SA, Mojcik CF, Brodsky RA, Bell L. Discovery and development of the complement inhibitor eculizumab for the treatment of paroxysmal nocturnal hemoglobinuria. *Nat Biotechnol.* 2007 Nov;25(11):1256–64.
75. Nusinow SR, Zuraw BL, Curd JG. The hereditary and acquired deficiencies of complement. *Med Clin North Am.* 1985 May;69(3):487–504.
76. Ricklin D, Lambris JD. Complement-targeted therapeutics. *Nat Biotechnol.* 2007 Nov;25(11):1265–75.
77. Langer-Gould A, Atlas SW, Bollen AW, Pelletier D. Progressive multifocal leukoencephalopathy in a patient treated with natalizumab. *N Engl J Med.* 2005.
78. Jones DS, Barstad PA, Feild MJ, Hachmann JP, Hayag MS, Hill KW, et al. Immunospecific reduction of antioligonucleotide antibody-forming cells with a tetrakis-oligonucleotide conjugate (LJP 394), a therapeutic candidate for the treatment of lupus nephritis. *J Med Chem.* 1995 Jun 9;38(12):2138–44.

79. Weisman MH, Bluestein HG, Berner CM, de Haan HA. Reduction in circulating dsDNA antibody titer after administration of LJP 394. *J Rheumatol.* 1997 Feb;24(2):314–18.
80. Furie R. Abetimus sodium (riquent) for the prevention of nephritic flares in patients with systemic lupus erythematosus. *Rheum Dis Clin North Am.* 2006 Feb;32(1):149–56, x.
81. Alarcon-Segovia D, Tumlin JA, Furie RA, McKay JD, Cardiel MH, Strand V, et al. LJP 394 for the prevention of renal flare in patients with systemic lupus erythematosus: Results from a randomized, double-blind, placebo-controlled study. *Arthritis Rheum.* 2003 Feb;48(2):442–54.
82. Vyse TJ, Drake CG, Rozzo SJ, Roper E, Izui S, Kotzin BL. Genetic linkage of IgG autoantibody production in relation to lupus nephritis in New Zealand hybrid mice. *J Clin Invest.* 1996;98(8):1762–72.
83. Izui S, McConahey PJ, Clark JP, Hang LM, Hara I, Dixon FJ. Retroviral gp70 immune complexes in NZB x NZW F2 mice with murine lupus nephritis. *J Exp Med.* 1981 Aug 1;154(2):517–28.
84. Waisman A, Shoenfeld Y, Blank M, Ruiz PJ, Mozes E. The pathogenic human monoclonal anti-DNA that induces experimental systemic lupus erythematosus in mice is encoded by a VH4 gene segment. *Int Immunol.* 1995 Apr;7(4):689–96.
85. Waisman A, Ruiz PJ, Israeli E, Eilat E, Konen-Waisman S, Zinger H, et al. Modulation of murine systemic lupus erythematosus with peptides based on complementarity determining regions of a pathogenic anti-DNA monoclonal antibody. *Proc Natl Acad Sci USA.* 1997 Apr 29;94(9):4620–25.
86. Shlomchik M, Mascelli M, Shan H, Radic MZ, Pisetsky D, Marshak-Rothstein A, et al. Anti-DNA antibodies from autoimmune mice arise by clonal expansion and somatic mutation. *J Exp Med.* 1990;171(1):265–92.
87. Crow MK. Type I interferon in systemic lupus erythematosus. *Curr Top Microbiol Immunol.* 2007;316:359–86.
88. Feng X, Wu H, Grossman JM, Hanvivadhanakul P, FitzGerald JD, Park GS, et al. Association of increased interferon-inducible gene expression with disease activity and lupus nephritis in patients with systemic lupus erythematosus. *Arthritis Rheum.* 2006 Sep;54(9):2951–62.
89. Hron JD, Peng SL. Type I IFN protects against murine lupus. *J Immunol.* 2004 Aug 1;173(3):2134–42.
90. Santiago-Raber ML, Baccala R, Haraldsson KM, Choubey D, Stewart TA, Kono DH, et al. Type-I interferon receptor deficiency reduces lupus-like disease in NZB mice. *J Exp Med.* 2003 Mar 17;197(6):777–88.
91. Lawson BR, Prud'homme GJ, Chang Y, Gardner HA, Kuan J, Kono DH, et al. Treatment of murine lupus with cDNA encoding IFN-gammaR/Fc. *J Clin Invest.* 2000 Jul;106(2):207–15.
92. Davignon JL, Cohen PL, Eisenberg RA. Immunological effects of recombinant interferon gamma in vivo in normal mice: Failure to induce autoantibodies. *Int J Immunopharmacol.* 1990;12(6):691–98.
93. Peng SL, Moslehi J, Craft J. Roles of interferon-gamma and interleukin-4 in murine lupus. *J Clin Invest.* 1997 Apr 15;99(8):1936–46.
94. Griffiths B, Mosca M, Gordon C. Assessment of patients with systemic lupus erythematosus and the use of lupus disease activity indices. *Best Pract Res Clin Rheumatol.* 2005 Oct;19(5):685–708.

5 Biologic Therapies for Inflammatory Bowel Disease

Martin H. Holtmann and Markus F. Neurath

CONTENTS

KEY WORDS: IBD, Crohn's disease, immunotherapy, T cells, cytokines, TNF

ABSTRACT

The past decades have dramatically increased our understanding of the pathophysiology of chronic inflammatory bowel disease (IBD). This has mainly been facilitated by the advent of molecular biologic and recombinant techniques. The focus on immunological aspects, in particular on T-cell immunology with the analysis of regulatory cytokine signaling pathways has provided the basis for innovative treatment strategies. Many of these novel strategies are still experimental and subject to clinical studies, while others have already been incorporated into the treatment armamentarium of routine clinical management in IBD. The availability of an increasing number of recombinant proteins in the treatment of IBD is a unique example for translational research from the bedside to the bench and back to the bedside. Interestingly, while the elucidation of the immunoregulatory pathomechanisms provided the basis for the development of these novel strategies, ongoing studies on the mechanisms of action of these substances in the clinical setting further increase our pathophysiological understanding considerably. This will further promote the development of more effective and more specific treatment strategies. It is the purpose of this chapter to outline our current pathoimmunological understanding of inflammatory bowel disease with special focus on the characterization of those proinflammatory players which have become targets for the rational development of recombinant antagonistic proteins. In particular, in trying to understand the mechanisms of action of the various different anti-TNF-strategies, important insights in TNF signaling pathways in inflammatory bowel disease and other diseases have been made.

INFLAMMATORY BOWEL DISEASE

Crohn's disease (CD) and ulcerative colitis (UC) are the two most common forms of inflammatory bowel disease (IBD) and have gained increasing medical and health economic importance during the past few decades especially in western countries due to their early manifestation in life between 15 and 35 years of age and their chronic character. The etiology of IBD is still unclear and should be considered as multifactorial according to recent studies.[1] Genetic factors seem to play a pathogenetic role, as do environmental, infectious, and immunological factors (Figure 5.1).[2]

Mucosal surfaces such as the intestinal mucosa are special interfaces for the interaction between the organism and the environment and possess an especially adapted immune system (gut-associated lymphoid tissue, or GALT).[3] Due to its large surface the gut can be regarded as the biggest immune organ of the human body. In the gut the organism is physiologically exposed to a large amount of antigens from the natural flora and the food. Whereas a systemic immune response to these foreign antigens should be prevented in order not to damage the organism, potentially pathogenic antigens have to be recognized and eliminated at a local level. The intestinal epithelium with the mucus represents a mechanic barrier, but it is mainly the accomplishment of the GALT to keep this fragile immunological balance between hyporesponsiveness and efficient immune defense.

Substantial progress has been made in the understanding of the pathogenesis of IBD in past years pursuing the view that IBD could result from disturbances of the intestinal barrier and a pathologic activation of the intestinal immune response toward luminal bacterial and nutritional antigens in a genetically susceptible individual. The main focus of immunologic research in IBD has been on the role and behavior of the T cell and its interaction with other cell populations.[4]

PATHOPHYSIOLOGICAL ROLE OF T CELLS

The acute and chronic inflammation of the bowel goes along with an activation of lamina propria T cells leading to an increased cytokine production.[5] Cytokine patterns of CD and UC show significant differences. In CD, $CD4^+$ T cells produce

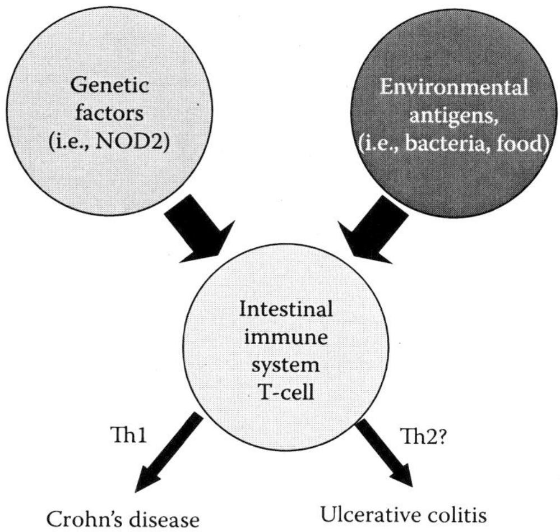

FIGURE 5.1 Multifactorial pathogenesis of IBD. According to the current paradigm the etiology of inflammatory bowel disease is multifactorial. IBD evolves in genetically susceptible patients (i.e., NOD2/CARD15), who are exposed to yet unknown environmental antigens (i.e., bacterial or nutritional antigens). This leads to an activation of the intestinal immune system with a pivotal role of T-cell-mediated immune responses. While Crohn's disease is prototypic of a Th1-mediated disease, data supporting a Th2 driven pathogenesis of ulcerative colitis are less convincing.

increased amounts of interleukin 12 (IL-12), tumor necrosis factor alpha (TNFα), and interferon gamma (IFNγ), which would fit to a Th1 cytokine profile. The production of the Th2 cytokines IL-4 and IL-5 in CD is decreased. In UC, IL-5 production is increased whereas IFNγ production by anti-CD2/CD28 stimulated lamina propria T cells is unchanged compared to control patients.[6] IL-13 has recently been identified as a key mediator in experimental UC-like disease.[7] However, the cytokine profile in UC cannot be classified as a Th2 phenotype without restrictions, since the Th2 cytokine IL-4 is decreased in UC.

The relevance of the Th1–Th2 paradigm has been questioned additionally by the emergence of a third type of $CD4^+$ helper cells termed Th3 cells, mainly producing transforming growth factor beta (TGFβ). In an experimental model, blockade of endogenous TGFβ aggravates colitis suggesting an anti-inflammatory, protective role of TGFβ.[8] $CD4CD25^+$ regulatory T cells have also recently been identified as a distinct T-cell population characterized by the production of the anti-inflammatory cytokine IL-10.[9] An additional T-cell subpopulation is characterized by the production of IL-17.[10] The differentiation of naïve T cells toward this IL-17 phenotype is promoted by TGFβ, while expansion and maintenance is regulated by IL-23. Numerous experimental animal models, especially knockout or transgenic models, have documented the pathophysiological relevance of distinct cytokine dysregulations, that is, TNFα, IL-2, IL-6, IL-10, IL-12, and INFγ.[11]

TUMOR NECROSIS FACTOR-α

TNFα belongs to the best characterized cytokines in IBD.[12] The primary translational product of the human TNFα is the membrane-bound TNFα (mTNFα) of 233 amino acids' length (26 kD). Metalloproteinases such as TACE and ADAM 10 cleave the extracellular domain of mTNFα and thus release soluble TNFα (sTNFα) of 157 amino acids (17 kD). A probably crucial role of mTNFα signaling via TNF-receptor 2 in chronic inflammatory disease states has not been recognized until recently, as discussed below. Both mTNFα and sTNFα form noncovalently linked homotrimers, which occurs intracellularly. mTNFα is capable of inducing signaling in a receptor-independent manner (reverse signaling).[13]

The pathogenic role of TNFα in inflammatory bowel disease is well established by clinical and experimental studies.[12] Although serum levels of TNFα are not elevated significantly in patients with IBD,[14] lamina propria mononuclear cells from colon biopsies of untreated patients with CD and UC spontaneously produce more TNFα than cells from controls when cultured *in vitro*.[15]

An important pathogenic mechanism of TNFα in the mucosa seems to be the stimulation of a Th1 T-cell response. Lamina propria T lymphocytes from colon biopsies of patients with CD incubated with TNFα produce increased amounts of the Th1 cytokines IL-2, IFNγ, and TNFα itself.[16] The stimulation of TNFα secretion by TNFα itself suggests a possible positive feedback mechanism, which could potentially contribute to the perpetuation of inflammation. Yet another possible mechanism of TNFα action could be the activation of endogenous matrix metalloproteinases (MMP), which results in damage of the extracellular matrix of the mucosa.[17]

The critical role of TNFα for the development of colitis could be reproduced in various established experimental animal models of colitis including the TNBS-model (2,4,6-trinitrobenzene sulfonic acid), the DSS-model (dextrane sulfate sodium), the IL-10 knockout mouse, and the $CD4^+CD45RB^{high}$, respectively $CD4^+CD62L^+$adoptive transfer model.[18–21]

TNF RECEPTORS AND SIGNALING

TNFα exerts its effects via two TNF-specific membrane-bound receptors with a molecular weight of approximately 55 to 60 kDa (TNF-R1 = p55) and 75 to 80 kDa (TNF-R2 = p75), respectively, which belong to the TNF/nerve growth factor (NGF) receptor family (Figure 5.2).[22] Both TNF-R1 and TNF-R2 are composed of two identical subunits and are glycosylated. While TNF-R1 and TNF-R2 are quite similar in their extracellular regions—both receptors possess multiple cysteine rich motifs—their intracellular domains exhibit striking structural differences most likely reflecting different signaling pathways.

Extracellular ligand binding elicits complex intracellular signaling cascades. Important signaling cascades of TNF-R1 include the activation of kinases, the induction of apoptosis, and the activation of the proinflammatory transcription factor nuclear factor kappa B (NFκB). NFκB is found in the cytosol of nonactivated monocytes and T cells as an inactive heterodimer of a p50 and p65 subunit, bound to the inhibitory IκB protein. Upon stimulation of the cell, IκB is phosphorylated by

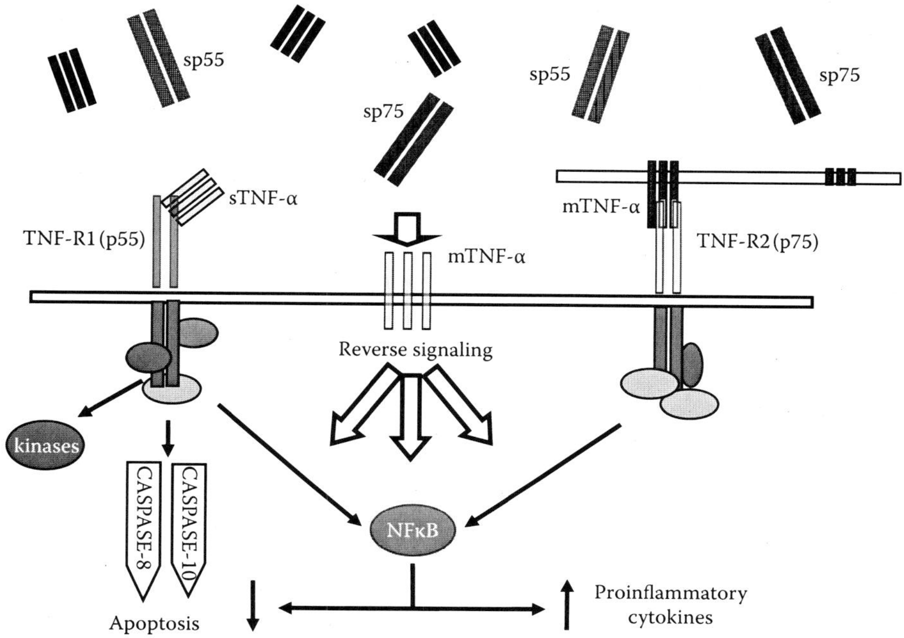

FIGURE 5.2 TNF signaling. TNFα mediates its effects via two specific cell surface receptors, TNF-R1 (p55) and TNF-R2 (p75). TNF-R1 is mainly activated via soluble TNFα (sTNFα), while membrane bound TNFα (mTNFα) is the principal ligand of TNF-R2. Recent data suggest that signaling via mTNFα/TNF-R2 plays a critical role in the pathogenesis of CD. Signaling via TNF-R2 leads to activation of NFκB, a strong proinflammatory transcription factor. Concurrent inhibition of proapoptotic factor may further promote inflammation. TNF receptors exist in a soluble form generated by shedding of the extracellular portions. These soluble receptors have preserved ligand binding capacity and might thus contribute to the regulation of ligand availability. The finding that mTNFα is capable of eliciting intracellular signaling without receptor interaction (reverse signaling) adds yet another level of complexity to TNF signaling. Novel recombinant protein-based anti-TNF strategies target different factors of this pathway, some by imitating the mechanism of action of endogenous components, such as soluble receptor p55.

IκB kinase and consecutively releases NFκB.[23] The active subunit of NFκB, p65, moves into the nucleus and directly interacts with the promoter region of several proinflammatory genes such as TNFα, IL-1, IL-6, and IL-12.[24] Inhibition of NFκB is the underlying mechanism of many established anti-inflammatory treatment strategies. Whereas salicylic acids inhibit IκB-kinase (IKK), corticosteroids stimulate IκB synthesis and additionally inhibit p65 in the nucleus by direct complexation.[25]

For a long time TNF-R1 was considered the principal mediator of TNF signal transduction. This view was mainly based on the fact that TNF-R2 binds sTNFα with a 20-fold lower binding affinity than TNF-R1.[26] However, it was then found that TNF-R2 is preferentially activated by mTNFα with high affinity in a paracrine fashion and probably in an autocrine loop, as well.[27] The main cellular response of

TNF-R2 is the activation of NFκB. These observations make it possible that the mTNFα/TNF-R2 system could play an important immunoregulatory role at the local level.

To assess the net cellular response of TNF signaling it is important to consider that the two principal effects of TNFα signal transduction, activation of NFκB and induction of apoptosis, can be antagonistic to each other. NFκB inhibits apoptosis through upregulation of antiapoptotic signaling factors.[28,29] It is conceivable that concurrent inhibition of apoptosis in addition to the upregulation of proinflammatory cytokines is an important proinflammatory mechanism of NFκB.

The hypothesis that TNFα signaling via TNF-R2 may play an independent pathogenic role could be demonstrated impressively in a transgenic mouse model overexpressing two alleles of the human TNF-R2, which can be activated by murine TNFα equally efficient as the mouse receptor.[30] These mice spontaneously developed a severe general inflammatory syndrome involving pancreas, liver, kidneys, and lungs. NFκB was constitutively increased in mononuclear cells of the peripheral blood.

Strong direct evidence for a crucial role of TNF-R2 signaling in the pathogenesis of CD has been provided by recent experimental studies.[31] In this work, it could be shown that TNF-R2 expression is significantly increased on mononuclear cells in peripheral blood and in the lamina propria of patients with active CD. In a murine model system of experimental colitis, overexpression of TNF-R2 led to severe aggravation of colitis. This was mediated by induction of a Th1-like cytokine profile of mononuclear cells and by inhibition of apoptosis in the lamina propria. Both pathomechanisms are relevant in CD, as well. These data contribute to understanding of the differential clinical efficacy of different anti-TNF-strategies (see below).

Elevated levels of soluble TNF-R1 and TNF-R2 can be detected in the urine of patients with CD and UC and correlate with high disease indices.[32] Elevated levels of soluble TNF-R2 can also be detected in various severe inflammatory or autoimmune disease states, that is, sepsis, chronic viral hepatitis, acute pancreatitis, systemic lupus erythematodes, rheumatoid arthritis, and AIDS.[33–37] It is unclear, however, how these elevated levels of soluble TNF-R2 can be explained. They could represent a regulatory mechanism to bind and inactivate soluble TNFα ligand.

TNF signaling thus represents a complex network of a ligand in a membrane-bound and soluble form, two receptors with different ligand affinities and intracellular signaling cascades, soluble forms of receptors that modify the availability of ligand and reverse signaling of the membrane-bound ligand without receptor involvement. These different components of TNF signaling are targeted by the anti-TNF-strategies available or are mimicked in order to block TNF signaling.

Currently there are four recombinant anti-TNF-antibodies and two recombinant fusion proteins available or under clinical testing in IBD: infliximab, adalimumab, certolizumab, CDP 571, etanercept, and onercept. Although sharing TNF-antagonizing properties, these drugs display striking differences in clinical efficacy in distinct inflammatory disorders. The elucidation of their different mechanisms of action in distinct clinical settings helps us to learn more about details and peculiarities of TNF signal transduction in distinct inflammatory disorders. The different features of TNFα antagonists are summarized in Table 5.1 and Figure 5.3.

TABLE 5.1
Biologics and Their Targets—A Synopsis of Biologics, Their Targets, and the Type of Substance

Substance	Construct	Binding Properties	Complement Fixation	Antibody-Dependent Cytotoxicity	Induction of Apoptosis	Efficacy in IBD	Efficacy in Rheumatological Disorders
Inflximab	Chimeric mouse-human IgG_1 mAb	sTNF mTNF	Yes	Yes	Yes	Yes	Yes
Adalimumab	Human IgG_1 mAb	sTNF mTNF	Yes	Yes	Yes	Yes	Yes
Certolizumab (CDP870)	Humanized Fab fragment linked to PEG	sTNF mTNF	No	No	?	(Yes)	(Yes)
CDP 571	Humanized IgG_4 mAb	sTNF mTNF	No	No	?	No	No
Etanercept	IgG_1 Fc fragment linked to 2 p75	sTNF	No	No	No	No	Yes
Onercept	Soluble p55	sTNF mTNF	No	No	?	?	?

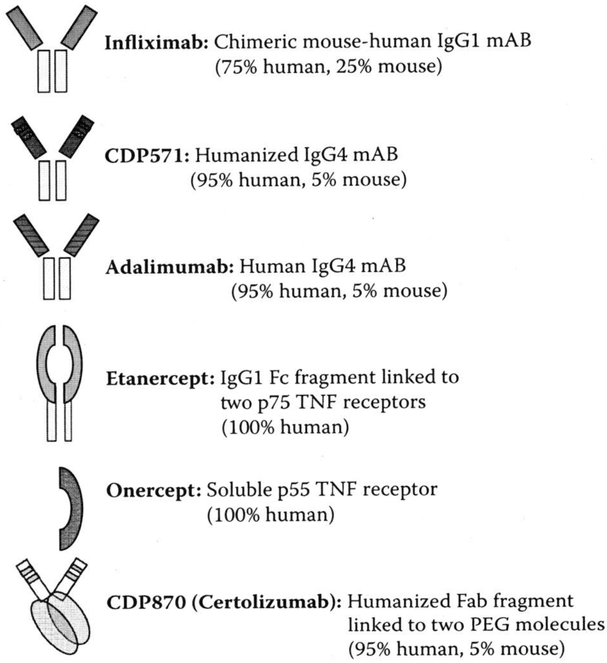

FIGURE 5.3 Recombinant anti-TNF strategies. Infliximab was the first recombinant anti-TNF-antibody. The residual murine proportions have been considered a major drawback in clinical practice because the mouse epitopes were ascribed to development of HACAs presumably responsible for allergic side effects. In an attempt to reduce immunogenicity, humanized antibodies were generated, further reducing the proportions of murine origin (CDP571). Adalimumab is a fully human monoclonal antibody. However, antibodies to CDP571 and adalimumab can still be detected (human antihuman antibodies). Etanercept and onercept represent extracellular portions of p75 and p55 with preserved ligand binding capacity, thus probably imitating the regulatory role of endogenous soluble receptor. CDP870 distinguished itself by the attachment of two polyethylene glycol molecules, ensuring prolonged continuous release after subcutaneous application.

Infliximab

Most clinical experience exists for infliximab (cA2). Infliximab is a genetically engineered immunoglobulin (IgG_1) murine–human monoclonal antibody with a constant region of human IgG1κ-Ig representing 75 percent of the molecule and a variable region of a monoclonal mouse antihuman antibody representing 25 percent. Infliximab binds both sTNFα and mTNFα and most likely blocks the interaction of TNFα with the TNF receptors this way.[38,39] Clinical studies were first performed in rheumatoid arthritis, where good efficacy could be shown.[40] Other rheumatological disorders followed including psoriatic arthritis,[41] juvenile chronic arthritis,[42] psoriasis,[43] and ankylosing spondarthritis.[44] The first clinical data on infliximab in CD were published more than 10 years ago. In a placebo-controlled trial in 108 patients with moderate to severe CD comparing dosages of 5, 10, and 20 mg/kg bodyweight with placebo, 81 percent of the patients given 5 mg/kg, 50 percent of the patients given 10 mg/kg, and 64 percent of the patients given 20 mg/kg bodyweight had a clinical response after 4 weeks versus 17 percent in the placebo group.[45] Overall, 33 percent of the patients treated with verum achieved clinical remission versus 4 percent in the placebo group. Additional studies followed suggesting efficacy in the maintenance of remission by a routine administration every 8 weeks.[46,47] Other indications for infliximab in CD are induction and maintenance therapy of fistulizing disease.[48,49] The observation that infliximab, as opposed to many other treatment regimens, was able to induce clinical remission accompanied by mucosal healing when surveyed endoscopically led to the concept of "mucosal healing" as an independent treatment goal and end point in studies.[50] This, however, is still a matter of debate. In addition, there still exists controversy with regard to the role of concomitant immunosuppressive therapy.[51] In case of loss of efficacy, the infliximab dose can be increased from 5 to 10 mg/kg bodyweight.[52]

In spite of the Th1/Th2 paradigm in the pathogenesis of CD and UC emphasizing the pivotal role of TNF in CD opposed to UD, more recent phase III trials have demonstrated an almost equal potency of infliximab in the treatment of UC.[53] The Acute Ulcerative Colitis Trials (ACT1 and ACT 2) each enrolled 364 patients with moderate to severe active disease who had been intolerant to or who had failed to improve with conventional therapy. Induction of remission was performed with infusions given at 0, 2, and 6 weeks; for maintenance, infliximab was given every 8 weeks until 46 weeks (ACT1) or 22 weeks (ACT2). Response rates were about 65 percent versus 33 percent of placebo at week 8; remission rates were about 35 percent versus 10 percent in the placebo group. Clinical response could be sustained by continued treatment in a significant proportion of the patients over 54 weeks. To date, 34 controlled studies including a total of 894 patients with UC have confirmed the efficacy of infliximab in this disease.[54] Of note, several trials have addressed the efficacy of infliximab as rescue therapy in fulminant disease and could show that infliximab is capable of preventing emergency proctocolectomy in the majority of cases.[55,56]

The general tolerability of infliximab is good. Most adverse side effects are probably related to the immunogenicity of infliximab that leads to the formation of antibodies to infliximab (ATI; human antichimeric antibodies = HACA) in 3 to 17 percent of the cases (Baert F NEJM 2993)[57] and include infusion reactions (in 4 to 16 percent), delayed hypersensititivity-like reactions (in 1 percent), and autoimmune phenomena. A rare, but severe side effect is the exacerbation of latent infections, especially tuberculosis.[58,59] This reflects the important role of TNFα in the antimicrobial defense of the immune system. Available long-term safety data provide no evidence for an increased risk for lymphoproliferative disorders or others malignancies.[60]

Adalimumab

Adalimumab is a recombinant human IgG_1 monoclonal antibody. Adalimumab binds to sTNFα and mTNFα and induces apoptosis in monocytes.[61] A large phase III study has shown that adalimumab is efficacious and well tolerated in the long-term treatment of rheumatoid arthritis.[62] Short-term efficacy could also be shown for psoriatic arthritis.[63] Adalimumab is now well established in the treatment of CD. In a large, randomized, controlled dose-ranging trial in 299 anti-TNF naïve patients with moderate to severe CD, adalimumab was given as subcutaneous injections on weeks 0 and 2 of 40 mg/20 mg, 80 mg/40 mg, or 160 mg/80 mg or placebo.[64] After 4 weeks, the dosage 160 mg/80 mg was able to achieve a remission rate of 36 percent versus 12 percent in placebo. Adalimumab is efficacious in the maintenance of remission, as well.[65] Interestingly, uncontrolled studies showed clinical efficacy of adalimumab in cases of infliximab intolerability and refractoriness.[66] The theoretical advantage of adalimumab with regard to adverse side effects is reduced immunogenicity due to the lack of mouse epitopes. However, development of human antihuman antibodies (anti-adalimumab antibodies = AAA) is still observed. Side effects include injection-site reactions and infections.

Certolizumab

Certolizumab (CDP870) is a Fab fragment of a humanized monoclonal anti-TNF antibody attached to polyethylene glycol molecules. Although binding to both sTNFα and mTNFα, certolizumab does not induce apoptosis in peripheral blood lymphocytes and monocytes.[67] The clinical efficacy of certolizumab has not yet been established definitively. In a phase II trial, certolizumab showed clinical response after 12 weeks only in a subgroup of patients with elevated C-reactive protein (CRP) level as sign of systemic inflammation.[68] In a large controlled study in 662 patients these observation could be confirmed.[69] In the subgroup of patients with elevated CRP, certolizumab achieved a clinical response beyond placebo, but not in the total group. Clinical remission was not achieved either. Data in terms of efficacy in maintenance treatment with certolizumab were equivocal, as well.[70] Adverse events include injection-site reactions, exacerbations of CD, and infections.

CDP571

CDP571 (nerelimomab) is a humanized anti-TNF antibody derived from a mouse antihuman TNFα monoclonal antibody. The complementary-determining region of this antibody has been linked to a human IgG_4 antibody. Binding occurs to both soluble and membrane-bound TNFα; no data have been reported regarding induction of apoptosis. CDP571 has shown only temporary marginal efficacy 2 weeks after treatment and in a subgroup of patients with elevated CRP.[71] Another study in 71 patients with steroid dependent CD utilized CDP571 in different doses versus placebo.[72] The primary end point of steroid freeness was achieved at week 16. However, another larger trial in 271 patients could not confirm these results.[72] Tolerability and safety data are limited and are comparable to infliximab.

Etanercept

Etanercept is a fusion protein consisting of two identical chains of recombinant human anti-TNF receptor p75 monomers fused to the Fc domain of human IgG_1. Etanercept binds only to sTNFα, but not to μTNFα. Unlike infliximab and adulimumab, etanercept fails to induce apoptosis in lamina propria T lymphocytes.[73] Etanercept has shown good efficacy in rheumatoid arthritis and the indication has been extended to many other rheumatoid disorders successively including psoriatic arthritis, juvenile chronic arthritis, plaque psoriasis, and ankylosing spondarthritis.[74–80] Interestingly, in CD etanercept has failed to show any efficacy.[81]

Onercept

Onercept is a recombinant form of human soluble TNF-R1 (p55). In an initial pilot study on 12 patients with CD, Onercept showed clinical response beyond placebo.[82] This result could not be confirmed in a phase II study.[83] More importantly, however, phase III trials on onercept in psoriasis had to be discontinued because of two cases of severe sepsis, one of which was lethal.

MECHANISMS OF ACTION

Soon after clinical efficacy of infliximab was shown, the mechanisms of action *in vivo* were investigated. In preclinical studies using a transfection model system with expression of uncleavable mTNFα, it had been shown that infliximab has a higher binding affinity to mTNFα than etanercept, while both agents bind sTNFα.[84] These results could be confirmed on activated lamina propria T lymphocytes.[73] The potential of infliximab and etanercept to neutralize biologically active sTNFα was similar.[85]

Because defective apoptosis of mononuclear cells seems to play a role in the pathogenesis of IBD,[86] it was soon hypothesized that the rapid clinical effect of infliximab might be due to induction of apoptosis. And, in fact, in patients with CD treated with infliximab, dose-dependent induction of peripheral blood monocyte apoptosis by a CD95/CD95L-dependent pathway within a few hours could be shown.[87] This effect could be imitated by F(ab)2 fragments of infliximab which lack the Fc domain. The finding that infliximab induces increased apoptosis of lamina propria T cells in patients with CD is probably even more relevant.[88]

The ability to bind mTNFα correlates with the inducibility of apoptosis, since only infliximab, but not etanercept, induces apoptosis in lamina propria T lymphocytes. Interestingly, induction of apoptosis in cultured peripheral blood monocytes could also be shown for adalimumab, which shares the isotype class IgG_1 with infliximab.[61] In this experimental setting, adalimumab, infliximab, and etanercept reduced the levels of sTNFα in the culture, but only adalimumab and infliximab reduced the production of IL-10 and IL-12.

For infliximab and adalimumab additional relevant mechanisms of actions may be the ability to fix complement and antibody-dependent cytotoxicity. These are features that may be referred to their common subtype class IgG_1.

However, binding to mTNFα does not always lead to blockade of TNF signaling, but can elicit reverse transsignaling. In an intriguing experimental approach, wild-type and serin-replaced mutant forms of mTNFα were stably transfected in human Jurkat T cells. Treatment with infliximab, but not with etanercept, led to IL-10 production, apoptosis, and G_0/G_1 cell cycle arrest. These effects were abolished by substitution of all three cytoplasmic serine residues of mTNFα by alanine residues.[89] This work not only reveals an additional mechanism of action of infliximab, but also provides insight into a so-far unrecognized role of mTNFα in chronic inflammation.

OTHER ANTI-TNF STRATEGIES

CNTO-148 (golimumab) is another fully human anti-TNF antibody that can be administered subcutaneously. Clinical testing has been restricted to rheumatoid arthritis so far and preliminary unpublished data suggest statistically significant reductions of signs and symptoms of rheumatoid arthritis. Afelimomab is a monoclonal anti-TNF antibody F(ab′)2 fragment. Published data exist from controlled clinical trials in patients with severe sepsis syndrome. Afelimomab slightly reduces 28-day all-cause mortality in a subgroup of patients with sepsis with elevated IL-6 levels.[90,91] No data or press releases, however, are available on these agents in IBD.

RDP58 is a decapeptide that inhibits synthesis of proinflammatory cytokines such as TNF and is thus a potentially interesting drug in IBD.[92] RDP58 was investigated in two studies in a total of 127 patients with mild to moderate active UC.[93] In doses of 200 and 300 mg/day more than 70 percent achieved improvement with a CAI of 3 or less at day 28 versus 43 percent treated with placebo ($p = 0.016$). RDP58 was well tolerated.

OTHER CYTOKINE-BASED AND ANTI-CD4$^+$ T-CELL APPROACHES

Many of the proinflammatory cytokines considered pathogenetically relevant in IBD have become targets of novel recombinant treatment approaches (Table 5.2). Fontolizumab is a humanized anti-IFNγ antibody developed by Protein Design Labs and has been investigated in two placebo-controlled studies in patients with CD comparing different doses and different numbers of injections.[94,95] In one study in which fontolizumab was given on days 0 and 28, fontolizumab showed significantly increased response over placebo at day 56.

The human IgG_1 monoclonal anti-IL-12p40 antibody ABT-874 from Abbott Laboratories has been successfully tested in a controlled phase II study in 79 patients with CD. Patients received seven weekly subcutaneous doses of 1 or 3 mg anti-IL12 per kilogram bodyweight or placebo and different nontreatment intervals. Anti-IL-12 showed statistically significant efficacy initially but not beyond 18 weeks.[96]

In a phase I/IIa dose escalating study, the orally administered small molecule STA 5326, inhibiting the dimer IL-12/23, showed clinical activity with a dosing of 28 mg qd and above.[97] IL-6 signaling is the specific target of MRA, a humanized monoclonal antibody to IL-6 receptor. MRA has been tested in a placebo-controlled trial in 36 patients with active CD.[98] Although well tolerated, the evidence for clinical efficacy is ambiguous. The anti-inflammatory IL-10 (from Schering Plough) has been utilized directly as recombinant protein delivered subcutaneously in three large phase III trials comprising 800 patients, but has failed to show any effect.[99–101] A very intriguing approach was pursued in an open-label study with transgenic bacteria expressing IL-10.[102] In the study, 10 patients with active CD were treated with

TABLE 5.2
Mechanisms of Action of Anti-TNF-Strategies—A Synopsis of the Available Recombinant Anti-TNF Agents with Their Distinct Properties

Target	Agent	Type
TNFα	Infliximab, adalimumab, CDP571, golimumab	Antibody F(ab)2-fragment
	Certolizumab, afelimomab	Fusion protein
	Onercept, etanercept	Small molecule
	RDP-58,	
IL-6	Atlizumab (MRA)	Antibody
IL-12	ABT-874	Antibody
IL-10	IL-10	Interleukin
	Transgenic *Lactococcus lactis*	Bacteria
IL-11	Oprelvekin	Interleukin
IL-12/IL-23	STA 5326, CNTO 1275	Small molecule
α4-Integrin	Natalizumab, MLN-02	Antibody
ICAM-1	Alicaofrsen	Antisense oligonucleotide
INFγ	Fontolizumab	Antibody
IL-2-R (CD25)	Basiliximab, daclizumab	Antibody
CD 3	Visilizumab	Antibodies
Monocytes	Sagramostim	Growth factor
MAP kinase	CNI-1493, BIRB 796	Small molecule

genetically modified *Lactococcus lactis* in which the thymidylate synthase gene was replaced with a synthetic sequence encoding mature human IL-10. The treatment was tolerated without adverse effects and a clinical effect could be seen. This study can be considered a proof of principle study for the feasibility of administration of genetically modified bacteria.

Subcutaneous IL-11 (from Genetics Institute) was tested in a placebo-controlled trial in patients with mild to moderate CD.[103] However, IL-11 failed to achieve statistically significant clinical efficacy beyond placebo. Steroid resistance is a major problem in the treatment of IBD. The *in vitro* observation that IL-2 renders lymphocytes steroid resistant provided the rationale for anti-IL-2 approaches in the treatment of UC. Basiliximab is an anti-IL-2 receptor (CD25) antibody. In an uncontrolled open-label study in 10 patients with steroid-resistant UC, a single bolus of 40 mg basiliximab was administered.[104] At week 8, 9 of 10 patients were in clinical remission; 8 of 9 responders relapsed, but remission could be re-achieved with steroids and immunosuppression, so that 7 of 10 patients were in full remission at week 24, 5 off all steroids.

However, these results could not be confirmed in a larger, placebo-controlled trial with daclizumab, another humanized monoclonal antibody to the IL-2 receptor (CD25). In a placebo-controlled trial in 159 patients with moderate UC, intravenous daclizumab was tested in different doses (1 mg/kg bodyweight at weeks 0 and 4, or 2 mg/kg at weeks 0, 2, 4, and 6) versus placebo.[105] There was no clinical effect beyond placebo.

Visilizumab is a humanized IgG_2 monoclonal antibody which binds to CD3 and induces T-cell apoptosis. In an open-label trial enrolling 32 patients with severe steroid refractory UC, a bolus of visilizumab on 2 consecutive days was administered. At day 30, 84 percent of patients demonstrated a clinical response; 41 percent achieved remission. However, in the first year of follow-up, 55 percent required either salvage therapy or colectomy.[106]

The pivotal role of the $CD4^+$ T-helper cell in the pathogenesis of IBD is also taken into account in treatment strategies that attempt to downregulate T-cell recruitment to the inflamed focus by blocking adhesion molecules. Early studies blocking the intercellular adhesion molecule 1 (ICAM-1) with antisense phosphorothioate oligonucleotide alicaforsen showed extremely diverging results between CD and UC, depending on the route of administration.

In 299 patients with active CD, alicaforsen was administered intravenously for 2 or 4 weeks versus placebo. The primary end point was steroid-free remission week 14. There was no clinically relevant effect of alicaforsen compared to placebo.[107] In a subsequent study in 331 subjects with CD treated with a modified protocol, these negative results were confirmed.[108] Alicaforsen given intravenously in patients with CD thus has no measurable clinical effect.

Two trials, however, studied the alicaforsen given in different doses as an enema formulation of 60 mL in patients with mild to moderately active UC. Alicaforsen 2 and 4 mg/mL once daily for 28 days improved the disease activity index significantly in about 70 percent versus 11.5 percent placebo at month 3.[109] A subsequent study, however, failed to show significant clinical efficacy at week 6, but confirmed a long-term effect from week 18 to 30.[110] When compared to the mesalamine in a trial with a treatment period limited to 6 weeks, alicaforsen showed equal short-term efficacy at week 6, but this effect lasted 128 and 146 days beyond discontinuation of treatment versus 54 days in the mesalamine group.[111] Alicaforsen may thus be a promising option for topical treatment in UC.

A monoclonal antibody approach was pursued with natalizumab, a humanized IgG_4 antibody to α4 integrin. While effective in multiple sclerosis, which is also T-cell mediated, natalizumab failed to show any clinically relevant efficacy in several large studies in active CD.[112–115] Of note, three cases of JC virus-associated progressive multifocal leukoencephalopathy under natalizumab treatment have been reported.[116,117]

MLN-02 is another humanized IgG_1 monoclonal anti-integrin antibody directed against the heterodimeric epitope a4b7 (Figure 5.4). This antibody showed efficacy in UC.[118] Concerns regarding the risk of PML under MLN-02 treatment have been allayed by the following considerations. Natalizumab blocks specifically the α4 subunit of integrins which is incorporated in both the α4β1 heterodimer that interacts with VCAM-1 in the brain and the α4β7 heterodimer that interacts with MAdCAM-1 specifically expressed in the gut. Natalizumab thus interferes with T-cell homing in both the brain and the gut. As it is directed against the heterodimer α4β7 as a whole, MLN-02 only blocks MAdCAM-1 mediated T-cell homing in the gut, thus preventing the negative impact on immune defense in the brain. It remains to be seen if these theoretical considerations really apply in clinical practice in the long run.

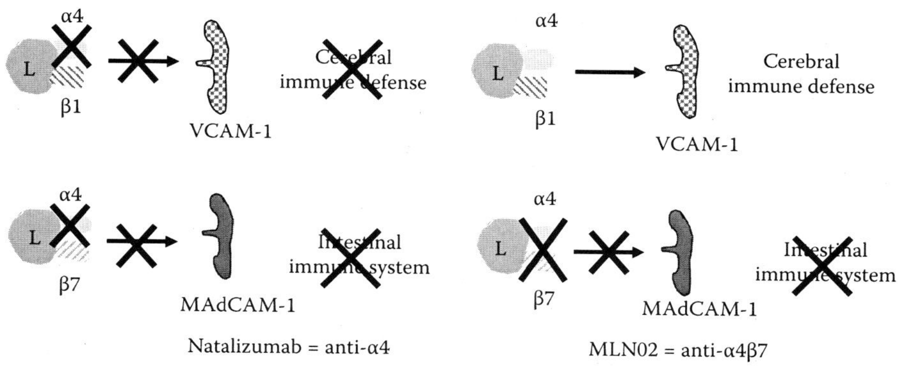

FIGURE 5.4 Differential mechanisms of action of anti-integrins. Inhibition of T-cell homing is a therapeutic approach to decrease inflammation. Migration of T cells is mediated by distinct adhesion molecules which interact specifically. T cells can pass the blood–brain barrier through interaction of the α4β1 integrin heterodimer with the adhesion molecule VCAM-1. For homing into the gut, T cells expressing the heterodimer α4β7 interact with the adhesion molecule MAdCAM-1, which is specifically expressed in the gut. The monoclonal anti-integrin antibody natalizumab blocks the α4 subunit and thus inhibits homing of T cells to both the brain and gut. The antigenic epitope of MLN-02 is the heterodimer α4β7 as a whole. MLN-02 thus blocks homing of T cells to the gut, but not to the brain.

The granulocyte–macrophage colony-stimulating factor sagramostim is a hematopoietic growth factor. The rationale for the utilization of sagramostim for treatment of IBD is that sagramostim stimulates immune cells of the innate immune system and thus strengthens the immediate unspecific response of the intestinal immune system toward detrimental microbial or nonmicrobial stimuli. In a randomized controlled study in 124 patients with moderate to severe CD sagramostim was given 6 μg/kg bodyweight subcutaneously for 56 days.[119] The primary end point of clinical response defined as a decrease of the CDAI of 70 points was missed, but secondary end points of decrease of CDAI of 100 points, clinical remission, and improvement of quality of life were achieved with statistical significance. Sagramostim may have a potential role in the treatment of CD.

RDP58 and STA 5326 are not the only immune-modulating small molecules potentially effective in IBD. CNI-1493 is a guanylhydrazone, which inhibits mitogen-activated protein kinases (MAPKs). In an open-label trial, 12 patients were given either 8 or 25 mg/m^2 body surface for 12 days.[120] At weeks 4 and 8, there was a significant decrease in clinical activity index CDAI of 120 points; the majority displayed mucosal and fistula healing as well as steroid tapering. Of note, in a large multicenter placebo-controlled trial in 284 patients with moderate to severe CD the MAPK-inhibitor BIRB 796 failed to show any clinical effect including activity index, CRP, quality of life, and endoscopic assessment.[121] These diverging results can either be due to differences in the biologic effect at the cellular level in spite of the common target or be due to study bias in the open-label design of the study of Hommes et al.[120] A large phase III trial on CNI-1493 should address this issue.

CONCLUSIONS

It is striking that only 50 to 70 percent of all patients respond clinically to distinct anticytokine strategies such as anti-TNF or anti-IL-12 antibodies given the pivotal role of these cytokines in murine models of colitis. In fact, there is increasing evidence that the classic Th1-Th2 paradigm derived from the focus on T-cell immunology may be too simplistic to explain IBD. Two considerations may be relevant. CD and UC may represent common final pathophysiological pathways of different etiological triggers. These different pathomechanisms may involve different signaling cascades. In addition, individual cytokines may play different roles in different phases of the disease. In particular, CD is characterized by two distinct phases: an initial, inductive phase and a chronic effector phase.

Increasing evidence suggests that for the inductive phase, cell components of the innate immune system, such as monocytes/macrophages play a predominant role.[122,123] Monocytes/macrophages are important antigen presentation cells and produce proinflammatory cytokines such as IL-1, TNFα, IL-6, and possibly IL-12 and IL-18. The first CD susceptibility gene codes for the intracellular protein NOD2/CARD15, which represents an intracellular receptor for bacterial products, in particular muramyl dipeptide, are strongly expressed in monocytes/macrophages and epithelial cells.[124–126] *In vitro* studies on loss of function mutations of NOD2 revealed that wild-type NOD2/CARD15 activates NFκB.

This apparently paradoxical finding has led to the view that in CD the initial response of the innate immune system to antigens of the physiological intestinal flora or the food may be defective, resulting in a detrimental activation of the acquired immune system with chronic inflammation. In this phase, dysregulation of T cells and T-cell-associated cytokine regulatory networks plays the crucial role. In fact, in a murine model of a loss of function mutation of NOD2, elevated NFκB activation in response to muramyl dipeptide could be shown with increased susceptibility to bacterial induced inflammation.[127,128]

It is thus conceivable that the blockade of a certain cytokine—although a key proinflammatory cytokine in colitis models—can be beneficial under certain circumstances, but ineffective or even detrimental otherwise. This is likely to apply not only to the targets of anticytokine strategies currently available, but also to all potential future targets such as IL-13, IL-18, IL-23, IL-27, and others. For medical and cost-effectiveness reasons it will be challenging to identify predictive markers for the responsiveness of an individual patient to a distinct cytokine-targeted treatment strategy. With regard to safety, the serious adverse events following natalizumab treatment and the risk of exacerbation of silent infections and sepsis under anti-TNF treatment should alert one to the two faces of Janus inherent to T cell and T-cell associated cytokine targeted strategies.

REFERENCES

1. Podolsky, D., Inflammatory bowel disease. *N Engl J Med* 2002. 347: 417–429.
2. Fiocchi, D., Inflammatory bowel disease: Etiology and pathogenesis. *Gastroenterology* 1998. 115: 182–205.

3. Nagler-Anderson, C., Tolerance and immunity in the intestinal immune system. *Crit Rev Immunol* 2000. 20: 103–120.
4. Neurath, M. F., Finotto, S. and Glimcher, L. H., The role of Th1/Th2 polarization in mucosal immunity. *Nat Med* 2002. 8: 567–573.
5. Strober, W., Ludviksson, B. R. and Fuss, I. J., The pathogenesis of mucosal inflammation in murine models of inflammatory bowel disease and Crohn disease. *Ann Intern Med* 1998. 128: 848–856.
6. Fuss, I. J., Neurath, M., Boirivant, M., Klein, J. S., de la Motte, C., Strong, S. A., Fiocchi, C. and Strober, W., Disparate CD4+ lamina propria (LP) lymphokine secretion profiles in inflammatory bowel disease. Crohn's disease LP cells manifest increased secretion of IFN-gamma, whereas ulcerative colitis LP cells manifest increased secretion of IL-5. *J Immunol* 1996. 157: 1261–1270.
7. Heller, F., Florian, P., Bojarski, C., Richter, J., Christ, M., Hillenbrand, B., Mankertz, J., Gitter, A. H., Burgel, N., Fromm, M., Zeitz, M., Fuss, I., Strober, W. and Schulzke, J. D., Interleukin-13 is the key effector Th2 cytokine in ulcerative colitis that affects epithelial tight junctions, apoptosis, and cell restitution. *Gastroenterology* 2005. 129: 550–564.
8. Fuss, I. J., Boirivant, M., Lacy, B. and Strober, W., The interrelated roles of TGF-beta and IL-10 in the regulation of experimental colitis. *J Immunol* 2002. 168: 900–908.
9. Maul, J., Loddenkemper, C., Mundt, P., Berg, E., Giese, T., Stallmach, A., Zeitz, M. and Duchmann, R., Peripheral and intestinal regulatory CD4+ CD25(high) T cells in inflammatory bowel disease. *Gastroenterology* 2005. 128: 1868–1878.
10. Bettelli, E., Oukka, M. and Kuchroo, V. K., T(H)-17 cells in the circle of immunity and autoimmunity. *Nat Immunol* 2007. 8: 345–350.
11. Wirtz, S. and Neurath, M. F., Animal models of intestinal inflammation: New insights into the molecular pathogenesis and immunotherapy of inflammatory bowel disease. *Int J Colorectal Dis* 2000. 15: 144–160.
12. Holtmann, M. H., Schutz, M., Galle, P. R. and Neurath, M. F., Functional relevance of soluble TNF-alpha, transmembrane TNF-alpha and TNF-signal transduction in gastrointestinal diseases with special reference to inflammatory bowel diseases. *Z Gastroenterol* 2002. 40: 587–600.
13. Eissner, G., Kirchner, S., Lindner, H., Kolch, W., Janosch, P., Grell, M., Scheurich, P., Andreesen, R. and Holler, E., Reverse signaling through transmembrane TNF confers resistance to lipopolysaccharide in human monocytes and macrophages. *J Immunol* 2000. 164: 6193–6198.
14. Nielsen, O. H., Vainer, B., Madsen, S. M., Seidelin, J. B. and Heegaard, N. H., Established and emerging biological activity markers of inflammatory bowel disease. *Am J Gastroenterol* 2000. 95: 359–367.
15. Reinecker, H. C., Steffen, M., Witthoeft, T., Pflueger, I., Schreiber, S., MacDermott, R. P. and Raedler, A., Enhanced secretion of tumour necrosis factor-alpha, IL-6, and IL-1 beta by isolated lamina propria mononuclear cells from patients with ulcerative colitis and Crohn's disease. *Clin Exp Immunol* 1993. 94: 174–181.
16. Plevy, S. E., Landers, C. J., Prehn, J., Carramanzana, N. M., Deem, R. L., Shealy, D. and Targan, S. R., A role for TNF-alpha and mucosal T helper-1 cytokines in the pathogenesis of Crohn's disease. *J Immunol* 1997. 159: 6276–6282.
17. Pender, S. L., Tickle, S. P., Docherty, A. J., Howie, D., Wathen, N. C. and MacDonald, T. T., A major role for matrix metalloproteinases in T cell injury in the gut. *J Immunol* 1997. 158: 1582–1590.
18. Neurath, M., Fuss, I. and Strober, W., TNBS-colitis. *Int Rev Immunol* 2000. 19: 51–62.
19. Kuhn, R., Lohler, J., Rennick, D., Rajewsky, K. and Muller, W., Interleukin-10-deficient mice develop chronic enterocolitis. *Cell* 1993. 75: 263–274.

20. Singh, B., Read, S., Asseman, C., Malmstrom, V., Mottet, C., Stephens, L. A., Stepankova, R., Tlaskalova, H. and Powrie, F., Control of intestinal inflammation by regulatory T cells. *Immunol Rev* 2001. 182: 190–200.
21. Atreya, R., Mudter, J., Finotto, S., Mullberg, J., Jostock, T., Wirtz, S., Schutz, M., Bartsch, B., Holtmann, M., Becker, C., Strand, D., Czaja, J., Schlaak, J. F., Lehr, H. A., Autschbach, F., Schurmann, G., Nishimoto, N., Yoshizaki, K., Ito, H., Kishimoto, T., Galle, P. R., Rose-John, S. and Neurath, M. F., Blockade of interleukin 6 trans signaling suppresses T-cell resistance against apoptosis in chronic intestinal inflammation: Evidence in Crohn disease and experimental colitis in vivo. *Nat Med* 2000. 6: 583–588.
22. Loetscher, H., Brockhaus, M., Dembic, Z., Gentz, R., Gubler, U., Hohmann, H. P., Lahm, H. W., Van Loon, A. P., Pan, Y. C., Schlaeger, E. J. et al., Two distinct tumour necrosis factor receptors—Members of a new cytokine receptor gene family. *Oxf Surv Eukaryot Genes* 1991. 7: 119–142.
23. Auphan, N., DiDonato, J. A., Rosette, C., Helmberg, A. and Karin, M., Immunosuppression by glucocorticoids: inhibition of NF-kappa B activity through induction of I kappa B synthesis. *Science* 1995. 270: 286–290.
24. Neurath, A. R., Strick, N., Li, Y. Y. and Debnath, A. K., *Punica granatum* (pomegranate) juice provides an HIV-1 entry inhibitor and candidate topical microbiocide. *Ann N Y Acad Sci* 2005. 1056: 311–327.
25. Yin, M. J., Yamamoto, Y. and Gaynor, R. B., The anti-inflammatory agents aspirin and salicylate inhibit the activity of I(kappa)B kinase-beta. *Nature* 1998. 396: 77–80.
26. Grell, M., Douni, E., Wajant, H., Lohden, M., Clauss, M., Maxeiner, B., Georgopoulos, S., Lesslauer, W., Kollias, G., Pfizenmaier, K. and Scheurich, P., The transmembrane form of tumor necrosis factor is the prime activating ligand of the 80 kDa tumor necrosis factor receptor. *Cell* 1995. 83: 793–802.
27. Haas, E., Grell, M., Wajant, H. and Scheurich, P., Continuous autotropic signaling by membrane-expressed tumor necrosis factor. *J Biol Chem* 1999. 274: 18107–18112.
28. Wang, C. Y., Mayo, M. W., Korneluk, R. G., Goeddel, D. V. and Baldwin, A. S., Jr., NF-kappaB antiapoptosis: Induction of TRAF1 and TRAF2 and c-IAP1 and c-IAP2 to suppress caspase-8 activation. *Science* 1998. 281: 1680–1683.
29. Rothe, M., Sarma, V., Dixit, V. M. and Goeddel, D. V., TRAF2-mediated activation of NF-kappa B by TNF receptor 2 and CD40. *Science* 1995. 269: 1424–1427.
30. Douni, E. and Kollias, G., A critical role of the p75 tumor necrosis factor receptor (p75TNF-R) in organ inflammation independent of TNF, lymphotoxin alpha, or the p55TNF-R. *J Exp Med* 1998. 188: 1343–1352.
31. Holtmann, M. H., Douni, E., Schutz, M., Zeller, G., Mudter, J., Lehr, H. A., Gerspach, J., Scheurich, P., Galle, P. R., Kollias, G. and Neurath, M. F., Tumor necrosis factor-receptor 2 is up-regulated on lamina propria T cells in Crohn's disease and promotes experimental colitis in vivo. *Eur J Immunol* 2002. 32: 3142–3151.
32. Hadziselimovic, F., Emmons, L. R. and Gallati, H., Soluble tumour necrosis factor receptors p55 and p75 in the urine monitor disease activity and the efficacy of treatment of inflammatory bowel disease. *Gut* 1995. 37: 260–263.
33. Schroder, J., Stuber, F., Gallati, H., Schade, F. U. and Kremer, B., Pattern of soluble TNF receptors I and II in sepsis. *Infection* 1995. 23: 143–148.
34. Marinos, G., Naoumov, N. V., Rossol, S., Torre, F., Wong, P. Y., Gallati, H., Portmann, B. and Williams, R., Tumor necrosis factor receptors in patients with chronic hepatitis B virus infection. *Gastroenterology* 1995. 108: 1453–1463.
35. de Beaux, A. C., Goldie, A. S., Ross, J. A., Carter, D. C. and Fearon, K. C., Serum concentrations of inflammatory mediators related to organ failure in patients with acute pancreatitis. *Br J Surg* 1996. 83: 349–353.

36. Gabay, C., Cakir, N., Moral, F., Roux-Lombard, P., Meyer, O., Dayer, J. M., Vischer, T., Yazici, H. and Guerne, P. A., Circulating levels of tumor necrosis factor soluble receptors in systemic lupus erythematosus are significantly higher than in other rheumatic diseases and correlate with disease activity. *J Rheumatol* 1997. 24: 303–308.
37. Godfried, M. H., van der Poll, T., Jansen, J., Romijin, J. A., Schattenkerk, J. K., Endert, E., van Deventer, S. J. and Sauerwein, H. P., Soluble receptors for tumour necrosis factor: A putative marker of disease progression in HIV infection. *AIDS* 1993. 7: 33–36.
38. Siegel, S. A., Shealy, D. J., Nakada, M. T., Le, J., Woulfe, D. S., Probert, L., Kollias, G., Ghrayeb, J., Vilcek, J. and Daddona, P. E., The mouse/human chimeric monoclonal antibody cA2 neutralizes TNF in vitro and protects transgenic mice from cachexia and TNF lethality in vivo. *Cytokine* 1995. 7: 15–25.
39. Agnholt, J., Dahlerup, J. F. and Kaltoft, K., The effect of etanercept and infliximab on the production of tumour necrosis factor alpha, interferon-gamma and GM-CSF in in vivo activated intestinal T lymphocyte cultures. *Cytokine* 2003. 23: 76–85.
40. Maini, S. R., Infliximab treatment of rheumatoid arthritis. *Rheum Dis Clin North Am* 2004. 30: 329–347, vii.
41. Antoni, C. E., Kavanaugh, A., Kirkham, B., Tutuncu, Z., Burmester, G. R., Schneider, U., Furst, D. E., Molitor, J., Keystone, E., Gladman, D., Manger, B., Wassenberg, S., Weier, R., Wallace, D. J., Weisman, M. H., Kalden, J. R. and Smolen, J., Sustained benefits of infliximab therapy for dermatologic and articular manifestations of psoriatic arthritis: Results from the infliximab multinational psoriatic arthritis controlled trial (IMPACT). *Arthritis Rheum* 2005. 52: 1227–1236.
42. Gerloni, V., Pontikaki, I., Gattinara, M., Desiati, F., Lupi, E., Lurati, A., Salmaso, A. and Fantini, F., Efficacy of repeated intravenous infusions of an anti-tumor necrosis factor alpha monoclonal antibody, infliximab, in persistently active, refractory juvenile idiopathic arthritis: Results of an open-label prospective study. *Arthritis Rheum* 2005. 52: 548–553.
43. Reich, K., Nestle, F. O., Papp, K., Ortonne, J. P., Evans, R., Guzzo, C., Li, S., Dooley, L. T. and Griffiths, C. E., Infliximab induction and maintenance therapy for moderate-to-severe psoriasis: A phase III, multicentre, double-blind trial. *Lancet* 2005. 366: 1367–1374.
44. Marzo-Ortega, H., McGonagle, D., Jarrett, S., Haugeberg, G., Hensor, E., O'Connor, P., Tan, A. L., Conaghan, P. G., Greenstein, A. and Emery, P., Infliximab in combination with methotrexate in active ankylosing spondylitis: A clinical and imaging study. *Ann Rheum Dis* 2005. 64: 1568–1575.
45. Targan, S. R., Hanauer, S. B., van Deventer, S. J., Mayer, L., Present, D. H., Braakman, T., DeWoody, K. L., Schaible, T. F. and Rutgeerts, P. J., A short-term study of chimeric monoclonal antibody cA2 to tumor necrosis factor alpha for Crohn's disease. Crohn's Disease cA2 Study Group. *N Engl J Med* 1997. 337: 1029–1035.
46. Rutgeerts, P., D'Haens, G., Targan, S., Vasiliauskas, E., Hanauer, S. B., Present, D. H., Mayer, L., Van Hogezand, R. A., Braakman, T., DeWoody, K. L., Schaible, T. F. and Van Deventer, S. J., Efficacy and safety of retreatment with anti-tumor necrosis factor antibody (infliximab) to maintain remission in Crohn's disease. *Gastroenterology* 1999. 117: 761–769.
47. Hanauer, S. B., Feagan, B. G., Lichtenstein, G. R., Mayer, L. F., Schreiber, S., Colombel, J. F., Rachmilewitz, D., Wolf, D. C., Olson, A., Bao, W. and Rutgeerts, P., Maintenance infliximab for Crohn's disease: The ACCENT I randomised trial. *Lancet* 2002. 359: 1541–1549.

48. Present, D. H., Rutgeerts, P., Targan, S., Hanauer, S. B., Mayer, L., van Hogezand, R. A., Podolsky, D. K., Sands, B. E., Braakman, T., DeWoody, K. L., Schaible, T. F. and van Deventer, S. J., Infliximab for the treatment of fistulas in patients with Crohn's disease. *N Engl J Med* 1999. 340: 1398–1405.
49. Sands, B. E., Anderson, F. H., Bernstein, C. N., Chey, W. Y., Feagan, B. G., Fedorak, R. N., Kamm, M. A., Korzenik, J. R., Lashner, B. A., Onken, J. E., Rachmilewitz, D., Rutgeerts, P., Wild, G., Wolf, D. C., Marsters, P. A., Travers, S. B., Blank, M. A. and van Deventer, S. J., Infliximab maintenance therapy for fistulizing Crohn's disease. *N Engl J Med* 2004. 350: 876–885.
50. D'Haens, G., Van Deventer, S., Van Hogezand, R., Chalmers, D., Kothe, C., Baert, F., Braakman, T., Schaible, T., Geboes, K. and Rutgeerts, P., Endoscopic and histological healing with infliximab anti-tumor necrosis factor antibodies in Crohn's disease: A European multicenter trial. *Gastroenterology* 1999. 116: 1029–1034.
51. Lemann, M., Mary, J. Y., Duclos, B., Veyrac, M., Dupas, J. L., Delchier, J. C., Laharie, D., Moreau, J., Cadiot, G., Picon, L., Bourreille, A., Sobahni, I. and Colombel, J. F., Infliximab plus azathioprine for steroid-dependent Crohn's disease patients: A randomized placebo-controlled trial. *Gastroenterology* 2006. 130: 1054–1061.
52. Kaplan, G. G., Hur, C., Korzenik, J. and Sands, B. E., Infliximab dose escalation vs. initiation of adalimumab for loss of response in Crohn's disease: A cost-effectiveness analysis. *Aliment Pharmacol Ther* 2007. 26: 1509–1520.
53. Rutgeerts, P., Sandborn, W. J., Feagan, B. G., Reinisch, W., Olson, A., Johanns, J., Travers, S., Rachmilewitz, D., Hanauer, S. B., Lichtenstein, G. R., de Villiers, W. J., Present, D., Sands, B. E. and Colombel, J. F., Infliximab for induction and maintenance therapy for ulcerative colitis. *N Engl J Med* 2005. 353: 2462–2476.
54. Gisbert, J. P., Gonzalez-Lama, Y. and Mate, J., Systematic review: Infliximab therapy in ulcerative colitis. *Aliment Pharmacol Ther* 2007. 25: 19–37.
55. Lindgren, S. C., Flood, L. M., Kilander, A. F., Lofberg, R., Persson, T. B. and Sjodahl, R. I., Early predictors of glucocorticosteroid treatment failure in severe and moderately severe attacks of ulcerative colitis. *Eur J Gastroenterol Hepatol* 1998. 10: 831–835.
56. Lees, C. W., Heys, D., Ho, G. T., Noble, C. L., Shand, A. G., Mowat, C., Boulton-Jones, R., Williams, A., Church, N., Satsangi, J. and Arnott, I. D., A retrospective analysis of the efficacy and safety of infliximab as rescue therapy in acute severe ulcerative colitis. *Aliment Pharmacol Ther* 2007. 26: 411–419.
57. Hanauer, S. B., Wagner, C. L., Bala, M., Mayer, L., Travers, S., Diamond, R. H., Olson, A., Bao, W. and Rutgeerts, P., Incidence and importance of antibody responses to infliximab after maintenance or episodic treatment in Crohn's disease. *Clin Gastroenterol Hepatol* 2004. 2: 542–553.
58. Keane, J., Gershon, S., Wise, R. P., Mirabile-Levens, E., Kasznica, J., Schwieterman, W. D., Siegel, J. N. and Braun, M. M., Tuberculosis associated with infliximab, a tumor necrosis factor alpha-neutralizing agent. *N Engl J Med* 2001. 345: 1098–1104.
59. Myers, A., Clark, J. and Foster, H., Tuberculosis and treatment with infliximab. *N Engl J Med* 2002. 346: 623–626.
60. Lichtenstein, G. R., Cohen, R. D., Feagan, B. G., Sandborn, W. J., Salzberg, B., Chen, D. M., and Diamond, R. H., Safety of infliximab and other Crohn's disease therapies—Updated treat registry data with over 10,000 patient-years of follow-up (Abstract). *Gastroenterology* 2005. 128: A-580.
61. Shen, C., Assche, G. V., Colpaert, S., Maerten, P., Geboes, K., Rutgeerts, P. and Ceuppens, J. L., Adalimumab induces apoptosis of human monocytes: A comparative study with infliximab and etanercept. *Aliment Pharmacol Ther* 2005. 21: 251–258.

62. Weinblatt, M. E., Keystone, E. C., Furst, D. E., Kavanaugh, A. F., Chartash, E. K. and Segurado, O. G., Long-term efficacy and safety of adalimumab plus methotrexate in patients with rheumatoid arthritis: ARMADA 4-year extended study. *Ann Rheum Dis* 2005.
63. Mease, P. J., Gladman, D. D., Ritchlin, C. T., Ruderman, E. M., Steinfeld, S. D., Choy, E. H., Sharp, J. T., Ory, P. A., Perdok, R. J. and Weinberg, M. A., Adalimumab for the treatment of patients with moderately to severely active psoriatic arthritis: Results of a double-blind, randomized, placebo-controlled trial. *Arthritis Rheum* 2005. 52: 3279–3289.
64. Hanauer, S. B., Sandborn, W. J., Rutgeerts, P., Fedorak, R. N., Lukas, M., MacIntosh, D., Panaccione, R., Wolf, D. and Pollack, P., Human anti-tumor necrosis factor monoclonal antibody (adalimumab) in Crohn's disease: The CLASSIC-I trial. *Gastroenterology* 2006. 130: 323–333; quiz 591.
65. Colombel, J. F., Sandborn, W. J., Rutgeerts, P., Enns, R., Hanauer, S. B., Panaccione, R., Schreiber, S., Byczkowski, D., Li, J., Kent, J. D. and Pollack, P. F., Adalimumab for maintenance of clinical response and remission in patients with Crohn's disease: The CHARM trial. *Gastroenterology* 2007. 132: 52–65.
66. Sandborn, W. J., Hanauer, S., Loftus, E. V., Jr., Tremaine, W. J., Kane, S., Cohen, R., Hanson, K., Johnson, T., Schmitt, D. and Jeche, R., An open-label study of the human anti-TNF monoclonal antibody adalimumab in subjects with prior loss of response or intolerance to infliximab for Crohn's disease. *Am J Gastroenterol* 2004. 99: 1984–1989.
67. Fossati, G. and Nesbitt, A. M., Effect of the anti-TNF agents, adalimumab, etanercept, infliximab, and certolizumab PEGOL (CDP870) on the induction of apoptosis in activated peripheral blood lymphocytes and monocytes (Abstract). *Am J Gastroenterol* 2005. 100: S298.
68. Schreiber, S., Rutgeerts, P., Fedorak, R. N., Khaliq-Kareemi, M., Kamm, M. A., Boivin, M., Bernstein, C. N., Staun, M., Thomsen, O. O. and Innes, A., A randomized, placebo-controlled trial of certolizumab pegol (CDP870) for treatment of Crohn's disease. *Gastroenterology* 2005. 129: 807–818.
69. Sandborn, W. J., Feagan, B. G., Stoinov, S., Honiball, P. J., Rutgeerts, P., Mason, D., Bloomfield, R. and Schreiber, S., Certolizumab pegol for the treatment of Crohn's disease. *N Engl J Med* 2007. 357: 228–238.
70. Schreiber, S., Khaliq-Kareemi, M., Lawrance, I. C., Thomsen, O. O., Hanauer, S. B., McColm, J., Bloomfield, R. and Sandborn, W. J., Maintenance therapy with certolizumab pegol for Crohn's disease. *N Engl J Med* 2007. 357: 239–250.
71. Sandborn, W. J., Feagan, B. G., Radford-Smith, G., Kovacs, A., Enns, R., Innes, A. and Patel, J., CDP571, a humanised monoclonal antibody to tumour necrosis factor alpha, for moderate to severe Crohn's disease: A randomised, double blind, placebo controlled trial. *Gut* 2004. 53: 1485–1493.
72. Feagan, B. G., Sandborn, W. J., Lichtenstein, G., Radford-Smith, G., Patel, J. and Innes, A., CDP571, a humanized monoclonal antibody to tumour necrosis factor-alpha, for steroid-dependent Crohn's disease: A randomized, double-blind, placebo-controlled trial. *Aliment Pharmacol Ther* 2006. 23: 617–628.
73. Van den Brande, J. M., Braat, H., van den Brink, G. R., Versteeg, H. H., Bauer, C. A., Hoedemaeker, I., van Montfrans, C., Hommes, D. W., Peppelenbosch, M. P. and van Deventer, S. J., Infliximab but not etanercept induces apoptosis in lamina propria T-lymphocytes from patients with Crohn's disease. *Gastroenterology* 2003. 124: 1774–1785.

74. Moreland, L. W., Schiff, M. H., Baumgartner, S. W., Tindall, E. A., Fleischmann, R. M., Bulpitt, K. J., Weaver, A. L., Keystone, E. C., Furst, D. E., Mease, P. J., Ruderman, E. M., Horwitz, D. A., Arkfeld, D. G., Garrison, L., Burge, D. J., Blosch, C. M., Lange, M. L., McDonnell, N. D. and Weinblatt, M. E., Etanercept therapy in rheumatoid arthritis. A randomized, controlled trial. *Ann Intern Med* 1999. 130: 478–486.
75. Bathon, J. M., Martin, R. W., Fleischmann, R. M., Tesser, J. R., Schiff, M. H., Keystone, E. C., Genovese, M. C., Wasko, M. C., Moreland, L. W., Weaver, A. L., Markenson, J. and Finck, B. K., A comparison of etanercept and methotrexate in patients with early rheumatoid arthritis. *N Engl J Med* 2000. 343: 1586–1593.
76. Mease, P. J., Kivitz, A. J., Burch, F. X., Siegel, E. L., Cohen, S. B., Ory, P., Salonen, D., Rubenstein, J., Sharp, J. T. and Tsuji, W., Etanercept treatment of psoriatic arthritis: Safety, efficacy, and effect on disease progression. *Arthritis Rheum* 2004. 50: 2264–2272.
77. Takei, S., Groh, D., Bernstein, B., Shaham, B., Gallagher, K. and Reiff, A., Safety and efficacy of high dose etanercept in treatment of juvenile rheumatoid arthritis. *J Rheumatol* 2001. 28: 1677–1680.
78. Tyring, S., Gottlieb, A., Papp, K., Gordon, K., Leonardi, C., Wang, A., Lalla, D., Woolley, M., Jahreis, A., Zitnik, R., Cella, D. and Krishnan, R., Etanercept and clinical outcomes, fatigue, and depression in psoriasis: Double-blind placebo-controlled randomised phase III trial. *Lancet* 2006. 367: 29–35.
79. Leonardi, C. L., Powers, J. L., Matheson, R. T., Goffe, B. S., Zitnik, R., Wang, A. and Gottlieb, A. B., Etanercept as monotherapy in patients with psoriasis. *N Engl J Med* 2003. 349: 2014–2022.
80. Davis, J. C., van der Heijde, D. M., Braun, J., Dougados, M., Cush, J., Clegg, D., Inman, R. D., Kivitz, A., Zhou, L., Solinger, A. and Tsuji, W., Sustained durability and tolerability of etanercept in ankylosing spondylitis for 96 weeks. *Ann Rheum Dis* 2005. 64: 1557–1562.
81. Sandborn, W. J., Hanauer, S. B., Katz, S., Safdi, M., Wolf, D. G., Baerg, R. D., Tremaine, W. J., Johnson, T., Diehl, N. N. and Zinsmeister, A. R., Etanercept for active Crohn's disease: A randomized, double-blind, placebo-controlled trial. *Gastroenterology* 2001. 121: 1088–1094.
82. Rutgeerts, P., Lemmens, L., Van Assche, G., Noman, M., Borghini-Fuhrer, I. and Goedkoop, R., Treatment of active Crohn's disease with onercept (recombinant human soluble p55 tumour necrosis factor receptor): Results of a randomized, open-label, pilot study. *Aliment Pharmacol Ther* 2003. 17: 185–192.
83. Rutgeerts, P., Sandborn, W. J., Fedorak, R. N., Rachmilewitz, D., Tarabar, D., Gibson, P., Haagen Nielsen, O., Wild, G., Schreiber, S., Pena Rossi, C. and Zignani, M., Onercept for moderate-to-severe Crohn's disease: A randomized, double-blind, placebo-controlled trial. *Clin Gastroenterol Hepatol* 2006. 4: 888–893.
84. Scallon, B. J., Moore, M. A., Trinh, H., Knight, D. M. and Ghrayeb, J., Chimeric anti-TNF-alpha monoclonal antibody cA2 binds recombinant transmembrane TNF-alpha and activates immune effector functions. *Cytokine* 1995. 7: 251–259.
85. van den Brande, J., Hommes, D. W. and Peppelenbosch, M. P., Infliximab induced T lymphocyte apoptosis in Crohn's disease. *J Rheumatol Suppl* 2005. 74: 26–30.
86. Boirivant, M., Marini, M., Di Felice, G., Pronio, A. M., Montesani, C., Tersigni, R. and Strober, W., Lamina propria T cells in Crohn's disease and other gastrointestinal inflammation show defective CD2 pathway-induced apoptosis. *Gastroenterology* 1999. 116: 557–565.
87. Lugering, A., Schmidt, M., Lugering, N., Pauels, H. G., Domschke, W. and Kucharzik, T., Infliximab induces apoptosis in monocytes from patients with chronic active Crohn's disease by using a caspase-dependent pathway. *Gastroenterology* 2001. 121: 1145–1157.

88. ten Hove, T., van Montfrans, C., Peppelenbosch, M. P. and van Deventer, S. J., Infliximab treatment induces apoptosis of lamina propria T lymphocytes in Crohn's disease. *Gut* 2002. 50: 206–211.
89. Mitoma, H., Horiuchi, T., Hatta, N., Tsukamoto, H., Harashima, S., Kikuchi, Y., Otsuka, J., Okamura, S., Fujita, S. and Harada, M., Infliximab induces potent anti-inflammatory responses by outside-to-inside signals through transmembrane TNF-alpha. *Gastroenterology* 2005. 128: 376–392.
90. Reinhart, K., Menges, T., Gardlund, B., Harm Zwaveling, J., Smithes, M., Vincent, J. L., Tellado, J. M., Salgado-Remigio, A., Zimlichman, R., Withington, S., Tschaikowsky, K., Brase, R., Damas, P., Kupper, H., Kempeni, J., Eiselstein, J. and Kaul, M., Randomized, placebo-controlled trial of the anti-tumor necrosis factor antibody fragment afelimomab in hyperinflammatory response during severe sepsis: The RAMSES Study. *Crit Care Med* 2001. 29: 765–769.
91. Panacek, E. A., Marshall, J. C., Albertson, T. E., Johnson, D. H., Johnson, S., MacArthur, R. D., Miller, M., Barchuk, W. T., Fischkoff, S., Kaul, M., Teoh, L., Van Meter, L., Daum, L., Lemeshow, S., Hicklin, G. and Doig, C., Efficacy and safety of the monoclonal anti-tumor necrosis factor antibody F(ab′)2 fragment afelimomab in patients with severe sepsis and elevated interleukin-6 levels. *Crit Care Med* 2004. 32: 2173–2182.
92. Holtmann, M. M., RDP-58 (SangStat Medical). *IDrugs* 2003. 6: 1188–1194.
93. Travis, S., Yap, L. M., Hawkey, C., Warren, B., Lazarov, M., Fong, T. and Tesi, R. J., RDP58 is a novel and potentially effective oral therapy for ulcerative colitis. *Inflamm Bowel Dis* 2005. 11: 713–719.
94. Hommes, D. W., Mikhajlova, T. L., Stoinov, S., Stimac, D., Vucelic, B., Lonovics, J., Zakuciova, M., D'Haens, G., Van Assche, G., Ba, S., Lee, S. and Pearce, T., Fontolizumab, a humanised anti-interferon gamma antibody, demonstrates safety and clinical activity in patients with moderate to severe Crohn's disease. *Gut* 2006. 55: 1131–1137.
95. Reinisch, W., Hommes, D. W., Van Assche, G., Colombel, J. F., Gendre, J. P., Oldenburg, B., Teml, A., Geboes, K., Ding, H., Zhang, L., Tang, M., Cheng, M., van Deventer, S. J., Rutgeerts, P. and Pearce, T., A dose escalating, placebo controlled, double blind, single dose and multidose, safety and tolerability study of fontolizumab, a humanised anti-interferon gamma antibody, in patients with moderate to severe Crohn's disease. *Gut* 2006. 55: 1138–1144.
96. Mannon, P. J., Fuss, I. J., Mayer, L., Elson, C. O., Sandborn, W. J., Present, D., Dolin, B., Goodman, N., Groden, C., Hornung, R. L., Quezado, M., Yang, Z., Neurath, M. F., Salfeld, J., Veldman, G. M., Schwertschlag, U. and Strober, W., Anti-interleukin-12 antibody for active Crohn's disease. *N Engl J Med* 2004. 351: 2069–2079.
97. Burakoff, R., Barish, C. F., Riff, D., Pruitt, R., Chey, W. Y., Farraye, F. A., Shafran, I., Katz, S., Krone, C. L., Vander Vliet, M., Stevens, C., Sherman, M. L., Jacobson, E. and Bleday, R., A phase 1/2A trial of STA 5326, an oral interleukin-12/23 inhibitor, in patients with active moderate to severe Crohn's disease. *Inflamm Bowel Dis* 2006. 12: 558–565.
98. Ito, H., Takazoe, M., Fukuda, Y., Hibi, T., Kusugami, K., Andoh, A., Matsumoto, T., Yamamura, T., Azuma, J., Nishimoto, N., Yoshizaki, K., Shimoyama, T. and Kishimoto, T., A pilot randomized trial of a human anti-interleukin-6 receptor monoclonal antibody in active Crohn's disease. *Gastroenterology* 2004. 126: 989–996; discussion 947.
99. Schreiber, S., Fedorak, R. N., Nielsen, O. H., Wild, G., Williams, C. N., Nikolaus, S., Jacyna, M., Lashner, B. A., Gangl, A., Rutgeerts, P., Isaacs, K., van Deventer, S. J., Koningsberger, J. C., Cohard, M., LeBeaut, A. and Hanauer, S. B., Safety and efficacy of recombinant human interleukin 10 in chronic active Crohn's disease. Crohn's Disease IL-10 Cooperative Study Group. *Gastroenterology* 2000. 119: 1461–1472.

100. Fedorak, R. N., Gangl, A., Elson, C. O., Rutgeerts, P., Schreiber, S., Wild, G., Hanauer, S. B., Kilian, A., Cohard, M., LeBeaut, A. and Feagan, B., Recombinant human interleukin 10 in the treatment of patients with mild to moderately active Crohn's disease. The Interleukin 10 Inflammatory Bowel Disease Cooperative Study Group. *Gastroenterology* 2000. 119: 1473–1482.
101. van Deventer, S. J., Elson, C. O. and Fedorak, R. N., Multiple doses of intravenous interleukin 10 in steroid-refractory Crohn's disease. Crohn's Disease Study Group. *Gastroenterology* 1997. 113: 383–389.
102. Braat, H., Rottiers, P., Hommes, D. W., Huyghebaert, N., Remaut, E., Remon, J. P., van Deventer, S. J., Neirynck, S., Peppelenbosch, M. P. and Steidler, L., A phase I trial with transgenic bacteria expressing interleukin-10 in Crohn's disease. *Clin Gastroenterol Hepatol* 2006. 4: 754–759.
103. Sands, B. E., Winston, B. D., Salzberg, B., Safdi, M., Barish, C., Wruble, L., Wilkins, R., Shapiro, M. and Schwertschlag, U. S., Randomized, controlled trial of recombinant human interleukin-11 in patients with active Crohn's disease. *Aliment Pharmacol Ther* 2002. 16: 399–406.
104. Creed, T. J., Norman, M. R., Probert, C. S., Harvey, R. F., Shaw, I. S., Smithson, J., Anderson, J., Moorghen, M., Gupta, J., Shepherd, N. A., Dayan, C. M. and Hearing, S. D., Basiliximab (anti-CD25) in combination with steroids may be an effective new treatment for steroid-resistant ulcerative colitis. *Aliment Pharmacol Ther* 2003. 18: 65–75.
105. Van Assche, G., Sandborn, W. J., Feagan, B. G., Salzberg, B. A., Silvers, D., Monroe, P. S., Pandak, W. M., Anderson, F. H., Valentine, J. F., Wild, G. E., Geenen, D. J., Sprague, R., Targan, S. R., Rutgeerts, P., Vexler, V., Young, D. and Shames, R. S., Daclizumab, a humanised monoclonal antibody to the interleukin 2 receptor (CD25), for the treatment of moderately to severely active ulcerative colitis: A randomised, double blind, placebo controlled, dose ranging trial. *Gut* 2006. 55: 1568–1574.
106. Plevy, S., Salzberg, B., Van Assche, G., Regueiro, M., Hommes, D., Sandborn, W., Hanauer, S., Targan, S., Mayer, L., Mahadevan, U., Frankel, M. and Lowder, J., A phase I study of visilizumab, a humanized anti-CD3 monoclonal antibody, in severe steroid-refractory ulcerative colitis. *Gastroenterology* 2007. 133: 1414–1422.
107. Yacyshyn, B. R., Chey, W. Y., Goff, J., Salzberg, B., Baerg, R., Buchman, A. L., Tami, J., Yu, R., Gibiansky, E. and Shanahan, W. R., Double blind, placebo controlled trial of the remission inducing and steroid sparing properties of an ICAM-1 antisense oligodeoxynucleotide, alicaforsen (ISIS 2302), in active steroid dependent Crohn's disease. *Gut* 2002. 51: 30–36.
108. Yacyshyn, B., Chey, W. Y., Wedel, M. K., Yu, R. Z., Paul, D. and Chuang, E., A randomized, double-masked, placebo-controlled study of alicaforsen, an antisense inhibitor of intercellular adhesion molecule 1, for the treatment of subjects with active Crohn's disease. *Clin Gastroenterol Hepatol* 2007. 5: 215–220.
109. van Deventer, S. J., Tami, J. A. and Wedel, M. K., A randomised, controlled, double blind, escalating dose study of alicaforsen enema in active ulcerative colitis. *Gut* 2004. 53: 1646–1651.
110. van Deventer, S. J., Wedel, M. K., Baker, B. F., Xia, S., Chuang, E. and Miner, P. B., Jr., A phase II dose ranging, double-blind, placebo-controlled study of alicaforsen enema in subjects with acute exacerbation of mild to moderate left-sided ulcerative colitis. *Aliment Pharmacol Ther* 2006. 23: 1415–1425.
111. Miner, P. B., Jr., Wedel, M. K., Xia, S. and Baker, B. F., Safety and efficacy of two dose formulations of alicaforsen enema compared with mesalazine enema for treatment of mild to moderate left-sided ulcerative colitis: A randomized, double-blind, active-controlled trial. *Aliment Pharmacol Ther* 2006. 23: 1403–1413.

112. Miller, D. H., Khan, O. A., Sheremata, W. A., Blumhardt, L. D., Rice, G. P., Libonati, M. A., Willmer-Hulme, A. J., Dalton, C. M., Miszkiel, K. A. and O'Connor, P. W., A controlled trial of natalizumab for relapsing multiple sclerosis. *N Engl J Med* 2003. 348: 15–23.
113. Gordon, F. H., Lai, C. W., Hamilton, M. I., Allison, M. C., Srivastava, E. D., Fouweather, M. G., Donoghue, S., Greenlees, C., Subhani, J., Amlot, P. L. and Pounder, R. E., A randomized placebo-controlled trial of a humanized monoclonal antibody to alpha4 integrin in active Crohn's disease. *Gastroenterology* 2001. 121: 268–274.
114. Ghosh, S., Goldin, E., Gordon, F. H., Malchow, H. A., Rask-Madsen, J., Rutgeerts, P., Vyhnalek, P., Zadorova, Z., Palmer, T. and Donoghue, S., Natalizumab for active Crohn's disease. *N Engl J Med* 2003. 348: 24–32.
115. Sandborn, W. J., Colombel, J. F., Enns, R., Feagan, B. G., Hanauer, S. B., Lawrance, I. C., Panaccione, R., Sanders, M., Schreiber, S., Targan, S., van Deventer, S., Goldblum, R., Despain, D., Hogge, G. S. and Rutgeerts, P., Natalizumab induction and maintenance therapy for Crohn's disease. *N Engl J Med* 2005. 353: 1912–1925.
116. Kleinschmidt-DeMasters, B. K. and Tyler, K. L., Progressive multifocal leukoencephalopathy complicating treatment with natalizumab and interferon beta-1a for multiple sclerosis. *N Engl J Med* 2005. 353: 369–374.
117. Van Assche, G., Van Ranst, M., Sciot, R., Dubois, B., Vermeire, S., Noman, M., Verbeeck, J., Geboes, K., Robberecht, W. and Rutgeerts, P., Progressive multifocal leukoencephalopathy after natalizumab therapy for Crohn's disease. *N Engl J Med* 2005. 353: 362–368.
118. Feagan, B. G., Greenberg, G. R., Wild, G., Fedorak, R. N., Pare, P., McDonald, J. W., Dube, R., Cohen, A., Steinhart, A. H., Landau, S., Aguzzi, R. A., Fox, I. H. and Vandervoort, M. K., Treatment of ulcerative colitis with a humanized antibody to the alpha4beta7 integrin. *N Engl J Med* 2005. 352: 2499–2507.
119. Korzenik, J. R., Dieckgraefe, B. K., Valentine, J. F., Hausman, D. F. and Gilbert, M. J., Sargramostim for active Crohn's disease. *N Engl J Med* 2005. 352: 2193–2201.
120. Hommes, D., van den Blink, B., Plasse, T., Bartelsman, J., Xu, C., Macpherson, B., Tytgat, G., Peppelenbosch, M. and Van Deventer, S., Inhibition of stress-activated MAP kinases induces clinical improvement in moderate to severe Crohn's disease. *Gastroenterology* 2002. 122: 7–14.
121. Schreiber, S., Feagan, B., D'Haens, G., Colombel, J. F., Geboes, K., Yurcov, M., Isakov, V., Golovenko, O., Bernstein, C. N., Ludwig, D., Winter, T., Meier, U., Yong, C. and Steffgen, J., Oral p38 mitogen-activated protein kinase inhibition with BIRB 796 for active Crohn's disease: A randomized, double-blind, placebo-controlled trial. *Clin Gastroenterol Hepatol* 2006. 4: 325–334.
122. Berrebi, D., Maudinas, R., Hugot, J. P., Chamaillard, M., Chareyre, F., De Lagausie, P., Yang, C., Desreumaux, P., Giovannini, M., Cezard, J. P., Zouali, H., Emilie, D. and Peuchmaur, M., Card15 gene overexpression in mononuclear and epithelial cells of the inflamed Crohn's disease colon. *Gut* 2003. 52: 840–846.
123. Kelsall, B. and Strober, W., *Gut-associated lymphoid tissue: antigen handling and T lymphocte responses*. Academic Press, San Diego: 1999.
124. Ogura, Y., Bonen, D. K., Inohara, N., Nicolae, D. L., Chen, F. F., Ramos, R., Britton, H., Moran, T., Karaliuskas, R., Duerr, R. H., Achkar, J. P., Brant, S. R., Bayless, T. M., Kirschner, B. S., Hanauer, S. B., Nunez, G. and Cho, J. H., A frameshift mutation in NOD2 associated with susceptibility to Crohn's disease. *Nature* 2001. 411: 603–606.

125. Hugot, J. P., Chamaillard, M., Zouali, H., Lesage, S., Cezard, J. P., Belaiche, J., Almer, S., Tysk, C., O'Morain, C. A., Gassull, M., Binder, V., Finkel, Y., Cortot, A., Modigliani, R., Laurent-Puig, P., Gower-Rousseau, C., Macry, J., Colombel, J. F., Sahbatou, M. and Thomas, G., Association of NOD2 leucine-rich repeat variants with susceptibility to Crohn's disease. *Nature* 2001. 411: 599–603.
126. Hisamatsu, T., Suzuki, M., Reinecker, H. C., Nadeau, W. J., McCormick, B. A. and Podolsky, D., CARD15/NOD2 functions as an anti-bacterial factor in human intestinal epithelial cells. *Gastroenterology* 2003. 124: 993–1000.
127. Maeda, S., Hsu, L. C., Liu, H., Bankston, L. A., Iimura, M., Kagnoff, M. F., Eckmann, L. and Karin, M., Nod2 mutation in Crohn's disease potentiates NF-kappaB activity and IL-1beta processing. *Science* 2005. 307: 734–738.
128. Kobayashi, K. S., Chamaillard, M., Ogura, Y., Henegariu, O., Inohara, N., Nunez, G. and Flavell, R. A., Nod2-dependent regulation of innate and adaptive immunity in the intestinal tract. *Science* 2005. 307: 731–734.

6 Biologic Therapies for Psoriasis

Rosita Saraceno and Sergio Chimenti

CONTENTS

KEY WORDS: psoriasis, biologics, TNFα, IL-12p40, T cells, innate immunity

ABSTRACT

Psoriasis is a lifelong, chronic, and immune-mediated skin disease characterized by erythematous scaly plaques, affecting ~1 to 3 percent of the worldwide population. Several systemic drugs are available for the treatment of moderate to severe plaque psoriasis, including photochemotherapy, retinoids, cyclosporine, methotrexate, and fumarates. Although these options have been shown to maintain efficacy, they produce significant subjective side effects. The realization that psoriasis is an immune system dysfunction rather than a primary skin disorder has led to the development of a new generation of therapies (named "biologicals") for psoriasis that target the immune system. These drugs are directed against key cytokines and immune cells—notably tumor necrosis factor-alpha (TNFα), interleukin (IL)-12, IL-23, and T cells—that have been identified as critical factors contributing to the immunopathogenesis of the disease. Biologics are a heterogeneous group of monoclonal antibodies, fusion proteins, and recombinant cytokines, designed to modify and regulate pivotal and specific mechanisms involved in the immunopathogenesis of psoriasis.

Several clinical trials have demonstrated the safety and efficacy of this new generation of drugs in short- and long-term treatment.

INTRODUCTION

Psoriasis is a lifelong, chronic, immune-mediated skin disease characterized by erythematous scaly plaques. Psoriasis affects ~1 to 3 percent of the Caucasian population and in the United States the prevalence has been estimated between 2.2 and 2.6 percent.[1] The most common clinical variant is plaque psoriasis and occurs in more than 80 percent of affected patients.[2] Erythematous scaly plaques, round or oval, variable in size, typically affect scalp, lower back, intergluteal cleft, umbilical area, elbows, and knees (Figure 6.1). Up to 40 percent of patients with psoriasis develop an inflammatory, deforming arthritis called psoriatic arthritis.[3]

PATHOGENESIS

Both genetic and environmental factors contribute to the pathogenesis of psoriasis, with immune-mediated mechanisms playing a crucial role. Psoriasis is driven by T

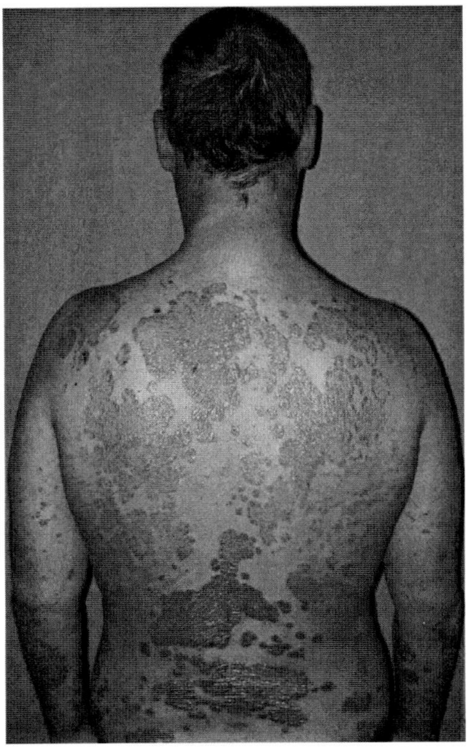

FIGURE 6.1 ***A color version of this figure follows page 176.*** Plaque-type psoriasis characterized by erythematous, scaly lesions.

cells which lead to skin inflammation, keratinocytes hyperproliferation, and vascular changes with dilated blood vessels in the papillary dermis. Type 1 and type 17 immune responses predominate in psoriatic lesions, and are responsible for the aforementioned epidermal and dermal changes.[4]

T lymphocytes and dendritic antigen-presenting cells (APCs) have been considered to be of major importance in the immunopathogenesis of psoriasis, and current evidence suggests that epidermal changes are caused by actions of these cells in skin lesions.[5,6] The role of T cells has been confirmed by the presence of activated T cells in psoriatic plaques where the majority of T lymphocytes that localize to the dermis are $CD4^+$ helper, while those that migrate to the epidermis are predominantly $CD8^+$ cytotoxic T cells. Activated T cells in psoriatic skin produce Th1 type cytokines, including interleukin-2 (IL-2), IL-1, interferon gamma (IFNγ) and tumor necrosis factor alpha (TNFα).[7] The most important evidence that psoriasis is an immune-mediated disease comes from the finding that psoriasis can be reversed by a lymphocyte-selective toxin (DAB389IL-2).[6] Psoriasis is now accepted as a T-cell-mediated disease thus targeting of T-cell function and/or trafficking is a logical approach to therapy. The original proof of concept for T-cell-targeted therapies in psoriasis came with the demonstration that anti-CD4 monoclonal antibodies were effective. Other proofs were represented by the cure of the disease by bone marrow transplantation from healthy persons or observing the transfer of the disease by transplantation of bone marrow from psoriatic patients.

The importance of T-cell activation, T-cell migration into the dermis, and T-cell binding to keratinocytes has been demonstrated as key steps in the pathogenesis of psoriasis. Two adhesion molecules are thought to play important roles in this process: lymphocyte function-associated antigen-1 (LFA-1) and intercellular adhesion molecule-1 (ICAM-1). LFA-1 expression is increased on memory T cells and ICAM-1 is expressed on vascular endothelial cells at sites of inflammation as well as on keratinocytes in a variety of T-cell–mediated disorders including psoriasis.[7]

Effector cells of innate immunity are also involved in the pathogenesis of psoriasis including neutrophils, plasmacitoid dendritic cells (PDCs), and $CD11c^+$ dendritic cells (DCs). Activated DCs, producing IL-23 and iNOS, are present in psoriasis plaques. Plasmacytoid DCs produce high levels of IFNα while $CD11c^+$ DCs express high levels of TNFα.[8] pDCs comprise a dendritic cell population highly specialized for sensing viral and certain microbial infections. In contrast to myeloid dendritic cells, pDCs uniquely express Toll-like receptor 7 (TLR 7) and TLR 9, intracellular receptors that recognize viral/microbial nucleic acids within the endosomal compartments. Endogenous antimicrobial peptides have been recently described in the pathogenesis of psoriasis; these peptides convert nonstimulatory self-DNA into a potent trigger of pDCs to produce IFNα. The antimicrobial peptides break innate tolerance to self-DNA and, in a similar way to viral responses, elicit a rapid and robust induction of IFNs that initiate innate and adaptive immunity.[9]

Moreover, neutrophils, natural killer cells, and DCs are also crucially involved in a complex network producing and releasing proinflammatory cytokines, including TNFα, IFNγ, IL-17, IL-22, and IL-23.[10–14] Intrinsic and genetically determined keratinocyte alterations are also relevant to psoriasis pathogenesis. In particular, psoriatic keratinocytes have an abnormal sensitivity to growth factors and cytokines, resistance to apoptosis, and an abnormal differentiation program. Our improved understanding of immune system involvement in the etiology of psoriasis has led to a number of treatments targeting specific steps in the immunopathogenesis of the disease.

FROM TRADITIONAL THERAPIES TO BIOLOGICS

Systemic drugs available for the treatment of moderate to severe psoriasis include photochemotherapy, retinoids, cyclosporine, methotrexate, and fumarates. Traditional treatments, although effective in mild to moderate plaque-type psoriasis, have numerous shortcomings and side-effect profiles precluding long-term use. Among these drugs, those that clear psoriasis most quickly and conveniently are among the most toxic. Those that have the safest therapeutic profile tend to work relatively slowly and incompletely or inconveniently. The realization that psoriasis is an immune system dysfunction rather than a primary skin disorder has led to the development of a new generation of therapies for psoriasis that target the immune system. Many of these medications are termed "biological therapies" because they are protein drugs produced in living organisms. Biologics are a set of bioengineered proteins used to modify immune reactions. They are designed to modify defined pathophysiological pathways that regulate pivotal immunological processes such as lymphocyte activation, interaction with APCs and endothelial cells (adhesion and migration), and production and action of cytokines and chemokines. They interfere with specific components of the autoimmune response and, in contrast to general immunosuppressants, do not affect the entire immune system nor do they have significant end-organ toxicity. Antibodies, fusion proteins, and recombinant cytokines are examples of these biologics.

These medications have been created using recombinant DNA technology and are produced in living tissue cultures. Biologics can be divided into three main groups: monoclonal antibodies, fusion proteins, and cytokines. A standardized nomenclature has been set to classify these medications. Generic names of chimeric monoclonals end with *-ximab,* humanized monoclonals end with *-zumab,* and human monoclonal antibodies end with *-umab.* Receptor antibody fusion proteins end with *-cept.* These drugs target the immune system at multiple levels: cell-adhesion and T-cell activation, production of cytokines, chemokines, and angiogenetic factors. Many of these medications are designed to modify specific pathophysiological pathways regulating pivotal immunological processes such as the T-cell activation and cytokine production. Our knowledge of the distinct cytokine profile of psoriatic lesions and our increasing expertise in manipulating genetic material have resulted in the development of targeted systemic agents opening a new field of promising future treatments.

Monoclonal antibody technology can be developed by fusing immortal tumor cells (myeloma cells) with antibody-producing B lymphocyte cells to produce "hybridomas," that continuously synthesize identical (or "monoclonal") antibodies. Monoclonal antibodies are preparations of a single antibody-secreting B cell (from the spleen of a mouse immunized with the target protein) fused with myeloma cells (cancerous B cells with unlimited growth that have lost the ability to make their own antibodies).

In accordance with the available data, biologics have a more favorable side-effect profile than conventional treatments. Numerous biologics are approved by the U.S. Food and Drug Administration (FDA) or European Medicines Agency (EMEA) for the treatment of psoriasis, including an anti-CD11a antibody, namely, efalizumab, and an antileukocyte function-associated molecule 3 (LFA3)/CD2 receptor, alefacept.[15] In addition, three anti-TNFα agents (adalimumab, etanercept, and infliximab) are now commercially available (Table 6.1).

TNFα AND TNF ANTAGONISTS

Recent advances in the understanding of the immunopathogenesis of psoriasis and psoriatic arthritis, combined with the developments in biotechnology, have led to the introduction of novel therapeutic options.[16] Indeed, several studies have suggested that proinflammatory cytokines, in particular TNFα, play a key role in intensifying inflammatory responses associated with both psoriasis and psoriatic arthritis. TNFα is a member of a growing family of cytokines referred to as the TNF superfamily, comprising at least 20 peptides which include common mediators such as lymphotoxin (LT), Fas ligand (FasL), and CD40 ligand (CD40L). Higher concentrations of TNFα are found in psoriatic plaques than in either the uninvolved skin of patients with psoriasis or normal individuals as a result of the overexpression by T lymphocytes, keratinocytes, and DCs.[17–22]

TABLE 6.1
Biologic Therapies for the Treatment of Psoriasis and Psoriatic Arthritis

Agent Target	Agent (Brand Name)	Status
T-cell modulator	Alefacept (Amevive)	Approved in U.S. in the treatment of plaque type psoriasis
T-cell modulator	Efalizumab (Raptiva)	Approved in EU and U.S. in the treatment of plaque-type psoriasis
Anti-TNF monoclonal antibody	Infliximab (Remicade)	Approved in EU and U.S. in the treatment of plaque-type psoriasis and psoriatic arthritis
Anti-TNF monoclonal antibody	Adalimumab (Humira)	Approved in EU and U.S. in the treatment of plaque-type psoriasis and psoriatic arthritis
Soluble-TNF receptor	Etanercept (Enbrel)	Approved in EU and U.S. in the treatment of plaque-type psoriasis and psoriatic arthritis
Anti-IL-12p40	ABT-874 STA-5326 CNTO-1275	Phase III and II ongoing trials for psoriasis and psoriatic arthritis

TNFα is a proinflammatory cytokine produced by many different cell types, including activated T cells, keratinocytes, Langerhans cells, and other dendritic APCs. TNFα is inactive as a monomer, but it forms active trimers, composed of three monomers which exert their biologic activity by binding to specific high-affinity cell surface receptors: p75 and p55. Both receptors can be shed from the cell surface and give rise to soluble receptors able to compete with the cell-bound receptors for binding to TNFα . This cytokine has numerous effects on the immune response, driving activation and recruitment of other inflammatory cells, amplifying production of IL-1, IL-6, and IL-8, and activating nuclear transcription factors such as NF-kB to propagate and maintain the inflammatory response.[23] In psoriasis, levels of both TNFα and TNFα receptors are higher and more widely expressed in lesional skin compared to normal skin.[18,19] TNFα induces Langerhans cells maturation, creating more efficient APCs with increased T-cell costimulatory molecules and their migration from the skin to the lymph nodes, where antigen presentation and T-cell activation take place. Moreover, TNFα leads to the production of transforming growth factor-α (TGFα), which may drive keratinocyte proliferation in psoriasis.[24]

TNFα induces keratinocytes to produce cytokines, including IL-6, IL-7, IL-15, IL-18, TNFα itself, chemokines, adhesion molecules, and growth factors.[20] Overexpressed TNFα thus acts in a self-reinforcing, positive feedback loop to maintain and to enhance local inflammation. TNFα is a very important regulator of skin inflammation, but it is also the key initiator of innate immunity. Psoriatic disease activity and severity have been found to correlate with circulating levels of TNFα.[19] However, TNFα is also crucial for skin homeostasis. In particular, excessive TNFα suppression is likely to interfere with local immunity, proper cell growth and differentiation, and programmed cell death (apoptosis). TNF antagonists modify the inflammatory burden of psoriasis, with T cells in the skin and synovial fluid and macrophages in the synovial sublining environment significantly reduced. Prevention of leukocyte infiltration is likely to be related to the down-regulation of cellular and vascular adhesion molecules and decreased production of cytokines and chemokines, and to the interference with T-cell activation.[21,22] Interruption of the self-sustaining loop mediated by overexpressed TNFα and reversal of cutaneous inflammation promote plaque clearing.[17–21] Increased levels of TNFα have been observed in skin, synovial fluid, and synovial tissues from affected patients,[19,20] and allelic polymorphisms in the promoter region for TNFα have been shown to correlate with certain features of the disease.[18] Currently available anti-TNFα agents designed specifically to inhibit TNFα include etanercept, infliximab, and, more recently, adalimumab (Figure 6.2).

The inclusive eligibility criteria for the use of biologics, especially the anti-TNFα inhibitors, are >18 years of age, moderate to severe plaque-type psoriasis or psoriatic arthritis, and no response or ineligibility to conventional current systemic treatments. Exclusion criteria are live vaccinations, history of chronic or recurrent serious infectious disease or opportunistic infection, evidence of internal malignancies or serious infections within 3 months prior to the first infusion, signs or symptoms of systemic lupus erythematosus, demyelinating diseases, lymphoproliferative diseases, noncutaneous malignancies within the previous 5 years, and serious congestive heart failure.

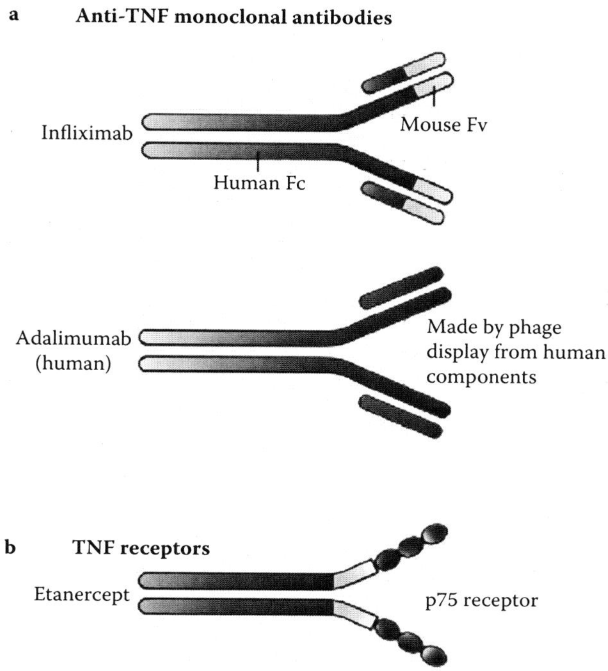

FIGURE 6.2 Structure of TNFα inhibitors approved for the treatment of psoriasis.

INFLIXIMAB

Infliximab is a biologic therapy developed for the treatment of chronic inflammatory diseases mediated by TNFα, and is approved in the United States and Europe for the treatment of gastroenterological and rheumatological diseases, namely, Crohn's disease, ulcerative colitis, psoriatic arthritis, ankylosing spondylitis, and rheumatoid arthritis. Infliximab has been approved by the FDA and EMEA for the treatment of plaque psoriasis. The approval was based on data obtained from previous clinical trials.

This agent is a chimeric anti-TNFα monoclonal IgG_1 antibody, with high affinity and avidity for soluble and cell-surface transmembrane forms of TNFα (Figure 6.2 and Figure 6.3). Detection of increased levels of TNFα in psoriatic plaque suggests a role of this factor on keratinocyte proliferation, expression of adhesion molecules on keratinocytes and endothelial cells, cell trafficking, and enhancement of neutrophil functions.[21] In addition, increased serum TNFα levels are related with a higher disease activity and reduced by effective therapies.[17] By blocking both soluble and cell-surface transmembrane TNFα, infliximab induces rapid decrease of epidermal T-cell infiltration, downregulates adhesion molecules expression, and normalizes keratinocyte proliferation.[22]

Infliximab has a molecular weight of 149,100 Da and comprises a mouse variable region and a human IgG_1/α constant region.[25] Each infliximab molecule can bind to two TNFα molecules and to TNFα already bound to the TNF receptor. Up to three infliximab molecules can bind and block a single trimer of TNFα (Figure 6.3).

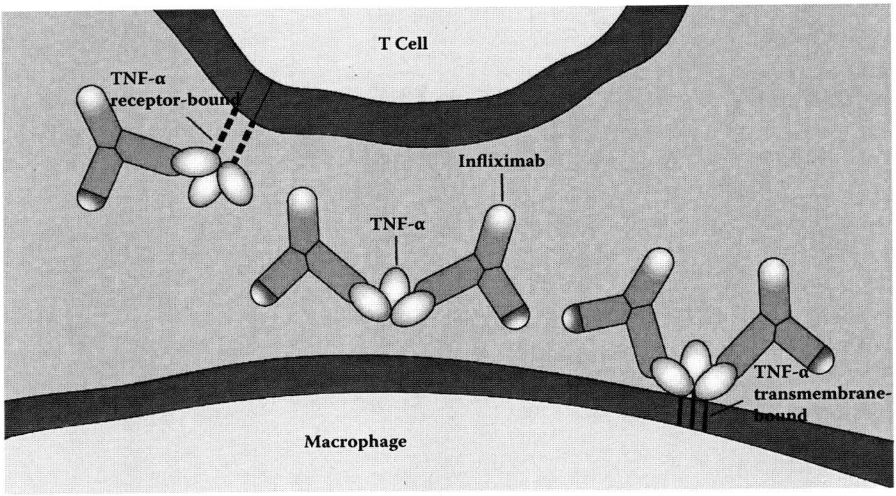

FIGURE 6.3 Mechanism of action of infliximab.

Binding to soluble and transmembrane TNFα molecules in the plasma and the disease tissue, infliximab directly neutralizes TNFα and retards or prevents monomeric association from forming the biologically active trimer.[22] Infliximab-bound TNFα cannot bind to or activate the TNFα receptor on target cells. In the skin, TNF receptors bearing cells include keratinocytes, Langerhans cells, T cells, natural killer (NK) cells, and endothelial cells.

The recommended dose is an infusion of 5 mg/kg followed by additional doses at 2 and 6 weeks after the first infusion and then every 8 weeks thereafter. It is administered to patients via intravenous infusion over 2 to 3 hours. In general, higher infliximab blood levels are associated with higher clinical efficacy. Patients with undetectable serum infliximab concentrations (<0.1 μg/mL) are less likely to maintain response. The median elimination half-life for infliximab is 7.62 (interquartile range, 6.62 to 10.15) days for the 5 mg/kg-dose.[22]

In clinical trials for patients with moderate to severe psoriasis, infliximab infusions were given at variable dosages of 3 to 5 to 10 mg/kg of bodyweight suggesting 5 mg/kg as the recommended dose. These studies showed rapid improvement from baseline and infliximab was generally well tolerated.[26–35] In the "EXPRESS" phase III study a total of 378 adult patients were enrolled in a 4:1 ratio to receive either infliximab 5 mg/kg or placebo.[28] The infliximab group received intravenous infusions of infliximab 5 mg/kg at weeks 0, 2, 6 and every 8 weeks until week 46. The placebo group were infused at weeks 0, 2, 6, 14, and 22, crossing over in a double-blind manner to infliximab 5 mg/kg at weeks 24, 26, and 30 and every 8 weeks up to week 46. About two-thirds of patients achieved an almost complete clinical PASI 75 response. In this study a significantly greater improvement of nail psoriasis in infliximab-treated compared with placebo-treated patients was also observed.

Etanercept

Etanercept is a recombinant human TNFα receptor fusion protein that acts as a competitive inhibitor of TNFα by binding to and inactivating endogenous TNFα, thereby preventing its interactions with cell surface receptors.[36] Etanercept has been approved by the FDA and the EMEA for the control of signs and symptoms, inhibiting radiographic progression, and improving function in adults with psoriatic arthritis; it is also approved for moderate to severe plaque-type psoriasis and its safety and efficacy have been demonstrated in several randomized clinical trials. The safety and efficacy of etanercept have been widely demonstrated in both psoriatic arthritis and moderate to severe plaque-type psoriasis.[37–44]

Etanercept is a fully human dimeric fusion protein produced by Chinese hamster ovary cells and is a recombinant molecule consisting of the extracellular domain of the TNFα p75 receptor fused to the inert FC portion of IgG_1 molecule (Figure 6.2). Its molecular weight is 150 kDa consisting of 934 amino acids.[36] The Fc portion of IgG_1 serves to stabilize etanercept and prolongs the median half-life which is 4.8 days. The dimeric nature of etanercept permits binding of the protein to two free or receptor bound molecules of TNFα at an affinity 50 to 1,000 times that of soluble monomeric forms of the TNFα receptor. Etanercept binds to both TNFα and LTα (formerly known as TNFβ). In addition, unlike the monoclonal antibodies, etanercept inhibits the activity of LTα, another member of the TNF superfamily. This is in contrast to the monoclonal antibodies that specifically bind TNF and have no effect on LTα.

In contrast to anti-TNF antibodies, etanercept does not induce complement-mediated TNFα cell lysis *in vitro*[36,37] and modulates biologic responses that are induced or regulated by TNFα. Two phase III randomized clinical trials on etanercept as monotherapy in moderate to severe plaque-type psoriasis have shown a substantial clinical benefit, expressed as improvement in PASI, global assessment by physicians and patients, and improvement in quality of life measures of functional status.[40,41] These clinical trials showed efficacy, tolerability, and a long-term clinical response of subcutaneously administered etanercept.

The efficacy of re-treatment with etanercept has also been reported. The re-treatment response appears to be similar to that of initial treatment; after 12 weeks of re-treatment with etanercept, patients' mean PASI scores were found to be similar to their original scores after the initial 12-week treatment period.[43]

Furthermore, a randomized, placebo-controlled trial recently demonstrated that etanercept is also effective in children and adolescents with moderate to severe plaque psoriasis. This phase III study showed a statistically significant and clinically meaningful reduction in disease severity as early as week 2 of weekly treatment with etanercept at 0.8 mg/kg (to a maximum of 50 mg) in children and adolescents with moderate to severe plaque psoriasis.[44]

The etanercept label states a recommended dose of etanercept of 25 mg twice weekly (BIW). Alternatively, 50 mg BIW may be used for up to 12 weeks, followed, if necessary, by 25 mg BIW. Treatment should be given until remission is achieved, for up to 24 weeks. However, use of the higher dose of etanercept has been shown to produce a more rapid response, thus maximizing patient benefit within the initial treatment period. The current license for etanercept recommends that the agent should be initiated at 25 mg BIW, or 50 mg BIW for plaque psoriasis. However, the improved efficacy of higher-dose etanercept was confirmed by 3-month results from an ongoing, 2-year, phase III study, indicating that patients' chances of responding to treatment are greater with etancercept 50 mg BIW. This is an important consideration as initial treatment with etanercept is given for a limited duration. On the basis of these data, it was recommended that, ideally, patients with moderate to severe disease would be started on 50 mg BIW.[45]

Adalimumab

Adalimumab is the third TNF antagonist, after infliximab and etanercept, to be approved in the United States and Europe. Like infliximab and etanercept, adalimumab binds to TNFα, preventing it from activating TNF receptors; adalimumab is a fully human monoclonal antibody, while infliximab is a mouse–human chimeric antibody and etanercept is a TNF receptor-IgG fusion protein. TNFα inactivation has proved to be important in downregulating the inflammatory reactions associated with autoimmune diseases. Adalimumab has been approved for the treatment of rheumatoid arthritis, psoriatic arthritis, ankylosing spondylitis, and Crohn's disease, and by 2008 was approved by the FDA for the treatment of moderate to severe chronic psoriasis and juvenile idiopathic arthritis.[46]

Adalimumab is a recombinant, fully human IgG_1 monoclonal antibody specific for human TNFα and blocks the interactions with the p55 and p75 cell surface TNF receptors (Figure 6.2). It was developed by harvesting antibody-producing B cells from many human donors. The resultant fully human protein, adalimumab, consists of 1330 amino acids and has a molecular weight of approximately 148 kDa.[46] This monoclonal antibody demonstrates a high specificity and affinity for TNFα but not other cytokines, such as LTβ. The half-life of adalimumab (2 weeks) is comparable to that of human IgG_1. It is indistinguishable in structure and function from naturally occurring human IgG_1 and is therefore theoretically less immunogenic than its predecessors and may therefore show a more prolonged duration of efficacy in chronic disease.

Adalimumab is administered as a subcutaneous injection. Initial data in psoriasis showed the efficacy and safety of adalimumab in patients with chronic plaque-type psoriasis.[47,48] The results of these studies demonstrated that adalimumab is effective in the treatment of moderate to severe psoriasis, as compared to placebo.[47] Adalimumab has also been shown, through sequential lesional biopsies obtained during treatment, to be able to reverse the decrease of Langerhans cells observed in psoriatic plaques, most probably through the inhibition of their TNFα-induced migration.[47] Adalimumab was initially approved for the treatment of rheumatoid arthritis and studies on patients affected by rheumatic disease treated with adalimumab showed that therapeutic response is influenced by genetic factors, in particular by the A-to-G polymorphism at position-308 in the promoter of the *TNF* α gene (subjects with a G/G genotype respond better than those with an A/G or A/A genotype),[48] and by the polymorphism at position-158 of the Fc-γ receptor type IIIA, a receptor for the invariable part of IgG antibodies (subjects with an F/F phenotype show a better response than those with a V/F or V/V phenotype).[50] It is unknown, however, if these polymorphisms also influence clinical response in psoriatic subjects. Adalimumab is administered subcutaneously, at a standard dose of 40 mg every other week in patients with rheumatoid arthritis or psoriatic arthritis. The recommended dosing of adalimumab for the treatment of adult patients with plaque-type psoriasis is an initial dose of 80 mg, followed by 40 mg given every other week starting 1 week after the initial dose.[47]

Several trials have reported the short-and long-term clinical efficacy, safety, and tolerability of adalimumab in the treatment of psoriasis and psoriatic arthritis. One of the trials directly compared the clinical efficacy, safety, and tolerability of adalimumab versus a traditional systemic agent (MTX) and placebo in the treatment of moderate to severe chronic plaque psoriasis. In this comparative study, adalimumab was more efficacious than either methotrexate or placebo for the treatment of moderate to severe plaque psoriasis.[47]

Efalizumab

Efalizumab is a recombinant, humanized monoclonal antibody that targets the α subunit (CD11a) of leukocyte function associated antigen-1 (LFA-1). By blocking the interaction of CD11a and the intercellular adhesion molecule-1 (ICAM-1), efalizumab interferes at multiple levels with the pathogenesis of psoriasis, inhibiting leukocyte trafficking into the skin as well as the continuous activation of T-lymphocytes by APCs in the dermis and epidermis. Adhesion molecules are essential in the process of T-cell rolling on the inner surface of cutaneous microvessels and consequently in their extravasation and migration to the skin.[51] T cells also express transmembrane glycoproteins bearing Sialyl-LewisX determinants, which act as ligands for both E- and P-selectin, including CLA and glycosylated P-selectin glycoprotein ligand 1 (CD162). Efomycine M, an inhibitor of both E- and P-selectins, inhibited the development of psoriasis in a SCID mouse model.[52] However, an anti-E-selectin inhibitor failed in psoriasis clinical trials.[53]

LFA-1 and VLA-4 are integrins on T cells binding to endothelial cell-expressed VCAM-1 and ICAM-1 mediating rolling and facilitating diapedis. Efalizumab, targeting T-cell adhesion and migration by the inhibition of LFA-1 (CD11a), has shown efficacy and safety in psoriasis.[54]

Efalizumab is indicated for use in adults with moderate to severe plaque psoriasis who are eligible for systemic therapy and considered as "high need" and is self-administered by the patient once weekly by subcutaneous injection at the recommended dose (1 mg/kg/week).[56] Several clinical trials have been performed with efalizumab, which validated its efficacy in short-term (12 weeks), mid-term (24 weeks), or long-term (36 or more months) treatment.[55–57] There are several phase III studies that confirm the efficacy and safety of efalizumab in patients with moderate to severe disease and that report the optimal dosing regimen.

Results from additional ongoing trials are anticipated to further support the efficacy and safety observed to date and to characterize the optimal duration of efalizumab treatment for patients with moderate to severe disease. In particular, 2-year experience in treating more than 100 patients with moderate to severe psoriasis with efalizumab at two dermatology centers in Italy confirmed the efficacy and safety of efalizumab. Furthermore, in this long-term study individual cases provided additional information about efalizumab re-treatment, the administration of efalizumab in patients with a history of heart failure, and the management of patients who became pregnant or conceived while receiving efalizumab therapy.[56]

Alefacept

Alefacept is a recombinant fusion protein, which is composed of the first extracellular domain of leukocyte function-associated antigen 3 (LFA-3) fused to the constant fraction (Fc) portion of the human immunoglobulin (Ig)G_1.[58] LFA-3, the ligand for CD2, is expressed on APCs' surface. CD2 is expressed on mature T lymphocytes and NK cells. LFA-3 is a costimulatory signal necessary for T-cell proliferation and activation. Alefacept selectively binds to unique targets: the LFA-3 portion binds to CD2 on T-cells and the Fc portion binds to FRγRIII receptors on NK cells and macrophages. The LFA-3 domain of this biologic agent inhibits the activation of T cells by means of costimulatory molecules occurring on APCs. Hence alefacept inhibits T-cell activation. Moreover, alefacept exerts an immunosuppressive action, binding its Fc portion to FRγRIII receptors on accessory cells and inducing T-cell apoptosis by releasing of granzyme by NK cells. As CD2 expression is higher on memory effector T cells ($CD4^+CD45RO^+$ and $CD8^+CD45RO^+$) than naïve ($CD45RA^+$) T cells, alefacept selectively eliminates activated T lymphocytes, which are predominant in the psoriatic skin lesions.[59–62]

Alefacept represents the first biologic agent approved in the United States for the treatment of chronic plaque psoriasis.[63] The dual mechanism of action involves both the inhibition of T-cell activation and the selective induction of apoptosis of memory T cells. The safety and efficacy of alefacept as monotherapy for the treatment of moderate to severe chronic plaque psoriasis has been demonstrated in multiple clinical trials.[64–69] A second course of alefacept has been demonstrated to be well tolerated and beneficial with respect to prolonging the duration of response and increasing the proportion of responders. Disease rebound and flare have not been noted after alefacept treatment discontinuation.[70] The safety and efficacy of multiple treatment courses with alefacept have been well documented, suggesting that alefacept is worthy of consideration for the treatment of psoriasis.[71]

Alefacept can be administered intramuscularly or intravenously, at dosages of 15 and 7.5 mg/week, respectively. Efficacy was first evaluated in two randomized, double-blind, placebo-controlled phase III trials by Krueger et al.[72] and Lebwohl et al.,[69] who used intravenous and intramuscular administration, respectively.

IL-12 AND IL-12 P40 ANTAGONISM

Two cytokines that are thought to be important in the development of Th1 immune responses in psoriasis are IL-12 and IL-23.[73] Both cytokines are produced by macrophages and dendritic cells, and function by activating T cells and NK cells. IL-12 and IL-23 are members of a heterodimeric family of soluble cytokines that comprise p35/p40 and p19/p40 protein subunits, respectively. The IL-12p40 subunit of both cytokines binds to the transmembrane IL-12 receptor b1 (IL-12Rb1) that is found on the surface of immune cells. Thus, interruption of the IL-12p40/IL-12Rb1 interaction will prevent the biologic activity of both IL-12 and IL-23.[74] The functions of IL-12 have been well characterized and include induction of IFNγ, differentiation of Th1 cells, and bridging between innate and adaptive immunity.[75] IL-23 has been proposed to have functions that are similar, but not identical, to those of IL-12.[76] IL-12, which was first discovered in 1989, plays an important role in the cell-mediated immune response, which is responsible for defense against certain intracellular bacterial and parasitic infections. IL-23 involves the stimulation of a subset of CD4+ T-cells (sometimes called TH-17 T cells) to produce IL-17 (Figure 6.4). IL-17 is a critical component in the establishment and perpetuation of autoimmune inflammation.[4] IL-17 induces the production of proinflammatory cytokines, predominately by endothelial cells and macrophages. It is believed that IL-17 and IFNγ synergize to increase production of proinflammatory cytokines by keratinocytes, which is likely important for the development of inflammation in the skin seen in psoriasis.

Monoclonal antibodies against a subunit shared by IL-12 and IL-23 have recently been developed and clinical trials targeting IL-12 and IL-23 performed.[77] These antibodies target IL-12 and IL-23 signaling pathways and by targeting the p40 subunit that is shared by both the IL-12 and IL-23 receptors. Investigators have reported results of a double-blind, placebo-controlled phase II and III study of subcutaneously administered Centocor compound (CNTO 1275).[78] Patients were randomized to placebo or one of four doses of CNTO 1275: one 45-mg dose, one 90-mg dose, 4 weekly 45-mg doses, and 4 weekly 90-mg doses. Two other inhibitors of the IL-12/IL-23 pathway are in clinical trials. Synta Pharmaceuticals (Lexington, MA) has a small molecule inhibitor of transcription of IL-12/IL-23, STA 5326. The results of placebo-controlled clinical studies of both molecules were very promising.[79]

The third compound that is undergoing testing in psoriasis is ABT-874 (Abbott Laboratories, Abbott Park, IL). A phase II, multicenter, randomized, double-blind, placebo-controlled trial demonstrated statistically and clinically significant efficacy of ABT-874 in the treatment of moderate to severe chronic plaque psoriasis. Most of the patients (90 percent) in the ABT-874 multiple-dose groups achieved a PASI 75 or greater response by week 12, compared with 3 percent of placebo-treated patients. Treatment with ABT-874 was well tolerated, and most adverse events were mild.[80]

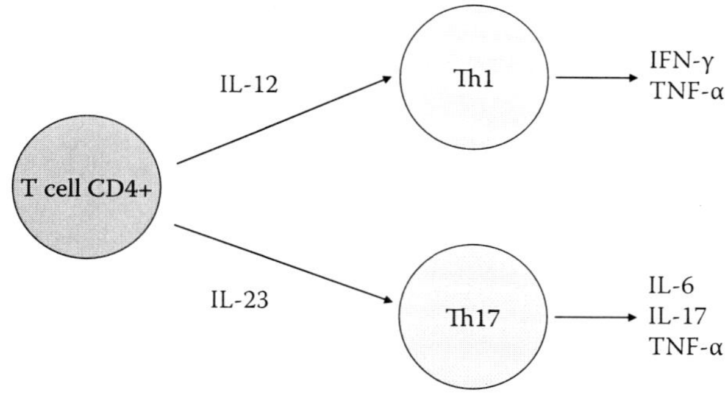

FIGURE 6.4 Pathway engaged by IL-12 and IL-23.

CONCLUSIONS

Despite that a large majority of patients are able to adequately control psoriasis using current systemic treatments or biologic agents, some of them are not eligible for systemic therapies because of contraindications due either to their side effects or to their associated diseases. Furthermore, some patients are not responsive at all to available systemic therapies, both traditional and biologic. The deeper our knowledge goes to understanding the importance of the immune system in the pathogenesis of psoriasis, the faster more specific and effective agents will be developed. At the moment the clinical research is focused mainly on anti-TNFα and the new anti-IL-12p40.

These biologic response modifiers, like any other therapeutics, carry potential risks of adverse events such as infections. As with any immunosuppressant, biologic therapies might reduce resistance to infection. For example, TNFα and IL-12p40 are important in host defense against infection, and inhibitors of this cytokine have been associated with cases of tuberculosis reactivation. Likewise, T cells are crucial to the immune response; thus, their reduction or inhibition could theoretically increase the risk for infection.

REFERENCES

1. Lowes MA, Bowcock AM, Krueger JG. Pathogenesis and therapy of psoriasis. *Nature* 2007; 22; 866–873.
2. Lebwohl M. Psoriasis. *Lancet* 2003; 361, 1197–1204.
3. Setty AR, Choi HK. Psoriatic arthritis epidemiology. *Curr Rheumatol Rep* 2007; 9, 449–454.
4. Lowes MA, Kikuchi T, Fuentes-Duculan J, Cardinale I, Zaba LC, Haider AS, Bowman EP, Krueger JG. Psoriasis vulgaris lesions contain discrete populations of Th1 and Th17 T Cells. *J Invest Dermatol* 2008; 128, 1207–1211.
5. Wang H, Peters T, Kess D, Sindrilaru A, Oreshkova T, Van Rooijen N, Stratis A, Renkl AC, Sunderkötter C, Wlaschek M, Haase I, Scharffetter-Kochanek K. Activated macrophages are essential in a murine model for T cell-mediated chronic psoriasiform skin inflammation. *J Clin Invest* 2006;116, 2105–2114.
6. Gottlieb SL, Gilleaudeau P, Johnson R, Estes L, Woodworth TG, Gottlieb AB, Krueger JG. Response of psoriasis to a lymphocyte-selective toxin (DAB389IL-2) suggests a primary immune, but not keratinocyte, pathogenic basis. *Nat Med* 1995; 442–447.
7. Bowcock AM, Krueger JG. Getting under the skin: the immunogenetics of psoriasis. *Nat Rev Immunol* 2005; 699–711.
8. Büchau AS, Gallo RL. Innate immunity and antimicrobial defense systems in psoriasis. *Clin Dermatol* 2007; 25, 616–624.
9. Lande R, Gregorio J, Facchinetti V, Chatterjee B, Wang YH, Homey B, Cao W, Wang YH, Su B, Nestle FO, Zal T, Mellman I, Schröder JM, Liu YJ, Gilliet M. Plasmacytoid dendritic cells sense self-DNA coupled with antimicrobial peptide. *Nature* 2007; 449, 564–569.
10. Liu YJ. IPC: professional type 1 interferon-producing cells and plasmacytoid dendritic cell precursors. *Annu Rev Immunol* 2005; 23, 275–306.
11. Colonna M, Trinchieri G, Liu YJ. Plasmacytoid dendritic cells in immunity. *Nature Immunol* 2004; 5, 1219–1226.

12. Kadowaki N, Ho S, Antonenko S, Malefyt RW, Kastelein RA, Bazan F, Liu YJ. Subsets of human dendritic cell precursors express different Toll-like receptors and respond to different microbial antigens. *J Exp Med* 2001; 194, 863–869.
13. Latz E, Schoenemeyer A, Visintin A, Fitzgerald KA, Monks BG, Knetter CF, Lien E, Nilsen NJ, Espevik T, Golenbock DT. TLR9 signals after translocating from the ER to CpG DNA in the lysosome. *Nature Immunol* 2004; 5, 190–198.
14. Diebold SS, Kaisho T, Hemmi H, Akira S, Reis E, Sousa C. Innate antiviral responses by means of TLR7-mediated recognition of single-stranded RNA. *Science* 2004; 303, 1529–1531.
15. Sterry W, Barker J, Boehncke WH, et al. Biologicals in psoriasis consensus. *Br J Dermatol* 2004; 151, 3–17.
16. Boss JD, de Rie MA. The pathogenesis of psoriasis: immunological facts and speculations. *Immunol Today* 1999; 20, 40–46.
17. Ettehadi P, Greaves MW, Wallach D, Aderka D, Camp RD. Elevated tumor necrosis factor-α (TNF-α) biological activity in psoriatic skin lesions. *Clin Exp Immunol* 1994; 96, 146–151.
18. Hohler T, Kruger A, Schneider PM, Schopf RE, Knop J, Rittner C, Meyer zum Büschenfelde KH, Märker-Hermann E. A TNF-α promoter polymorphism is associated with juvenile onset of psoriasis and psoriatic arthritis. *J Invest Dermatol* 1997; 109, 562–565.
19. Partsch G, Steiner G, Leeb BF, et al. Highly increased levels of tumor necrosis factor-α and other proinflammatory cytokines in psoriatic arthritis synovial fluid. *J Rheumatol* 1997; 24, 518–523.
20. Ritchlin C, Haas-Smith SA, Hicks D, Cappuccio J, Osterland CK, Looney RJ. Patterns of cytokine production in psoriatic synovium. *J Rheumatol* 1998; 25, 1544–1552.
21. Gottlieb AB. Infliximab in psoriasis. *J Am Acad Dermatol* 2003; 49, S112–117.
22. Gottlieb AB, Masud S, Ramamurthi R, Abdulghani A, Romano P, Chaudhari U, Dooley LT, Fasanmade AA, Wagner CL. Pharmacodynamic and pharmacokinetic response to anti-tumor necrosis factor monoclonal antibody (Infliximab) treatment of moderate to severe psoriasis vulgaris. *J Am Acad Dermatol* 2003; 48, 68–75.
23. Krueger JG. The immunologic basis for the treatment of psoriasis with new biologic agents. *J Am Acad Dermatol* 2002; 46, 1–23.
24. Gottlieb AB, Chang CK, Posnett DN, Fanelli B, Tam JP. Detection of transforming growth factor alpha in normal, malignant, and hyperproliferative human keratinocytes. *J Exp Med* 1988; 167, 670–675.
25. Knight DM, Trinh H, Le J, Le J, Siegel S, Shealy D, McDonough M, Scallon B, Moore MA, Vilcek J, Daddona P, et al. Construction and initial characterization of A mouse-human chimeric anti-TNF antibody. *Mol Immunol* 1993 ; 30, 1443–1453.
26. Chaudhari U, Romano P, Mulcahy LD, Dooley LT, Baker DG, Gottlieb AB. Efficacy and safety of infliximab monotherapy for plaque-type psoriasis: A randomized trial. *Lancet* 2001; 357, 1842–1847.
27. Gottlieb AB, Chaudari U, Mulcahy LD, Li S, Dooley LT, Baker DG. Infliximab monotherapy provides rapid and substantial benefit for plaque-type psoriasis. *J Am Acad Dermatol* 2003; 48, 829–835.
28. Reich K, Nestle FO, Papp K, Ortonne JP, Evans R, Guzzo C, Li S, Dooley LT, Griffiths CE; EXPRESS study investigators. Infliximab induction and maintenance therapy for moderate-to-severe psoriasis: A phase III, multicentre, double-blind trial. *Lancet* 2005; 366, 1367–1374.
29. Kalb RE, Gurske J. Infliximab for the treatment of psoriasis: Clinical experience at the State University of New York at Buffalo. *J Am Acad Dermatol* 2005; 53, 616–622.
30. Smith CH, Jackson K, Bashir SJ. Infliximab for severe, treatment-resistant psoriasis: A prospective, open-label study. *Br J Dermatol* 2006; 155,160–169.

31. Ahmad K, Rogers S. Three years' experience with infliximab in recalcitrant psoriasis. *Clin Exp Dermatol* 2006; 31, 630–633.
32. Schopf RE, Aust H, Knop J. Treatment of psoriasis with the chimeric monoclonal antibody against tumor necrosis factor alpha, infliximab. *J Am Acad Dermatol* 2002; 46, 886–891.
33. Gottlieb AB, Evans R, Li S, Dooley LT, Guzzo CA, Baker D, Bala M, Marano CW, Menter A. Infliximab induction therapy for patients with severe plaque-type psoriasis: a randomized, double-blind, placebo-controlled trial. *J Am Acad Dermatol* 2004; 51, 534–542.
34. Oh CJ, Das KM, Gottlieb AB. Treatment with anti-tumor necrosis factor α (TNF-α) monoclonal antibody dramatically decreases the clinical activity of psoriasis lesions. *J Am Acad Dermatol* 2000; 42, 829–830.
35. Antoni CE, Kavanaugh A, Kirkham B, Tutuncu Z, Burmester GR, Schneider U, Furst DE, Molitor J, Keystone E, Gladman D, Manger B, Wassenberg S, Weier R, Wallace DJ, Weisman MH, Kalden JR, Smolen J. Sustained benefits of infliximab therapy for dermatologic and articular manifestations of psoriatic arthritis: results from the infliximab multinational psoriatic arthritis controlled trial (IMPACT). *Arthritis Rheum* 2005; 52, 1227–1236.
36. Goffe B, Clay CJ. Etanercept: An overview. *J Am Acad Dermatol* 2003; 49: S105–111.
37. Gordon K, Korman N, Frankel E, Wang H, Jahreis A, Zitnik R, Chang T. Efficacy of etanercept in an integrated multistudy database of patients with psoriasis. *J Am Acad Dermatol* 2006; 54: S101–111.
38. Leonardi CL, Powers JL, Matheson RT, Goffe BS, Zitnik R, Wang A, Gottlieb AB; Etanercept Psoriasis Study Group. Etanercept as monotherapy in patients with psoriasis. *N Engl J Med* 2003; 349, 2014–2022.
39. Gottlieb AB, Matheson RT, Lowe N, Krueger GG, Kang S, Goffe BS, Gaspari AA, Ling M, Weinstein GD, Nayak A, Gordon KB, Zitnik R. A randomized trial of etanercept as monotherapy for psoriasis. *Arch Dermatol* 2003; 139, 1627–1632.
40. Papp KA, Tyring S, Lahfa M, Prinz J, Griffiths CE, Nakanishi AM, Zitnik R, van de Kerkhof PC, Melvin L; Etanercept Psoriasis Study Group. A global phase III randomized controlled trial of etanercept in psoriasis: Safety, efficacy, and effect of dose reduction. *Br J Dermatol* 2005; 152, 1304–1312.
41. Mease P, Goffe B, Metz J, VanderStoep A, Finck B, Burge DJ. Etanercept in the treatment of psoriatic arthritis and psoriasis: A randomized trial. *Lancet* 2000; 356, 385–390.
42. Costanzo A, Mazzotta A, Papoutsaki M, Nisticò S, Chimenti S. Safety and efficacy study on etanercept in patients with plaque psoriasis. *Br J Dermatol* 2005; 152, 187–189.
43. Leonardi CL, Elewski B, Camisa C, Zitnik R. The efficacy and safety of etanercept in the retreatment of psoriasis after relapse (Abstract P568). *J Acad Dermatol* 2004; 50, 146.
44. Paller AS, Siegfried EC, Langley RG, Gottlieb AB, Pariser D, Landells I, Hebert AA, Eichenfield LF, Patel V, Creamer K, Jahreis A; Etanercept Pediatric Psoriasis Study Group. Etanercept treatment for children and adolescents with plaque psoriasis. *N Engl J Med* 2008; 358, 241–251.
45. Boehncke WH, Brasie RA, Barker J, Chimenti S, Daudén E, de Rie M, Dubertret L, Giannetti A, Katsambas A, Kragballe K, Naeyaert JM, Ortonne JP, Peyrí J, Prinz JC, Saurat JH, Strohal R, van de Kerkhof P, Sterry W; European Dermatology Expert Group. Recommendations for the use of etanercept in psoriasis: A European dermatology expert group consensus. *J Eur Acad Dermatol Venereol* 2006; 20, 988–998.
46. Humira package insert. Abbott Park, IL: Abbott.
47. Papoutsaki M, Costanzo A, Chimenti MS, Chimenti S. Adalimumab for the treatment of severe psoriasis and psoriatic arthritis. *Expert Opin Biol Ther* 2008; 8, 363–370.

48. Papoutsaki M, Chimenti MS, Costanzo A, Talamonti M, Zangrilli A, Giunta A, Bianchi L, Chimenti S. Adalimumab for severe psoriasis and psoriatic arthritis: An open-label study in 30 patients previously treated with other biologics. *J Am Acad Dermatol* 2007; 57, 269–275.
49. Gordon KB, Bonish BK, Patel T, Leonardi CL, Nickoloff BJ. The tumour necrosis factor-alpha inhibitor adalimumab rapidly reverses the decrease in epidermal Langerhans cell density in psoriatic plaques. *Br J Dermatol* 2005; 153, 945–953.
50. Seitz M, Wirthmuller U, Moller B, Villiger PM. The -308 tumour necrosis factor-alpha gene polymorphism predicts therapeutic response to TNFalpha-blockers in rheumatoid arthritis and spondyloarthritis patients. *Rheumatology* (Oxford) 2007; 46, 93–96.
51. Schon MP, Zollner TM and Boehncke HW. The molecular basis of lymphocyte recruitment to the skin: Clues for pathogenesis and selective therapies of inflammatory disorders. *J Invest Dermatol* 2002; 121, 951–962.
52. Schon MP, Krahn T, Schon M, Rodriguez ML, Antonicek H, Schultz JE, Ludwig RJ, Zollner TM, Bischoff E, Bremm KD, Schramm M, Henninger K, Kaufmann R, Gollnick HP, Parker CM, Boehncke WH. Efomycine M, a new specific inhibitor of selectin, impairs leukocyte adhesion and alleviates cutaneous inflammation. *Nat Med* 2002; 8, 366–372.
53. Bhushan M, Bleiker TO, Ballsdon AE, Allen MH, Sopwith M, Robinson MK, Clarke C, Weller RP, Graham-Brown RA, Keefe M, Barker JN, Griffiths CE. Anti-E-selectin is ineffective in the treatment of psoriasis: A randomized trial. *Br J Dermatol* 2002; 146, 824–831.
54. Jullien D, Prinz JC, Langley RG, et al. T-cell modulation for the treatment of chronic plaque psoriasis with efalizumab (Raptiva): Mechanisms of action. *Dermatology* 2004; 208, 297–306.
55. Leonardi C, Menter A, Hamilton T, Caro I, Xing B, Gottlieb AB. Efalizumab: Results of a 3-year continuous dosing study for the long-term control of psoriasis. *Br J Dermatol* 2008; 158(5): 1107–1116.
56. Costanzo A, Peris K, Talamonti M, Di Cesare A, Fargnoli MC, Botti E, Chimenti S. Long-term treatment of plaque psoriasis with efalizumab: An Italian experience. *Br J Dermatol* 2007; 156, 17–23.
57. Dubertret L, Sterry W, Bos JD, Chimenti S, Shumack S, Larsen CG, Shear NH, Papp KA; CLEAR Multinational Study Group. Clinical experience acquired with the efalizumab (Raptiva) (CLEAR) trial in patients with moderate-to-severe plaque psoriasis: Results from a phase III international randomized, placebo-controlled trial. *Br J Dermatol* 2006; 155, 170–181.
58. Kormeili T, Lowe NJ, Yamauchi PS. Psoriasis: Immunopathogenesis and evolving immunomodulators and systemic therapies; U.S. experiences. *Br J Dermatol* 2004; 151, 3–15.
59. Prinz JC. The role of T cells in psoriasis. *JEADV*, 2003; 17, 257–270.
60. Mehlis LS, Gordon KB. The immunology of psoriasis and biologic immunotherapy. *J Am Acad Dermatol* 2003; 49, s44–49.
61. Krueger GG, Callis KP. Development and use of alefacept to treat psoriasis. *J Am Acad Dermatol* 2003; 49, s87–97.
62. Gottlieb AB, Casale TB, Frankel E, Goffe B, Lowe N, Ochs HD, Roberts JL, Washenik K, Vaishnaw AK, Gordon KB. $CD4^+$ T cell-directed antibody responses are maintained in patients with psoriasis receiving alefacept: Results of a randomized study. *J Am Acad Dermatol* 2003; 49, 816–825.
63. Piascik P. Alefacept, first biologic agent approved for treatment of psoriasis. *J Am Pharm Assoc* 2003; 43, 649–650.

64. Ortonne JP. Clinical response to alefacept: Results of a phase 3 study of intramuscular administration of alefacept in patients with chronic plaque psoriasis. *JEADV*, 2003; 17, 12–16.
65. Gordon KB, Vaishnaw AK, O'Gorman J, Haney J, Menter A; Alefacept Clinical Study Group. Treatment of psoriasis with alefacept: Correlation of clinical improvement with reductions of memory T-cell counts. *Arch Dermatol* 2003; 139, 1563–1570.
66. Gade JN. Clinical update on alefacept: Consideration for use in patients with psoriasis. *Suppl J Manag Care Pharmacy* 2004; 10, s33–s37.
67. Krueger GG, Papp KA, Stough DB, Loven KH, Gulliver WP, Ellis CN; Alefacept Clinical Study Group. A randomized, double-blind, placebo-controlled phase III study evaluating efficacy and tolerability of 2 courses of alefacept in patients with chronic plaque psoriasis. *J Am Acad Dermatol* 2002; 47, 821–833.
68. Gordon KB, Langley RG. Remittive effects of intramuscular alefacept in psoriasis. *J Drugs Dermatol* 2003; 2, 624–628.
69. Lebwohl M, Christophers E, Langley R, Ortonne JP, Roberts J, Griffiths CE; Alefacept Clinical Study Group. An international, randomized, double-blind, placebo-controlled phase III trial of intramuscular alefacept in patients with chronic plaque psoriasis. *Arch Dermatol* 2003; 139, 719–727.
70. Hodak E, David M. Alefacept: A review of the literature and practical guidelines for management. *Dermatol Ther* 2004; 17, 383–392.
71. Krueger GG. Current concepts and review of alefacept in the treatment of psoriasis. *Dermatol Clin* 2004; 22, 407–426.
72. Krueger GG, Papp KA, Stough DB, Loven KH, Gulliver WP, Ellis CN; Alefacept Clinical Study Group. A randomized, double-blind, placebo controlled phase III study evaluating efficacy and tolerability of 2 courses of alefacept in patients with chronic plaque psoriasis. *J Am Acad Dermatol* 2002; 47, 821–833.
73. Lee E, Trepicchio WL, Oestreicher JL, et al. Increased expression of interleukin 23 p19 and p40 in lesional skin of patients with psoriasis vulgaris. *J Exp Med* 2004; 199, 125–130.
74. Kang BY, Kim TS. Targeting cytokines of the interleukin-12 family in autoimmunity. *Curr Med Chem* 2006; 13, 1149–1156.
75. Trinchieri G, Pflanz S, Kastelein RA. The IL-12 family of heterodimeric cytokines: New players in the regulation of T cell responses. *Immunity* 2003; 19, 641–644.
76. Oppmann B, Lesley R, Blom B, et al. Novel p19 protein engages IL-12p40 to form a cytokine, IL-23, with biological activities similar as well as distinct from IL-12. *Immunity* 2000; 13, 715–725.
77. Kauffman CL, Aria N, Toichi E, McCormick TS, Cooper KD, Gottlieb AB, et al. A phase I study evaluating the safety, pharmacokinetics, and clinical response of a human IL-12 p40 antibody in subjects with plaque psoriasis. *J Invest Dermatol* 2004; 123, 1037–1044.
78. Leonardi C, Yeilding N, Guzzo C, Wang Y, et al. A human interleukin-12/23 monoclonal antibody for the treatment of psoriasis. *N Engl J Med* 2007; 356, 580–592.
79. Gordon K, Tschen E, Korman N, Cather J, Gold M, Hampel F, et al. A phase II study of STA-5326, an oral IL-12 and IL-23 inhibitor, in patients with moderate to severe chronic plaque psoriasis. Paper presented at 4th International Congress, Royal College of Physicians; Dec 1–3, 2005; London.
80. Kimball AB, Gordon KB, Langley RG, Menter A, Chartash EK, Valdes J; ABT-874 Psoriasis Study Investigators. Safety and efficacy of ABT-874, a fully human interleukin 12/23 monoclonal antibody, in the treatment of moderate to severe chronic plaque psoriasis: Results of a randomized, placebo-controlled, phase 2 trial. *Arch Dermatol* 2008; 144, 200–207.

7 TNFα Inhibitors

Emerging Biologic Therapies for the Treatment of Asthma

Sudha Natarajan, Louis J. Vaickus, Jiyoun Kim, and Daniel G. Remick

CONTENTS

KEY WORDS: asthma, TNFα, bronchial hyperresponsiveness, eosinophil, infliximab, etanercept

ABSTRACT

Asthma is emerging as an increasingly prevalent disease in the developed world with incidences on the rise in developed countries. It is a complex inflammatory disease characterized by airways obstruction, bronchial hyperresponsiveness, accompanied by eosinophilic and neutrophilic inflammation. Upon activation, local immune and inflammatory cells release several immunomodulator molecules that serve to drive the inflammatory response. One such molecule, tumor necrosis factor-α (TNFα) has been shown to induce airway smooth muscle spasm and the downstream production and release of numerous proinflammatory agents. Inhibition of TNFα has shown marked improvement of symptoms in chronic inflammatory diseases such as rheumatoid arthritis, psoriasis, and Crohn's disease. Effective treatment in these inflammatory diseases suggests that blockade of TNFα may be a promising therapeutic option in the treatment of asthma. Several clinical trials have already shown improvement in asthmatic symptoms upon treatment with various TNFα inhibitors. Here, we highlight the role of TNFα in the induction and progression of the asthmatic response, and discuss the efficacy of existing TNFα antagonists, as well as newly emerging agents that might serve as treatment in the future.

ASTHMA AS A GROWING PUBLIC HEALTH CONCERN

Asthma has emerged as one of the most prevalent chronic conditions in modern industrial societies resulting in a significant strain on health care expenditures. The prevalence of asthma increases by 50 percent every 10 years worldwide. Interestingly, the prevalence of asthma shows a geographic and economic dependence with the highest rates of prevalence occurring in Europe and the United States. Asthma incidence in developing countries such as India and China is significantly lower, but rises as migration from rural areas to urban centers increases.[1] The high mortality rates associated with asthma make it a serious public health concern as roughly 250,000 individuals succumb to the disease each year.[1] In 1997, 26.7 million Americans had been diagnosed with asthma during their lifetime (96.6/1,000 persons).[2] As rates of asthma increase, the burdens brought on the health care system, from both direct and indirect costs, will continue to grow. From 1994 to 1996 complications of asthma were responsible for an estimated 14 million missed days of school, and 14.5 million missed days of work. In addition, asthmatic attacks caused close to 2 million emergency department visits and nearly 5,000 deaths in 1999.[2] Considering health care costs and lost productivity, asthma is responsible for an estimated annual loss of 14 billion dollars in national spending.[3]

CURRENT ASTHMA TREATMENT OPTIONS

Current treatment strategies are aimed at halting acute exacerbations, preventing/decreasing exacerbations, and slowing long-term progression of the disease. Short acting/relief medications can be split into two main groups: anticholinergics and short-acting bronchodilators. The most commonly employed anticholinergic to treat asthma is ipratropium bromide (Atrovent), which causes a decrease in smooth muscle tone in the airway by antagonism of muscarinic cholinergic receptors. Its action is thus limited to cholinergic bronchoconstriction and has no effect on antigen-induced, hypersensitivity-type attacks.[4]

Short-acting β2 agonists are the agents of choice for the treatment of acute asthmatic exacerbations. They stimulate β2 adrenergic receptors on airway smooth muscle (ASM) cells to cause rapid bronchodilation with limited systemic side effects (as are seen with less specific adrenergic agonists such as epinephrine). However, some of these agents may cause down-regulation of β2 receptors over time, necessitating a higher dose for relief.

Long-term control medications include corticosteroids, cromolyn compounds, anti-immunoglobulin E (anti-IgE) antibodies, leukotriene receptor antagonists, 5-lipoxygenase inhibitors, long-acting β2 agonists, and methylxanthines. These agents are used for symptom prevention, but are not used to treat asthma attacks.

Corticosteroids (prednisone, methyl prednisone, budesonide) are the single most powerful and effective class of compounds in the resolution of asthmatic inflammation and the long-term prevention of symptoms.[4,5] Inhaled corticosteroids are included in the regimen of patients with moderate to severe asthma as a periodic treatment to prevent exacerbations. By using inhalation as the route of administration, many of the systemic side effects are avoided. However, users are subject to cough, dysphonia, oral candidiasis, and other opportunistic infections of the respiratory tract.[4,5] Systemic corticosteroids are reserved for chronic treatment of patients with severe persistent asthma. They are sometimes used to quickly resolve acute symptoms in less severe cases before beginning conventional therapy. Unfortunately, several drawbacks exist with the prolonged use of systemic administration of corticosteroids. Some of these side effects may manifest as early as a few weeks into treatment, including slow wound healing and osteoporosis.[4]

Omalizumab is a monoclonal antibody specific for human IgE used in patients with severe, persistent asthma with high circulating IgE levels. It binds to circulating IgE and prevents it from binding to high-affinity Fc receptors on mast cells and basophils. This prevents the release of inflammatory mediators in the event of antigen exposure and binding to IgE. Although treatment is both safe and effective, the required frequent subcutaneous administration results in injection-site reactions in a large proportion of users. Depending on the required dose for an individual patient, treatment can range from $5,400 to $35,000 per year, making this treatment cost-prohibitive for many individuals with asthma.[6]

Montelukast and zafirlukast are selective inhibitors of the $CysLT_1$ leukotriene receptor. They are indicated for long-term control of mild persistent asthma and achieve modest results. Montelukast has been associated with infrequent cases of Churg-Straus syndrome. Zafirlukast inhibits hepatic p450 enzymes and thus may interfere with the metabolism of certain drugs. It has also been associated with cases of reversible hepatitis and rare cases of irreversible, fatal hepatitis.[4]

Given the heterogeneous nature of the asthmatic response, and the inability of current treatments to provide safe and long-term effectiveness in controlling asthma symptoms, considerable resources are allocated to the search for superior alternatives. Inflammatory modulators have been an area of intense study because the profile of asthmatic airways is inflammatory, even between attacks. Chronic airways inflammation causes progressive structural and functional damage leading to progressive worsening of the disease over time. One such inflammatory cytokine that has garnered much interest is tumor necrosis factor-α (TNFα). It has become of particular interest not only because of the central role it seems to play in asthma exacerbations, but also because anti-TNFα compounds have had great success in allaying other inflammatory conditions such as rheumatoid arthritis (RA), Crohn's disease, systemic sclerosis, and psoriasis.

BIOLOGY OF TUMOR NECROSIS FACTOR-α

TNFα (lymphotoxin, cachetin) is a powerful proinflammatory and immunomodulatory molecule that is elevated in many inflammatory conditions. It was originally discovered in 1968 as a potent lymphotoxic compound, and in 1975 was discovered to be potently stimulated by lipopolysaccharide (LPS) and an effective mediator of tumor dysregulation and necrosis.[7,8] It is now recognized to be a highly pleiotropic cytokine produced primarily by activated macrophages.[9,10] The functions of TNF range from intracellular pathogen containment to apoptosis induction to tissue regeneration to robust inflammation.[11]

Macrophage activation and TNFα release is classically considered to be dependent on two stimuli: interferon gamma (IFNγ) priming of macrophages and Toll-like receptor (TLR) activation-induced TNFα, which acts as its own stimulus for release on IFNγ primed macrophages.[9,10] IFNγ is produced in response to a wide variety of insults, especially intracellular pathogens such as viruses and certain bacteria. TLR activation is a part of an innate immune cascade designed to recognize highly conserved pathogenic motifs such as endotoxin, and certain polysaccharides, peptides, and nucleic acids.

TNFα is synthesized as monomer subunits and assembles to form type II transmembrane trimers, constitutively expressed on the cell surface.[11–13] Transmembrane TNFα has been shown to be a biologically active form of the molecule.[14] Upon activation, a metalloproteinase known as disintegrin/TACE/ADAM17 cleaves TNFα from the membrane to form a soluble homotrimer.[11,15] TNF signal transduction is accomplished via two membrane proteins, TNF-R1/p55 and TNF-R2/p75.[11,16–18] TNF-R1 is expressed in most tissues and its activation has a number of downstream effects that influence cell survival and the inflammatory state.[11,16,17] TNF-R1 contains the death domain (DD) which can activate the caspase system and is proapoptotic. It also may activate NF-κB and cIAP which promote cell survival.[11,16,17] Signaling through TNF-R1 also mediates most proinflammatory effects of TNFα including shock and endotoxin-induced tissue injury.[19] The factors that determine the fate of a TNF-R1-stimulated cell under the influence of these antagonistic signals are not well understood.

TNF-R2 is found mostly on immune cells and can only be fully activated by interaction with membrane-bound TNF. It seems to play an important role in the regulation and development of the lymphoid system and may also initiate an alternate apoptotic signal.[11] Specifically, TNF-R2 causes the proliferation of T and B cells, the induction of NF-κB, and the stimulation of cytotoxic T cells.[17]

The ratios of the two TNF receptors differ from tissue to tissue. The disparate signaling pathways and functions imply a highly pleiotropic response to TNF depending on the cell type receiving the stimulus and on the local and systemic environment. In addition, there is evidence for extensive cross-talk between TNF-R1 and TNF-R2. One such example is the factor TRAF2, which is recruited by TNF-R1 and TNF-R2 and allows for significant signal exchange between the two receptors.[11,16]

Interestingly, TNF has a higher affinity binding for TNF-R2, but dissociates from the receptor more quickly than it does from TNF-R1. This implies that p75 may bind TNF and then pass it off to p55 to cause a greater accumulation of the cytokine and thus a stronger activation of the receptor than would be possible if p55 alone were present on the membrane.[11,17,18] p55 and p75 may also exist as soluble receptors in order to bind free TNF and blunt its effect on target cells.[20–24] These circulating TNF receptors are shed at a significant level under basal conditions implying that they serve an important function in the control of TNF signaling.[24–26] Soluble receptors are also released in response to various stimuli including LPS, phorbol myristate (PMA), interleukin 10 (IL-10), and TNF itself.[11,23,24,27] In addition, soluble TNF receptors can be released by resting and stimulated monocytes; activated neutrophils, and T cells.[11,17,23,24,26,27] p55 is cleaved from the membrane by TACE/ADAM17; the mechanism by which p75 is solubilized has yet to be elucidated.[11]

Depending on the type of stimulus, and mediator(s) employed, differing ratios of soluble p55 and p75 will be released. The interplay of these two soluble receptors with free TNF, cross-talk between the membrane-bound receptors and soluble and membrane TNF, and the overall inflammatory state determine what the final result of TNF simulation will be. From granuloma formation, to apoptosis, from septic shock to tissue regeneration, the complex nature of the TNF signaling and control pathways renders outcome prediction very difficult.

Involvement of TNFα in Inflammatory Disease

TNFα has been shown to play a central role in the initiation and progression of various inflammatory diseases such as RA, Crohn's disease, and psoriasis, and has also been implicated in contributing to insulin resistance resulting in type 2 diabetes mellitus.[28,29] As is the case in many chronic inflammatory diseases, TNFα production is increased at the site of inflammation. As such, TNFα production is seen as a logical therapeutic target for treatment of these diseases. To date, immunobiologics targeted toward inhibition of TNFα have been tremendously successful in the treatment of RA and much effort is being put into the development of new antagonists and additional strategies aimed at TNFα inhibition.[30]

Several different mechanisms can be employed to neutralize the biologic effects of TNFα, including antisense oligonucleotides, inhibition of synthesis or secretion of the cytokine, soluble receptors that bind and inactivate the molecule, antibody-mediated binding, and receptor antagonists that bind TNFα receptors, thereby blocking binding of the molecule itself (Figure 7.1). Currently, three inhibitors of TNFα have been approved by the Food and Drug Administration (FDA) for treatment of RA, each approaching inhibition by a different mechanism. Etanerecept (Enbrel) is a soluble TNFα receptor whose extracellular domain consists of two identical segments of the human TNFα p75 receptor and also consists of the Fc portion of human IgG_1.[31] Infliximab (Remicade) is a chimeric human–mouse monoclonal antibody directed against TNFα, and adalimumab (Humira) is a human monoclonal anti-TNFα antibody.[32–34] Several other therapeutics such as TNFα converting enzyme

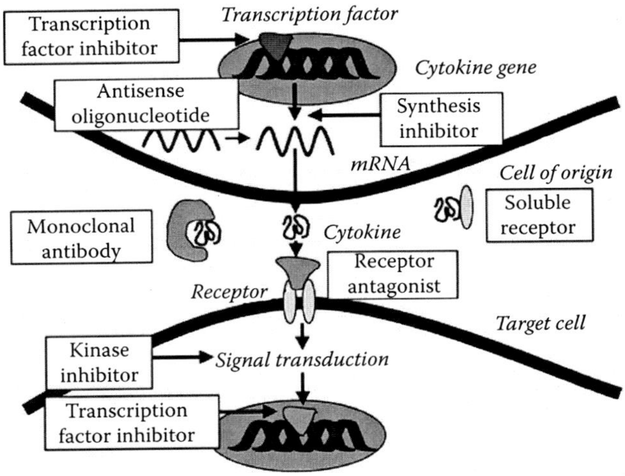

FIGURE 7.1 Methods of TNF inhibition. Currently, a soluble TNFα receptor and two different monoclonal antibodies have been approved by the FDA for treatment of RA. (From Barnes, P.J., *Annu Rev Pharmacol Toxicol*, 2002. 42: 81–98. With permission from Annual Reviews, www.annualreviews.org.)

(TACE) and matrix metalloproteinase (MMP) inhibitors are currently being developed and tested.[35,36]

Role of TNFα in Allergic Airways Disease

Both animal models and human research have shown an increase in TNFα in the airways of allergic subjects. In a mouse model of allergic asthma induced by a house dust extract containing large amounts of cockroach allergens and endotoxin, TNFα levels within the lung peaked 2 hours post allergen challenge and subsided by 24 hours (Figure 7.2). This production probably results, in part, because of IgE cross-linking and mast cell degranulation which leads to the release of pre-formed TNFα.[37] Alveolar macrophages also secrete large amounts of TNFα within the first hours after allergen encounter.[38] If unchecked, these inflammatory processes can result in long-term structural modifications and tissue remodeling in the lung resulting from the growth-promoting effects of TNFα on fibroblasts, as well as its ability to induce the release of MMP-9 by eosinophils.[39,40] TNFα can also induce the expression of ICAM-1 and VCAM-1 on epithelial cells, inducing the binding of T lymphocytes to airway smooth muscle cells resulting in migration of inflammatory cells into the airway interstitium.[41,42]

Another cardinal feature of asthma initiated by TNFα production is the onset of airways hyperresponsiveness (AHR). Although the mechanisms of AHR are not completely understood in either humans or animal models, TNFα seems to play a key role in its onset and maintenance. Inhalation of TNFα in healthy subjects is able to induce airways responsiveness and neutrophil recruitment.[43] Upon onset of the asthmatic response, TNFα is responsible for initiating many downstream steps

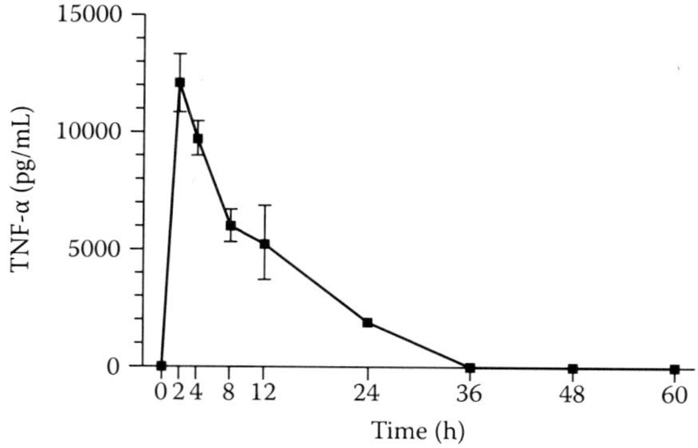

FIGURE 7.2 Kinetics of TNFα production. Allergen sensitization and challenge results in peak production of TNFα at 2 hours post final challenge, which subsides by 24 hours as measured in bronchoalveolar lavage (BAL) fluid. (From Kim, J., et al., *Clin Exp Allergy*, 2006. 36(1): 122–32. With permission.)

involved in the pathophysiology of asthma. In individuals with mild asthma, TNFα inhalation has been shown to induce increased sensitivity to methacholine challenge as well as a rise in neutrophil and eosinophil numbers in the lung.[44] TNFα may act directly on ASM or may induce the release of histamine and cysteinyl leukotrienes by mast cells.[45,46] Some evidence suggests that TNFα acts directly to alter the contractile properties of ASM. In an animal model, ASM exposed to TNFα became hyperresponsive to contractile agonists.[47] Recent evidence implicates both the p55 and p75 receptors of TNFα in AHR as serum levels of both receptors show inverse correlation to levels of AHR in both adults and children with asthma.[48] This clear relationship between TNFα and the hallmark indications of asthmatic inflammation make it an obvious target for therapeutic intervention for the treatment of asthma.

TNFα and Corticosteroid-Resistant Asthma

On the severe end of the spectrum, the morbidity and mortality associated with severe or corticosteroid-dependent asthma account for a disproportionate amount of the health care burden of the asthmatic population. In the United States, patients with severe asthma represent roughly 10 percent of patients, but are allocated 50 percent of health care resources for disease management. In contrast, patients with mild asthma represent about 70 percent of the asthmatic population, but require only 20 percent of treatment costs.[49] Currently, there is no universally accepted definition of steroid-resistant asthma, but it can be described by cases in which patients respond poorly to inhaled corticosteroids, necessitating the periodic or continuous use of high-dose oral corticosteroids.[50] It is also defined as failure to improve morning forced expiratory volume (FEV_1) by 15 percent following sustained treatment with oral prednisone. Most patients do not exhibit complete corticosteroid resistance, but the high doses and prolonged periods of treatment required for effectiveness are undesirable.[51] It should also be noted that TNF inhibitors are typically used for the treatment of RA or Crohn's disease after failed steroid therapy, so a medical precedent exists for the concept of treating steroid-resistant diseases with TNF inhibitors.

The pathology of severe asthma is also quite diverse. Although asthma is generally considered an eosinophilic disease, steroid-dependent asthma involves a more heterogeneous inflammatory response, with a significant population of patients with severe asthma having few eosinophils and lower FEV_1.[52] Patients with severe asthma also show increased involvement of neutrophils as well as increased airway remodeling.[53] The magnitude by which the asthmatic phenotype varies from patient to patient highlights the difficulties involved in developing treatments that are effective for a wide range of individuals.

A substantial link between steroid resistance and TNFα has been established. The presence of TNFα diminishes the ability of glucocoritcoids to suppress IL-6 production in whole blood in response to LPS.[54] Expression of TNFα in bronchoalveolar lavage fluid is significantly elevated in patients with steroid-treated severe asthma compared to those with steroid-treated mild asthma.[55] In contrast to patients with moderate asthma, patients with severe asthma showed increased expression of TACE, membrane-bound TNFα, and TNFα receptor 1 in peripheral blood mononuclear cells. Endobronchial biopsies of patients with severe asthma also show increased TNFα gene expression as well as increased numbers of TNFα immunoreactive cells.[56] In contrast, TNFα in individuals with mild asthma is generally localized to mast cell granules.[57] Administration of formoterol to patients with atopic asthma for 9 weeks resulted in a reduction in all cytokines examined with the exception of TNFα. The increased levels of TNFα taken together with the inability of corticosteroids to reduce TNFα levels in patients with severe asthma may suggest why these drugs are ineffective in treating severe asthma.[58]

TNFα INHIBITORS IN ANIMAL MODELS OF ASTHMA

The use of TNFα inhibitors in various animal models has met with significant success and has led the way for the use of these inhibitors in various clinical trials aimed at treating patients with asthma. Early studies using a TNFα receptor fusion protein to treat ovalbumin-induced asthma in both rats and guinea pigs showed a reduction in inflammatory cell recruitment into the bronchoalveolar lavage fluid,[32] including reduced eosinophilia and reduced microvascular leakage. Treatment with this inhibitor also reduced airway reactivity in response to substance P.[59] Mice deficient in both the p55 and p75 receptors showed markedly reduced eosinophilia, indicating that receptor targeting may also be an effective method of treatment.[60] More recent studies using a neutralizing antibody against TNFα showed marked reduction in the hallmark signs of asthmatic inflammation. In this murine model, in which asthma-like inflammation was induced using a house dust extract containing cockroach allergens, treatment with a TNF-neutralizing antibody resulted in a significant decrease in airways hyperresponsiveness, eosinophil recruitment, as well as total inflammatory cells recruited to the lung (Figure 7.3). Another mouse model of OVA-induced asthma treated with etanercept in models mimicking both mild/moderate asthma and severe asthma. In the severe asthma model, treatment with etanercept resulted

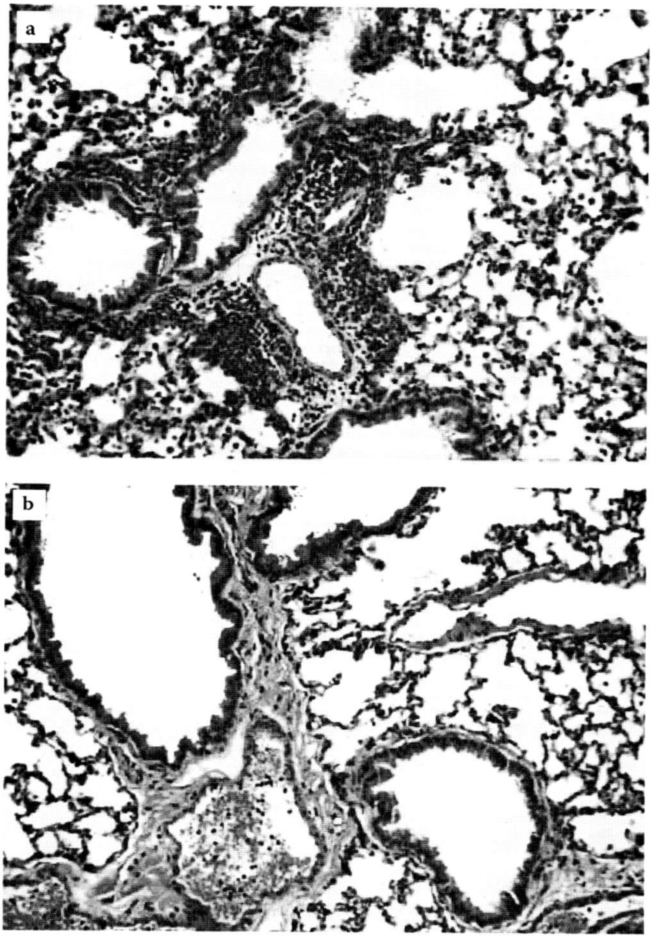

FIGURE 7.3 H&E staining of control antibody and anti-TNFα-treated mouse lung. (a) Representative section from a rat IgG-treated mouse showing clear perivascular and peribronchial infiltration of eosinophils and lymphocytes. (b) Representative section from anti-TNFα antibody treated lung showing marked reduction in inflammatory cell infiltrate. (From Kim, J., et al., *Clin Exp Allergy*, 2006. 36(1): 122–32. With permission.)[86]

in decreased eosinophilia in the severe asthmatic mice, while this reduction was less marked in mild/moderate mice. The abrogated eosinophilia seen in this model correlated with a lung environment rich in neutrophilic inflammation, suggesting that the reduced eosinophilia seen in patients with severe asthma is dependent on the presence of neutrophils.[61] This is in agreement with the success seen in clinical trials of TNFα inhibitors in patients with severe asthma, where neutrophilic inflammation is often found. Continuation of the use of TNFα inhibitors in these animal models is important given that the mechanism by which TNFα blockade works to treat asthma is still elusive.

TNFα INHIBITORS IN THE TREATMENT OF PATIENTS WITH ASTHMA

Several clinical trials employing various TNFα inhibitors have met with promising success, especially in the treatment of patients with severe asthma. In a randomized, double-blind, placebo-controlled study of mild/moderate asthma, patients were treated with etanercept, a soluble TNFα receptor two times per week for 2 weeks. Treatment did not attenuate pulmonary eosinophilia and actually resulted in increased IL-4 levels in epithelial lining fluid. This study also resulted in no noticeable change in methacholine responsiveness.[55] These results could indicate that TNFα antagonism will provide no beneficial effects to those with mild/moderate asthma; however, the duration of treatment in this study was quite short. It is possible that effectiveness of TNFα antagonism in these patients requires extended treatment.

Another double-blind placebo-controlled study involved treatment of patients with moderate asthma that was effectively treated with inhaled corticosteroids. These patients were treated with infliximab (a chimeric mouse/human monoclonal anti-TNFα antibody), by intravenous infusion at week 0, 2, and 6. There was no difference seen in morning peak expiratory flow (PEF), or sputum eosinophils, but the number of patients who experience asthma attacks was decreased in those treated with infliximab compared to placebo.[62] This study suggests that TNFα antagonism may be effective in moderate asthma; however, more studies are needed to determine whether prolonged treatment is beneficial. Additionally, a small pilot study has been conducted using marimastat, a synthetic matrix metalloprotease inhibitor on 12 patients with atopic asthma. Treated patients had reduced bronchial hyperresponsiveness, but no reduction in sputum inflammatory cells.[63]

To date, one study has been published comparing the use of etanercept in both patients with mild/moderate asthma and those with severe asthma. In this patient subset, levels of membrane-bound TNFα on peripheral blood mononuclear cells in patients with severe asthma was significantly elevated above those seen in patients with moderate asthma, and was accompanied by increases in TNFα receptor 1 and TACE. These patients were treated with etancercept two times per week for 10 weeks. Patients with severe asthma treated with etanercept showed a significant increase in the concentration of methacholine required to induce a 20 percent decrease in FEV_1 as well as increased quality of life score. Sputum histamine levels were also significantly decreased in treated patients with severe asthma compared to placebo; however, there was no change in other inflammatory indices such as sputum eosinophils or neutrophils. This suggests that etancercept is acting directly on airway smooth muscle cells to reduce bronchial hyperresponsiveness.[56]

It is important to note the specific features modulated by these antagonists if long-term treatment is to be considered. For example, no patient trials so far have demonstrated a decrease in inflammatory cell recruitment to the lung when TNFα antagonists are administered. This may suggest that TNFα treatment alone may not be sufficient as these inflammatory cells may continue to cause long-term airway remodeling and lung damage seen in severe asthma. More large-scale controlled studies will provide a clearer picture of the efficacy of these antagonists in various subpopulations of patients with asthma.

Side Effects Associated with TNFα Antagonism

Although TNFα inhibitors have demonstrated significant promise for asthma treatment, the side effects associated with their use must be carefully considered. Many of these side effects have been reported in the context of treatment of RA and some have been observed in the clinical trials for asthma as well. One of the most prevalent side effects is the reactivation of latent tuberculosis (TB) as well as increased incidences of malignancies. A meta-analysis of previously conducted clinical trials using infliximab and adalimumab for the treatment of RA determined that a dose-dependent risk of malignancies and serious infections exists in patients treated with the TNFα antagonists.[64] Given this risk, careful assessment and close monitoring of administered doses will be needed. Treatment of RAs using infliximab has also resulted in reported cases of TB both in Europe and the United States. Although activation of TB infection was not seen during RA clinical trials, after infliximab was approved for widespread use, 70 cases of TB were reported, most during the first 12 weeks of treatment. Indeed, once these TB cases were reported, screening for latent TB was administered in order to assess the risk of infection.[65,66] Interestingly, the use of another anti-TNFα antibody, adalimumab, has not to date been used in clinical trials for asthma treatment and has in fact been reported in one case to cause asthma in one patient being treated for RA. The asthma was not severe enough for the treatment to be stopped; however, this unexpected side effect highlights the need to elucidate the mechanisms by which each of these TNFα inhibitors function.[67] While assessment and interpretation of these risks is of great importance, one must not overlook the immense contribution to the improvement of quality of life that these TNFα antagonists have provided for patients with RA. Therefore, as with RA treatment it is imperative to find the balance between risk of adverse affects and benefit to the general asthmatic population.

The Future of TNFα Antagonists for the Treatment of Asthma

Given that TNFα is a complex and multifunctional cytokine that has roles in regulating allergy, cancer, infection, and autoimmunity, the deleterious side effects noted after its inhibition are expected and necessitate further research into more specific inhibition in the context of allergy, rather than global inhibition which may affect the body's ability to fight infection. Biologically active TNFα can exist in both soluble and membrane-bound forms. Recent evidence has suggested that soluble TNFα is associated with chronic inflammation,[68,69] while membrane-bound TNFα plays a principal role in fighting infection.[14,70,71] The current TNFα antagonists used to treat RA inhibit both transmembrane and soluble forms of TNFα, and this global inhibition may be responsible for increased susceptibility to infection. New studies have emerged in which dominant negative inhibitors are used to selectively inhibit soluble TNFα. These inhibitors form inactive mixed heterotrimers, eliminating native TNFα. They have so far been shown to reduce the severity of Parkinson's disease in a rat model.[72] These inhibitors are able to block cytokine release in human whole blood and attenuated arthritis severity score in a mouse model, while leaving intact the host response to a *Listeria* infection.[73] This method of inhibition may also be achieved using inhibitors of TACE to reduce soluble TNFα levels, leaving membrane-bound TNFα intact. Biologic precedence for this approach is apparent as levels of tissue inhibitor of metalloprotease 3, an inhibitor of TACE, are decreased in sputum samples of patients with asthma.[74]

OTHER EMERGING BIOTHERAPIES FOR THE TREATMENT OF ASTHMA

Several other exciting avenues for the development of asthma therapeutics are currently undergoing investigation (Table 7.1). These include diverse approaches such as modulation of innate immune signaling (TLR inhibition), the use of newly developed

TABLE 7.1
Summary of Emerging Biologics for the Treatment of Asthma

Molecule	Target	Comments
Etanercept[56]	TNFα	Clinical trial: Increased quality of life score and FEV_1 Decreased histamine levels
Infliximab[62]	TNFα	Clinical trial: Decreased number of patients experiencing asthmatic attack
Adalimumab[67]	TNFα	Caused asthma as side effect of RA treatment
XENP346[73]	TNFα	Animal model: Attenuated arthritis and Parkinson's disease severity score
Marimastat[63]	MMP inhibitor	Clinical trial: Reduced AHR No change in airway inflammatory cells
CRX-675[75]	TLR4 agonist	Clinical trial: Improved nasal symptom scores in allergic rhinitis
Pam3CSK4[76]	TLR2 agonist	Animal model: Attenuated IgE and inflammatory cell recruitment
CpG motifs[78]	TLR9 agonist	Clinical trial: Reduced Th2 response in favor of Th1
EPI-2010[80]	Adenosine A_1 receptor	Animal model: Attenuated bronchial obstruction Clinical trial: Modest improvement in asthmatic symptoms
Roflumilast[82]	PDE4 inhibitor	Clinical trial: Increased FEV_1 Improvements in exercise-induced asthma
LY294002[84]	PI3K inhibitor	Animal model: Attenuated inflammatory cell recruitment and mucus production
IC87114[85]	PI3K inhibitor p110δ-specific	Animal model: Attenuated serum IgE, tissue eosinophilia, AHR, and leukotrienes production

technology such as RNA interference, and enzyme inhibition. Several of these emerging therapies and the studies conducted so far are summarized here.

Toll-Like Receptor Modulation

TLRs are key molecules in both innate and adaptive immunity and play a role in activation of antigen-presenting cells and immune skewing toward Th1 or Th2 responses. Specifically, agonists of TLR2, 4, and 9 have been studied for the treatment of allergic airway diseases. Ligation of TLR4 by LPS has been shown to induce Th1-type responses. The synthetic TLR4 agonist CRX-675 has recently been tested in clinical trials as a treatment for allergic rhinitis, and showed some improvement in nasal symptom scores.[75] Administration of the TLR2 agonist Pam3CSK4 after established ovalbumin sensitization in a mouse model significantly reduced lung inflammation as well as serum IgE levels.[76] CpG sequences which bind to TLR9 show promise for attenuation of symptoms of allergic airways diseases as well in humans, as conjugation of ragweed allergen to CpG motifs significantly reduced Th2 immune responses in favor of a Th1 profile.[77]

RNA-Interference-Based Therapy

Gene targeting is a new and emerging mechanism for disease treatment, which offers the advantage of targeting specific genes or gene families, as opposed to global immunosuppression characteristic of steroid treatment. By acting at the level of translation, these molecules can inhibit expression of a single target gene. This inhibition can be achieved using antisense oligonucleotides, siRNA, and ribozymes. Genome-wide analysis has recently highlighted the loci of several asthma susceptibility genes potentially contributing to increased circulating IgE and airways hyperresponsiveness, which may be targeted by this therapy.[78] Preclinical trials targeting the adenosine A_1 receptor showed attenuated bronchial obstruction; however, clinical trials revealed only modest amelioration of symptoms in patients with mild asthma.[79,80] Because RNA interference is a relatively new technology, and is only recently being applied to the study of allergic diseases, more genes must be investigated as potential therapeutic targets. This technology provides a powerful tool for treatment as it has the potential to reduce drug-associated complications given that therapy can be targeted to a single gene.

Phosphodiesterase Inhibitors

Inhibitors of phosphodiesterase type 4 (PDE4) have been used successfully for the treatment of chronic obstructive pulmonary disorder (COPD) and are also being assimilated into asthma therapy. These inhibitors function by increasing levels of cyclic AMP, which has downstream anti-inflammatory effects. Treatment of patients with mild and moderate asthma with the PDE4 inhibitor Roflumilast resulted in reductions in both early and late phase FEV_1 as compared to placebo.[81] Evidence also suggests that PDE4 inhibitors protect against exercise-induced asthma, as treatment resulted in significantly decreased FEV_1 in patients, as well as significantly decreased serum TNFα.[82] These results are promising and suggest the possibility of PDE4 inhibitors as future oral asthma therapies.

Phosphoinositide 3-Kinase Inhibitors

Various members of the phosphoinositide 3-kinase (PI3K) family have been implicated in inflammatory cell recruitment, expression of inflammatory mediators, and steroid refractoriness in chronic inflammatory diseases. Murine studies have reported that administration of the global PI3K inhibitor LY294002 significantly reduced inflammatory cell recruitment and mucus production.[83] One advantage in this area of study is that inhibitors specific to several PI3K isoforms are also available. Inhibition of the p110δ isoform significantly reduced serum IgE, tissue eosinophilia, leukotriene production, and AHR in a murine model.[84] Several clinical trials using PI3K inhibitors for treatment of cancers are currently under way, and their efficacy for the treatment of human asthma is an emerging field of study.

CONCLUSIONS

The use of TNFα antagonists for the treatment of asthma represents a new and exciting avenue in the treatment of this costly disease. It is likely that a combination of sustained use of TNFα inhibitors, along with the numerous other novel and promising therapies, will afford more effective asthma treatments. As new molecules are continuously being discovered, the clinical applications of TNFα inhibitors may provide the much needed alternative to harmful long-term oral steroid use.

REFERENCES

1. Braman, S.S., The global burden of asthma. *Chest*, 2006. 130(1 Suppl): 4S–12S.
2. Mannino, D.M., et al., Surveillance for asthma—United States, 1980–1999. *MMWR Surveill Summ*, 2002. 51(1): 1–13.
3. *Morbidity and Mortality: 2007* Chart Book on Cardiovascular, Lung and Blood Diseases. National Heart, Lung and Blood Institute. 2007.
4. *Guidelines for the diagnosis and management of asthma.* National Heart, Lung and Blood Institute. 2007.
5. Barnes, P.J., Inhaled glucocorticoids for asthma. *N Engl J Med*, 1995. 332(13): 868–75.
6. Hendeles, L. and C.A. Sorkness, Anti-immunoglobulin E therapy with omalizumab for asthma. *Ann Pharmacother*, 2007. 41(9): 1397–410.
7. Kolb, W.P. and G.A. Granger, Lymphocyte in vitro cytotoxicity: Characterization of human lymphotoxin. *Proc Natl Acad Sci USA*, 1968. 61(4): 1250–55.
8. Carswell, E.A., et al., An endotoxin-induced serum factor that causes necrosis of tumors. *Proc Natl Acad Sci USA*, 1975. 72(9): 3666–70.
9. Mosser, D.M., The many faces of macrophage activation. *J Leukoc Biol*, 2003. 73(2): 209–12.
10. Gordon, S., Alternative activation of macrophages. *Nat Rev Immunol*, 2003. 3(1): 23–35.
11. Wajant, H., K. Pfizenmaier, and P. Scheurich, Tumor necrosis factor signaling. *Cell Death Differ*, 2003. 10(1): 45–65.

12. Tang, P., M.C. Hung, and J. Klostergaard, Human pro-tumor necrosis factor is a homotrimer. *Biochemistry*, 1996. 35(25): 8216–25.
13. Kriegler, M., et al., A novel form of TNF/cachectin is a cell surface cytotoxic transmembrane protein: Ramifications for the complex physiology of TNF. *Cell*, 1988. 53(1): 45–53.
14. Saunders, B.M., et al., Transmembrane TNF is sufficient to initiate cell migration and granuloma formation and provide acute, but not long-term, control of *Mycobacterium tuberculosis* infection. *J Immunol*, 2005. 174(8): 4852–59.
15. Black, R.A., et al., A metalloproteinase disintegrin that releases tumour-necrosis factor-alpha from cells. *Nature*, 1997. 385(6618): 729–33.
16. Chen, G. and D.V. Goeddel, TNF-R1 signaling: a beautiful pathway. *Science*, 2002. 296(5573): 1634–35.
17. Porteu, F. and C. Nathan, Shedding of tumor necrosis factor receptors by activated human neutrophils. *J Exp Med*, 1990. 172(2): 599–607.
18. Medvedev, A.E., et al., Distinct roles of the two tumor necrosis factor (TNF) receptors in modulating TNF and lymphotoxin alpha effects. *J Biol Chem*, 1996. 271(16): 9778–84.
19. Skerrett, S.J., et al., Role of the type 1 TNF receptor in lung inflammation after inhalation of endotoxin or *Pseudomonas aeruginosa*. *Am J Physiol*, 1999. 276(5 Pt 1): L715–27.
20. Fernandez-Botran, R., Soluble cytokine receptors: their role in immunoregulation. *FASEB J*, 1991. 5(11): 2567–74.
21. Heaney, M.L. and D.W. Golde, Soluble hormone receptors. *Blood*, 1993. 82(7): 1945–48.
22. Lantz, M., et al., Characterization in vitro of a human tumor necrosis factor-binding protein. A soluble form of a tumor necrosis factor receptor. *J Clin Invest*, 1990. 86(5): 1396–1402.
23. Van Zee, K.J., et al., Tumor necrosis factor soluble receptors circulate during experimental and clinical inflammation and can protect against excessive tumor necrosis factor alpha in vitro and in vivo. *Proc Natl Acad Sci USA*, 1992. 89(11): 4845–49.
24. Xanthoulea, S., et al., Tumor necrosis factor (TNF) receptor shedding controls thresholds of innate immune activation that balance opposing TNF functions in infectious and inflammatory diseases. *J Exp Med*, 2004. 200(3): 367–76.
25. Pinckard, J.K., et al., Constitutive shedding of both p55 and p75 murine TNF receptors in vivo. *J Immunol*, 1997. 158(8): 3869–73.
26. Leeuwenberg, J.F., T.M. Jeunhomme, and W.A. Buurman, Slow release of soluble TNF receptors by monocytes in vitro. *J Immunol*, 1994. 152(8): 4036–43.
27. Leeuwenberg, J.F., M.A. Dentener, and W.A. Buurman, Lipopolysaccharide LPS-mediated soluble TNF receptor release and TNF receptor expression by monocytes. Role of CD14, LPS binding protein, and bactericidal/permeability-increasing protein. *J Immunol*, 1994. 152(10): 5070–76.
28. Cottet, S., et al., cFLIP protein prevents tumor necrosis factor-alpha-mediated induction of caspase-8-dependent apoptosis in insulin-secreting betaTc-Tet cells. *Diabetes*, 2002. 51(6): 1805–14.
29. Miyazaki, Y., et al., Tumor necrosis factor alpha and insulin resistance in obese type 2 diabetic patients. *Int J Obes Relat Metab Disord*, 2003. 27(1): 88–94.
30. Mousa, S.A., O. Goncharuk, and D. Miller, Recent advances of TNF-alpha antagonists in rheumatoid arthritis and chronic heart failure. *Expert Opin Biol Ther*, 2007. 7(5): 617–25.
31. Toussirot, E. and D. Wendling, The use of TNF-alpha blocking agents in rheumatoid arthritis: an update. *Expert Opin Pharmacother*, 2007. 8(13): 2089–107.
32. Smolen, J.S., et al., Infliximab treatment maintains employability in patients with early rheumatoid arthritis. *Arthritis Rheum*, 2006. 54(3): 716–22.

33. Machold, K.P. and J.S. Smolen, Adalimumab—A new TNF-alpha antibody for treatment of inflammatory joint disease. *Expert Opin Biol Ther*, 2003. 3(2): 351–60.
34. Maini, R.N., et al., Anti-tumour necrosis factor specific antibody (infliximab) treatment provides insights into the pathophysiology of rheumatoid arthritis. *Ann Rheum Dis*, 1999. 58 Suppl 1: I56–60.
35. Tsuji, F., et al., Differential effects between marimastat, a TNF-alpha converting enzyme inhibitor, and anti-TNF-alpha antibody on murine models for sepsis and arthritis. *Cytokine*, 2002. 17(6): 294–300.
36. Le, G.T. and G. Abbenante, Inhibitors of TACE and Caspase-1 as anti-inflammatory drugs. *Curr Med Chem*, 2005. 12(25): 2963–77.
37. Gordon, J.R. and S.J. Galli, Release of both preformed and newly synthesized tumor necrosis factor alpha (TNF-alpha)/cachectin by mouse mast cells stimulated via the Fc epsilon RI. A mechanism for the sustained action of mast cell-derived TNF-alpha during IgE-dependent biological responses. *J Exp Med*, 1991. 174(1): 103–7.
38. Marshall, B.G., et al., Tumour necrosis factor-alpha production in human alveolar macrophages: Modulation by inhaled corticosteroid. *Eur Respir J*, 2000. 15(4): 764–70.
39. Palombella, V.J., J. Mendelsohn, and J. Vilcek, Mitogenic action of tumor necrosis factor in human fibroblasts: Interaction with epidermal growth factor and platelet-derived growth factor. *J Cell Physiol*, 1988. 135(1): 23–31.
40. Schwingshackl, A., et al., Human eosinophils release matrix metalloproteinase-9 on stimulation with TNF-alpha. *J Allergy Clin Immunol*, 1999. 104(5): 983–39.
41. Godding, V., et al., Adhesion of activated eosinophils to respiratory epithelial cells is enhanced by tumor necrosis factor-alpha and interleukin-1 beta. *Am J Respir Cell Mol Biol*, 1995. 13(5): 555–62.
42. Panettieri, R.A., Jr., et al., Activation of cAMP-dependent pathways in human airway smooth muscle cells inhibits TNF-alpha-induced ICAM-1 and VCAM-1 expression and T lymphocyte adhesion. *J Immunol*, 1995. 154(5): 2358–65.
43. Thomas, P.S., D.H. Yates, and P.J. Barnes, Tumor necrosis factor-alpha increases airway responsiveness and sputum neutrophilia in normal human subjects. *Am J Respir Crit Care Med*, 1995. 152(1): 76–80.
44. Thomas, P.S. and G. Heywood, Effects of inhaled tumour necrosis factor alpha in subjects with mild asthma. *Thorax*, 2002. 57(9): 774–78.
45. Adner, M., et al., An assay to evaluate the long-term effects of inflammatory mediators on murine airway smooth muscle: Evidence that TNFalpha up-regulates 5-HT(2A)-mediated contraction. *Br J Pharmacol*, 2002. 137(7): 971–82.
46. Huber, M., B. Beutler, and D. Keppler, Tumor necrosis factor alpha stimulates leukotriene production in vivo. *Eur J Immunol*, 1988. 18(12): 2085–88.
47. Amrani, Y., et al., Mechanisms underlying TNF-alpha effects on agonist-mediated calcium homeostasis in human airway smooth muscle cells. *Am J Physiol*, 1997. 273(5 Pt 1): L1020–28.
48. Halasz, A., et al., Role of TNF-alpha and its 55 and 75 kDa receptors in bronchial hyperreactivity. *Respir Med*, 2002. 96(4): 262–67.
49. Beasley, R., The burden of asthma with specific reference to the United States. *J Allergy Clin Immunol*, 2002. 109(5 Suppl): S482–89.
50. Chung, K.F., et al., Difficult/therapy-resistant asthma: The need for an integrated approach to define clinical phenotypes, evaluate risk factors, understand pathophysiology and find novel therapies. ERS Task Force on Difficult/Therapy-Resistant Asthma. European Respiratory Society. *Eur Respir J*, 1999. 13(5): 1198–208.
51. Goleva, E., et al., Airway remodeling and lack of bronchodilator response in steroid-resistant asthma. *J Allergy Clin Immunol*, 2007 Nov; 120(5): 1065–72.

52. Wenzel, S.E., et al., Evidence that severe asthma can be divided pathologically into two inflammatory subtypes with distinct physiologic and clinical characteristics. *Am J Respir Crit Care Med*, 1999. 160(3): 1001–8.
53. Busse, W., et al., Airway remodeling and repair. *Am J Respir Crit Care Med*, 1999. 160(3): 1035–42.
54. Franchimont, D., et al., Tumor necrosis factor alpha decreases, and interleukin-10 increases, the sensitivity of human monocytes to dexamethasone: potential regulation of the glucocorticoid receptor. *J Clin Endocrinol Metab*, 1999. 84(8): 2834–39.
55. Howarth, P.H., et al., Tumour necrosis factor (TNFalpha) as a novel therapeutic target in symptomatic corticosteroid dependent asthma. *Thorax*, 2005. 60(12): 1012–18.
56. Berry, M.A., et al., Evidence of a role of tumor necrosis factor alpha in refractory asthma. *N Engl J Med*, 2006. 354(7): 697–708.
57. Bradding, P., et al., Interleukin-4, -5, and -6 and tumor necrosis factor-alpha in normal and asthmatic airways: Evidence for the human mast cell as a source of these cytokines. *Am J Respir Cell Mol Biol*, 1994. 10(5): 471–80.
58. Wilson, S.J., et al., Effects of budesonide and formoterol on NF-kappaB, adhesion molecules, and cytokines in asthma. *Am J Respir Crit Care Med*, 2001. 164(6): 1047–52.
59. Renzetti, L.M., et al., Pharmacological evidence for tumor necrosis factor as a mediator of allergic inflammation in the airways. *J Pharmacol Exp Ther*, 1996. 278(2): 847–53.
60. Broide, D.H., et al., Inhibition of eosinophilic inflammation in allergen-challenged TNF receptor p55/p75- and TNF receptor p55-deficient mice. *Am J Respir Cell Mol Biol*, 2001. 24(3): 304–11.
61. Hutchison, S., et al., Tumour necrosis factor-alpha blockade suppresses murine allergic airways inflammation. *Clin Exp Immunol*, 2008 Jan; 151(1): 114–22.
62. Erin, E.M., et al., The effects of a monoclonal antibody directed against tumor necrosis factor-alpha in asthma. *Am J Respir Crit Care Med*, 2006. 174(7): 753–62.
63. Bruce, C. and P.S. Thomas, The effect of marimastat, a metalloprotease inhibitor, on allergen-induced asthmatic hyper-reactivity. *Toxicol Appl Pharmacol*, 2005. 205(2): 126–32.
64. Bongartz, T., et al., Anti-TNF antibody therapy in rheumatoid arthritis and the risk of serious infections and malignancies: Systematic review and meta-analysis of rare harmful effects in randomized controlled trials. *JAMA*, 2006. 295(19): 2275–85.
65. Keane, J., et al., Tuberculosis associated with infliximab, a tumor necrosis factor alpha-neutralizing agent. *N Engl J Med*, 2001. 345(15): 1098–104.
66. Ellerin, T., R.H. Rubin, and M.E. Weinblatt, Infections and anti-tumor necrosis factor alpha therapy. *Arthritis Rheum*, 2003. 48(11): 3013–22.
67. Bennett, A.N., et al., Adalimumab-induced asthma. *Rheumatology* (Oxford), 2005. 44(9): 1199–200.
68. Holtmann, M.H. and M.F. Neurath, Differential TNF-signaling in chronic inflammatory disorders. *Curr Mol Med*, 2004. 4(4): 439–44.
69. Ruuls, S.R., et al., Membrane-bound TNF supports secondary lymphoid organ structure but is subservient to secreted TNF in driving autoimmune inflammation. *Immunity*, 2001. 15(4): 533–43.
70. Torres, D., et al., Membrane tumor necrosis factor confers partial protection to *Listeria* infection. *Am J Pathol*, 2005. 167(6): 1677–87.
71. Olleros, M.L., et al., Contribution of transmembrane tumor necrosis factor to host defense against *Mycobacterium bovis* bacillus Calmette-Guerin and *Mycobacterium tuberculosis* infections. *Am J Pathol*, 2005. 166(4): 1109–20.
72. McCoy, M.K., et al., Blocking soluble tumor necrosis factor signaling with dominant-negative tumor necrosis factor inhibitor attenuates loss of dopaminergic neurons in models of Parkinson's disease. *J Neurosci*, 2006. 26(37): 9365–75.

73. Zalevsky, J., et al., Dominant-negative inhibitors of soluble TNF attenuate experimental arthritis without suppressing innate immunity to infection. *J Immunol*, 2007. 179(3): 1872–83.
74. Paulissen, G., et al., Expression of ADAMs and their inhibitors in sputum from patients with asthma. *Mol Med*, 2006. 12(7–8): 171–79.
75. Casale, T.B., J. Kessler, and F.A. Romero, Safety of the intranasal toll-like receptor 4 agonist CRX-675 in allergic rhinitis. *Ann Allergy Asthma Immunol*, 2006. 97(4): 454–56.
76. Patel, M., et al., TLR2 agonist ameliorates established allergic airway inflammation by promoting Th1 response and not via regulatory T cells. *J Immunol*, 2005. 174(12): 7558–63.
77. Marshall, J.D., et al., Immunostimulatory sequence DNA linked to the Amb a 1 allergen promotes T(H)1 cytokine expression while downregulating T(H)2 cytokine expression in PBMCs from human patients with ragweed allergy. *J Allergy Clin Immunol*, 2001. 108(2): 191–97.
78. Howard, T.D., D.A. Meyers, and E.R. Bleecker, Mapping susceptibility genes for allergic diseases. *Chest*, 2003. 123(3 Suppl): 363S–8S.
79. Nyce, J., Respirable antisense oligonucleotides: A new, third drug class targeting respiratory disease. *Curr Opin Allergy Clin Immunol*, 2002. 2(6): 533–36.
80. Ball, H.A., M.R. Van Scott, and C.B. Robinson, Sense and antisense: Therapeutic potential of oligonucleotides and interference RNA in asthma and allergic disorders. *Clin Rev Allergy Immunol*, 2004. 27(3): 207–17.
81. van Schalkwyk, E., et al., Roflumilast, an oral, once-daily phosphodiesterase 4 inhibitor, attenuates allergen-induced asthmatic reactions. *J Allergy Clin Immunol*, 2005. 116(2): 292–98.
82. Timmer, W., et al., The new phosphodiesterase 4 inhibitor roflumilast is efficacious in exercise-induced asthma and leads to suppression of LPS-stimulated TNF-alpha ex vivo. *J Clin Pharmacol*, 2002. 42(3): 297–303.
83. Duan, W., et al., An anti-inflammatory role for a phosphoinositide 3-kinase inhibitor LY294002 in a mouse asthma model. *Int Immunopharmacol*, 2005. 5(3): 495–502.
84. Lee, K.S., et al., Inhibition of phosphoinositide 3-kinase delta attenuates allergic airway inflammation and hyperresponsiveness in murine asthma model. *FASEB J*, 2006. 20(3): 455–65.
85. Barnes, P.J., Cytokine modulators as novel therapies for asthma. *Annu Rev Pharmacol Toxicol*, 2002. 42: 81–98.
86. Kim, J., et al., Anti-tumor necrosis factor-alpha antibody treatment reduces pulmonary inflammation and methacholine hyper-responsiveness in a murine asthma model induced by house dust. *Clin Exp Allergy*, 2006. 36(1): 122–32.

8 Protein-Based Therapies in Experimental and Clinical Transplantation

Gilles Benichou, Stephanie de Haseth, A. Benedict Cosimi, and Alessandro Alessandrini

CONTENTS

KEY WORDS: MHC peptides, monoclonal antibodies, T-cell costimulation blockade, transplant tolerance

ABSTRACT

Rejection by the host's immune system represents the main barrier to successful transplantation of allogeneic organs and tissues. The development of immunosuppressive drugs in the 1960s has permitted the large-scale transplantation of some organs such as kidneys and livers by preventing their early acute rejection. However, such treatment is associated with increased risks for infection and neoplasia. In addition, it is often toxic and ineffective in achieving long-term graft survival. Chronic allograft rejection is also a critical barrier to successful long-term survival of allotransplanted tissues and organs in patients. It is a slow process involving perivascular inflammation, fibrosis, and intimal hyperplasia leading to lumenal occlusion of graft vessels or bronchioles in the case of lung transplants. At present, there is no treatment to prevent or block the chronic rejection process. These observations underscore the need for the design of novel and more selective treatments in transplantation. During the past two decades, a number of protein-based therapies have been developed in an effort to suppress donor-specific T cells involved in acute and chronic rejection of allografts. In experimental models, successful transplant tolerance induction has been achieved using soluble major histocompatibility complex (MHC) proteins and peptides, presumably via the induction of host's regulatory T cells capable of suppressing proinflammatory antidonor immunity. Recently, some studies have shown that immune deviation of the immune response to graft tissue-specific antigens can also prolong allotransplant survival. In addition, many attempts have been made to modulate alloreactive T-cell responses using monoclonal antibodies (mAbs) directed to different receptors and cytokines involved in T-cell activation. In rodent and nonhuman primate models, some success has been achieved using antibodies blocking signals delivered via the CD28 and CD40 pathways. In clinical transplantation, non-FcR-binding humanized anti-CD3 antibodies have been shown to suppress acute rejection episodes and are still regularly used in patients. Novel antibodies directed to T and B cells such as anti-CD2, anti-CD20, anti-CD25, and anti-CD154 antibodies are currently being evaluated in clinical trials for their ability to prolong the survival of solid organ allotransplants in patients.

INTRODUCTION

Transplantation of organs, tissues, and cells represents a life-saving procedure for a variety of patients affected with incurable diseases. During the past 30 years, progress in surgical techniques and the development of immunosuppressive drugs have rendered possible the allotransplantation (intraspecies grafts) of some organs and tissues in patients with minimal risks of failure due to early acute rejection. However, immunosuppressive therapy is associated with nephrotoxicity, increased risks for infection, and neoplasia, and it does not achieve indefinite graft survival. In addition, a large portion of organ transplants succumb to chronic rejection, a slow process involving perivascular inflammation, fibrosis, and intimal hyperplasia leading to lumenal occlusion of graft vessels or bronchioles in the case of lung transplants. At present, there is no treatment to prevent or block the chronic rejection process.

Following transplantation of allogeneic tissues, recognition by recipient lymphocytes of foreign proteins expressed by donor cells (allorecognition) initiates a cascade of immunological reactions resulting in rapid elimination of donor cells and rejection of the graft.[1] Allorecognition occurs in the recipient's lymph nodes and spleen where $CD4^+$ T cells interact either directly with donor MHC class II molecules present on infiltrating donor dendritic cells (passenger leukocytes) or indirectly

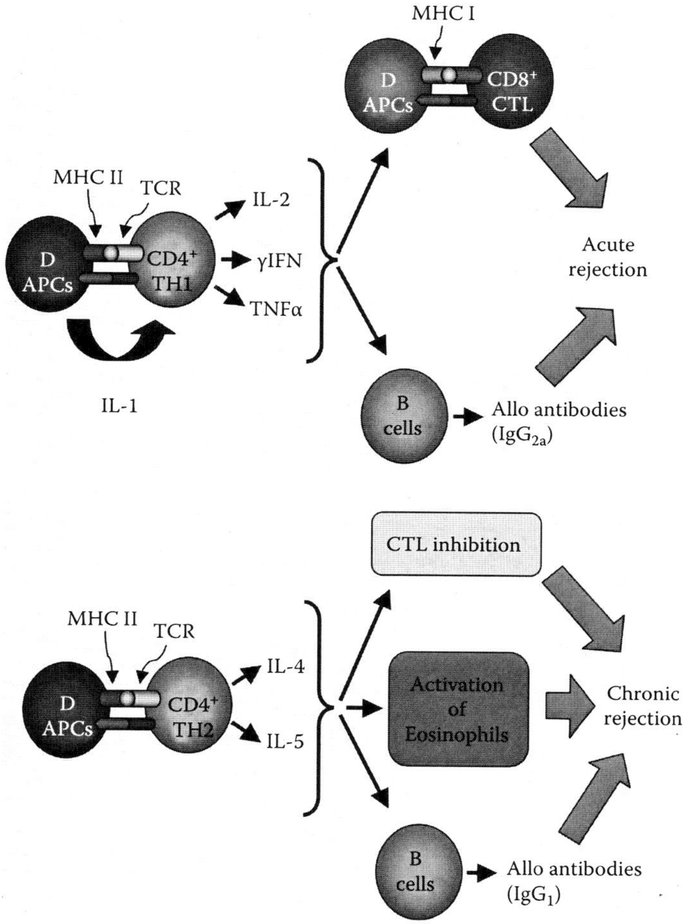

FIGURE 8.1 Cellular mechanisms involved in acute and chronic allograft rejection by Th1 and Th2 cells.

with processed donor peptides presented by recipient MHC class II$^+$ antigen-presenting cells (APCs)[2] (Figure 8.1). Both direct and indirect allorecognition pathways can trigger delayed type hypersensitivity (DTH) reactions, antibody production, and differentiation of CD8$^+$ cytotoxic T cells.[3] In direct alloreactivity, donor APCs present a multitude of peptides in MHC class II context thereby eliciting vigorous polyclonal CD4$^+$ T cell responses.[2] It is traditionally accepted that the direct alloresponse represents the driving force behind early acute graft rejection. Alternatively, the indirect T-cell alloresponse is oligoclonal in that it is mediated by a selected set of alloreactive T-cell clones recognizing a few dominant peptides presented by self-MHC.[4,5] Despite its lower frequency, the indirect alloresponse is sufficient to ensure acute rejection of skin grafts and presumably other transplants. In addition, there is accumulating evidence suggesting that indirect type of alloreactivity, via spreading of T-cell responses to cryptic determinants on self- and alloantigens, represents the driving force behind chronic rejection of allotransplants.[6–8]

During the past 20 years, many attempts have been made to achieve transplant tolerance defined as indefinite allograft survival without vasculopathy in the absence of continuous immunosuppressive treatment. Most efforts have focused on the selective inhibition of donor-specific T- and B-cell responses in recipients treated pre- and/or post-transplantation. In this chapter, we examine different aspects of protein-based therapies designed to prolong allotransplant survival and analyze their mechanisms of action and their impact on the rejection of allografts in experimental and clinical settings.

INDUCTION OF TRANSPLANT TOLERANCE USING DONOR MHC PROTEINS AND PEPTIDES

Transplantation Tolerance Using Intact Donor MHC Proteins

Allogeneic MHC proteins are regularly shed by cells and eventually processed into peptides by recipient APCs. This process can result in activation of recipient inflammatory T cells and allograft rejection. Alternatively, some studies show that, under appropriate conditions, soluble intact MHC proteins can suppress deleterious alloimmunity either by causing apoptosis of donor-reactive T cells or by activating regulatory T cells.[9] The probability that exposure to soluble MHC proteins might drive the immune response toward regulation rather than rejection may be related to the route by which the soluble antigens are released and the molecular forms they adopt. De Vito-Haynes et al. have reported the presence of high levels of donor soluble human leukocyte antigen (HLA) class I antigens in individuals with well-functioning lung transplants. Similarly, soluble MHC class I is produced abundantly by liver allografts, a feature associated with the tolerogenic properties of these organ transplants. Studies by Van Rood's group suggest that soluble MHC class I proteins can bind to CD8 receptors on T cells thus triggering release of soluble Fas ligand and subsequent T-cell apoptosis.[10,11] This is in partial agreement with other studies by Zavazava and Kabelitz showing that an interaction with the T-cell receptor (TCR) but not CD8 on $CD8^+$ T cells is required for apoptosis of alloreactive T cells.[12]

Transplantation Tolerance with MHC Peptides

Donor antigen-specific tolerization of T cells mediating a direct alloresponse may be difficult to design owing to the polyclonal and polyspecific nature of this response. In contrast, the indirect alloresponse is thought to involve a limited set of alloreactive T cells expressing selected TCR genes and recognizing a few dominant determinants on allogeneic MHC molecules.[4,13] This feature of indirect alloreactivity suggests that peptide-based strategies could be designed to manipulate this type of alloresponse *in* vivo similarly to what is being attempted in autoimmune disease models. Based on this principle, a number of laboratories have administered to transplant recipients synthetic peptides that were derived from MHC class I and II molecules in the effort to modulate the alloimmune response and subsequent allograft rejection.

Peptides Derived from Polymorphic Regions of MHC Class I Molecules

Significant prolongation of allograft survival has been accomplished upon intrathymic injection of MHC class I antigens[14] and allopeptides[15,16] in rats. This was associated with predominant expression of Th2 cytokines in the graft, in contrast to control animals that expressed Th1 cytokines, a result suggesting that this phenomenon was caused by immune deviation.[17] These peptides were more efficacious in promoting long-term allograft survival when coadministered along with a single dose of antilymphocyte serum (ALS). Donor-specific tolerance was confirmed by acceptance of a second cardiac transplant from the same donor, while third-party grafts were acutely rejected. In another study, ACI rats were injected intravenously with syngeneic dendritic cells (DCs) previously pulsed with a dominant allo-MHC class I (P5, RT1.A(u)) peptide of Wistar Furth (WF) origin. T cells from these rats were then inoculated intrathymically to a naïve ACI rat. This treatment resulted in a permanent and donor specific acceptance of pancreatic islets from WF donors.[18] In addition, oral feeding of peptide P5 in combination with a short course of cyclosporin A (CsA) prolonged the survival of DA cardiac allografts in Lewis recipient rats.[19] Recently, intratracheal delivery of a peptide derived from a variable region of the mouse MHC class I molecule K^b (region 54-68) was shown to induce reduced inflammatory responses to and long-term survival of allogeneic cardiac grafts, presumably via the generation of regulatory T cells.[20] In another set of studies, human MHC class I–derived peptides (HLA-B7.75-84 and HLA-B2702.75-84), referred to as ALLOTRAP, displayed some immunomodulatory effects in a rat model.[21] Apparently, these peptides produced their effect by preventing the differentiation of precursor $CD8^+$ T cells into effector cytotoxic T cells (CTLs), by inhibiting lysis by established CTLs and natural killer cells.[22] Finally, a study in our laboratory shows that intravenous injection of BALB/c-dm2 (L^d-) pregnant female mice with a dominant allogeneic MHC class I peptide (L^d 61-80) can tolerize corresponding allospecific T cells in her offspring and confer tolerance to MHC class I-mismatched BALB/c (L^d+) allotransplants.[23]

Peptides Derived from Polymorphic Regions of MHC Class II Molecules

Administration of peptides derived from MHC class II polymorphic sequences have also been shown to exert some suppressive effects on alloimmunity when administered via either the intrathymic or oral route.[24] Seminal studies by M. Sayegh et al. have shown that intrathymic injection of a single immunodominant MHC class II peptide could result in specific inhibition of primed T-cell proliferative responses to the corresponding determinant in vitro[25] as well as inhibition of delayed-type hypersensitivity (DTH) responses in vivo.[26] Here, oral tolerance was associated with a state of "immune deviation" characterized by a predominance of Th2 cytokine production.[27] Similarly, preoperative administration to LEW recipients of the WF-specific MHC class II peptides RT1.B2 in combination with low-dose CsA induced prolongation of allograft survival of WF small bowel transplants.[28]

Peptides Derived from Conserved Regions of MHC Molecules

Many naturally processed peptides found in the grooves of cell surface MHC class II molecules are derived from nonpolymorphic regions of both MHC class I and class

II amino acid sequences.[29] These peptides derived from a conserved region of the MHC class II chain inhibited rat MLR in a dose-dependent manner.[30,31] The human HLA-DQA1*0101 peptide was shown to inhibit both human and mouse mixed lymphocyte reactions (MLRs). Finally, MHC-derived peptides corresponding to CD4 and CD8 interacting regions have also been used to modulate immune functions. For example, peptides corresponding to the CD4 interacting regions of MHC class II molecules could directly interfere with CD4 interactions thereby suppressing T-helper responses.[32] Peptide analogues that mimic the putative interaction sites of CD8 and the MHC class I molecule prolonged skin allografts in an MHC class I-mismatched mouse model.[33] Additionally, a synthetic peptide with a sequence derived from the MHC class II-associated invariant chain peptide inhibited antigen-specific T-cell response in vitro and in vivo following immunization.[34] Apparently, this peptide inhibited the loading of antigenic peptides onto MHC class II molecules and the subsequent expression of the MHC molecule–peptide complexes at the cell surface.

MODULATION OF AUTOIMMUNITY TO TISSUE-SPECIFIC ANTIGENS

Activated autoreactive inflammatory T cells and autoantibodies specific to type V collagen (Col (V)) and cardiac myosin (CM) are consistently detected after lung and heart transplantation, respectively.[35–37] Clonal expansion of these T cells occurs only after an alloresponse, but once activated they can induce on their own rejection of allogeneic and even syngeneic transplants. Indirect rather than direct alloresponse triggers autoimmunity after transplantation presumably via antigen mimicry between autoantigen peptides and donor MHC peptides. It is plausible that inflammation and tissue damage associated with the initial alloresponse to donor MHC antigens causes the release of formerly sequestered autoantigens. The most compelling evidence for the role of autoimmunity in allograft rejection has been obtained through experiments designed to modulate the inflammatory T-cell responses to autoantigens. In the lung transplant model, Mares et al.[36] initially reported that WKY rats fed with 10 μg Col(V) every other day for 2 weeks prior to transplantation with an allogeneic F344 lung were tolerant to their allografts. In this model, prolongation of lung allograft survival was associated with a systemic production of transforming growth factor (TGF)-beta but not interleukin 4 (IL-4) and IL-10.[38] In another study, the influence of a type 2 anti-inflammatory T-cell response to CM on the rejection of cardiac allotransplants was investigated. A.TL mice were injected intraperitoneally with CM emulsified in incomplete Freund's adjuvant (IFA) and transplanted 10 days later with MHC class I-disparate A/J hearts. First, the anti-CM response in A.TL mice treated with CM-IFA was evaluated using ELISPOT. High frequencies of IL-4- and IL-5-producing T cells were detected in freshly isolated spleen cells from these mice. Remarkably, these mice retained donor hearts for more than 100 days, while control recipients rejected their transplants 10 days after grafting.[39] In contrast to acutely rejected hearts from untreated recipients exhibiting generalized interstitial inflammatory cell infiltration and myocyte damage, donor hearts from

CM/IFA-treated mice displayed well-preserved epicardium with minimal or no inflammatory cell infiltrates.[39]

Taken together, the results obtained in lung and heart transplant models suggest that the development of an autoimmune response after transplantation of allogeneic organs is an essential component of the rejection process. Most importantly, these studies show that tolerance induction to tissue-specific autoantigens can be effective in preventing acute rejection and delaying the onset and reducing the severity of chronic rejection of allografts.

PREVENTION OF ALLOTRANSPLANT REJECTION USING MONOCLONAL ANTIBODIES

In this section, we discuss some of the mAbs that have been used to modulate antidonor immunity and impair the rejection of allotransplants in experimental models.

Antibodies Directed to Surface Receptors on Immune Cells

Costimulation Blockade Using Antibodies

The concept of costimulation in lymphocyte activation was initially introduced by Lafferty and Cunningham who postulated that appropriate activation of T cells requires two distinct signals: signal 1 delivered via recognition of the antigen by the TCR and signal 2 delivered via interaction of costimulation receptors on T cells and their ligands on APCs[40] (Figure 8.2). Following antigen recognition in the absence of a second signal, T cells become anergic or undergo programmed cell death (apoptosis). In contrast, delivery of both signals results in T-cell proliferation, differentiation into effector cells, and cytokine release. Recently, other costimulation receptors on T cells that can deliver negative signals to T cells have been described. T-cell signaling via these receptors can lead to anergy, apoptosis, or the induction of regulatory cytokines. In addition, there is accumulating evidence that costimulatory signaling of T cells is also involved in the regulation of T–T-cell interactions and T-cell interactions with "non-professional APCs" such as endothelial and other parenchymal cells. Therefore, T-cell costimulation plays an essential role in the regulation of the T-cell response and the fate of the overall immune response to antigens. Based on this principle, transplant immunologists have attempted to modulate costimulation signals delivered to alloreactive T cells in an effort to suppress the rejection process and/or achieve transplant tolerance[41] (Table 8.1).

Many studies have focused on the modulation of signals delivered to T cells via CD28 and CTLA4 costimulation receptors. CD28 is constitutively expressed on naive T cells and it plays a critical role in T-cell activation via binding to B7-1 and B7-2 receptors on APCs.[42,43] CTLA4 is expressed only on activated T cells and binds to B7-1 and B7-2 with a much higher affinity than CD28. B7-1/B7-2 double knockout mice are unable to reject allogeneic cardiac transplants but reject skin allografts. CD28 KO mice reject both heart and skin allografts presumably via a process involving $CD8^+$ T cells and natural killer (NK) cells. Administration to rodents of anti-B7-2 but not anti-B7-1 prolongs allograft survival. In nonhuman primate models,

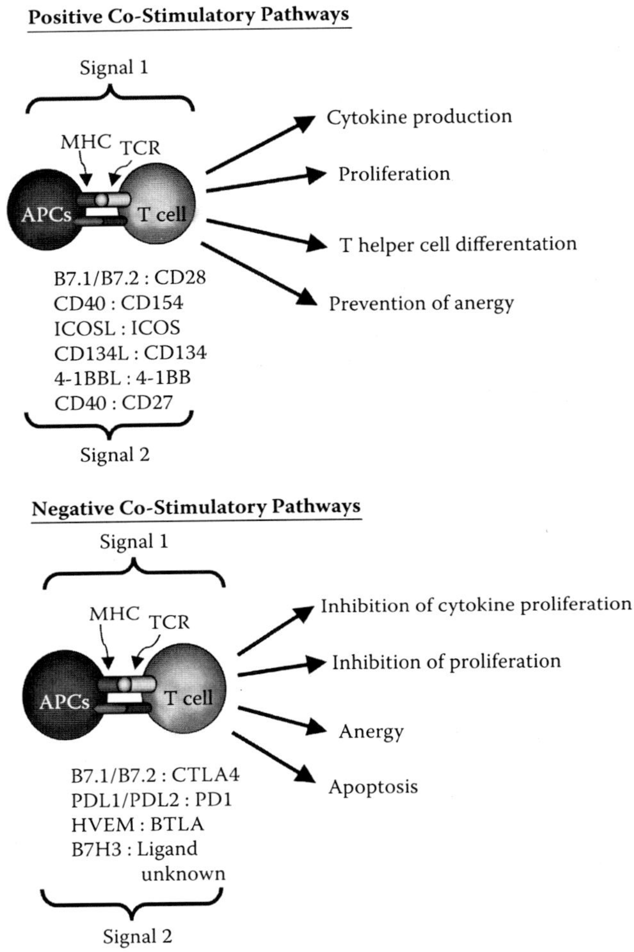

FIGURE 8.2 Cell surface receptors involved in the delivery of costimulation signals to T cells.

administration of either anti-B7-1 or B7-2 mAbs along with sirolimus or anti-CD40L mAbs also delayed the rejection of solid organ allotransplants.[44,45] However, the best results were obtained with CTLA4-Ig, a fusion protein which blocks the CD28-B7 pathway and prevents acute allograft rejection while inducing stable donor-specific tolerance in rodents.[46] In nonhuman primates, CTLA4-Ig was most tolerogenic in kidney- and islet-transplant models when injected along with anti-CD40L mAbs or rapamycin.[44,45]

CD40 and its ligand CD40L (CD154) are the most potent costimulation receptors belonging to the tumor necrosis factor receptor (TNFR)-TNF superfamily. CD40 is constitutively expressed on DCs and other professional APCs while its expression is inducible on endothelial cells and fibroblasts.[47] CD40L is abundant on activated

TABLE 8.1
T-Cell and B-Cell Surface Receptors, Experimental Treatments, and Transplant Rejection

Surface Molecule	Cell Type	Transplant Model	Effect on Transplant Rejection	Species	Ref.
CD2	T cell	Heart, skin	Tolerance (can be broken by adoptive transfer of naïve T cells)	Mouse	59
CD3	T cell	Kidney, islet, heart	Prolonged allograft survival	Mouse, human	90–93
CD4	T cell, APC	Heart, skin, kidney, bone marrow	Prolonged allograft survival	Mouse, human	60, 63
CD8	T cell, APC	Heart, skin, kidney, bone marrow	Prolonged allograft survival	Mouse, human	60–63
CD20	B cell	Hematopoietic stem cell transplantation (HSCT), kidney	Prolonged allograft survival	Human	100, 101
CD25	Activated T cells, regulatory T cells	Heart, liver	Prolonged allograft survival	Rat, monkey	64–66
CD28 and CD40L	T cell	Heart, islet, skin, kidney	Tolerance	Mouse, human	40–50
OX40 and ICOS	Activated and memory T cell		Prolonged allograft survival	Mouse	51–55
PD1	T cells	T cells	Prolonged allograft survival	Mouse	56–58

CD4⁺ T cells and it can also be found, although to a much lesser extent, on CD8⁺ T cells, B cells, and DCs. CD40 signaling is essential to the proliferation of B cells and their ability to produce antibodies and to differentiate into memory B cells. CD40–CD40L interaction provides a potent signal to APCs thus enhancing their ability to present alloantigens to T cells.[48] Likewise, administration of anti-CD40L mAbs has been shown to prevent acute rejection of skin, cardiac, and islet allografts in rodent models. However, costimulation blockade using anti-CD40L mAbs did not prevent the onset of chronic rejection.[49] In mice, coadministration of CTLA4-Ig and anti-CD40L mAbs resulted in long-term survival of fully allogeneic skin allografts and indefinite survival of cardiac allotransplants in the absence of chronic allograft vasculopathy.[46] In nonhuman primate models, mixed chimerism induction via infusion of recipients with donor bone marrow cells along with administration of anti-CD40L antibodies resulted in stable donor-specific tolerance and long-term survival of kidney allografts.[50] Although the mechanisms by which anti-CD40L mAb treatment contribute to tolerance induction are still unknown, there is accumulating evidence suggesting that treatment with these antibodies favors the expansion/activation of antigen specific regulatory T cells expressing FoxP3 molecules.

Two other molecules involved in the delivery of positive costimulatory signals to T cells have been shown to play important roles in transplant rejection: ICOS and CD134 (OX40).[51] ICOS is a CD28 homologue that is induced on activated T cells and resting memory T cells. ICOS-ICOSL (B7h) signaling results in heightened expression of CD40L on T cells, thus amplifying CD28 signaling and B7 expression of APCs.[51] ICOS blockade or ICOS deficiency has been shown to prolong the survival of cardiac allotransplants in mice without preventing cardiac allograft vasculopathy.[52] It is, however, noteworthy that early blockade of ICOS has been shown to accelerate allograft rejection, a result suggesting that ICOS may play different roles depending on the stage of the alloimmune response. CD134 is expressed primarily on activated $CD4^+$ T cells while its ligand CD134L or OX40L is present on activated B cells and DCs and to a lesser extent on endothelial cells and T cells.[53,54] Blockade of the CD134 signaling pathway combined with CTLA4-Ig has been shown to delay the rejection of skin and heart transplants in presensitized mice, presumably by altering the function of effector memory T cells.[53,54] Therefore, simultaneous blockade of CD134 and negative signaling of T cells via CTLA4-Ig may represent a promising strategy for tolerance induction in primates displaying significant numbers of allospecific memory T cells at the time of transplantation.[55]

Finally, PD1, a CD28 homologue, represents a key receptor on T cells capable of delivering negative costimulatory signals. PD1 is expressed on activated T cells but also on B cells and NK cells and macrophages.[56] PD1 interacts with PDL1 and PDL2 expressed on DCs and macrophages and some nonhematopoietic tissues in the case of PDL1. Anti-PDL1 antibody treatment has been shown to delay allograft rejection in rodents. It is noteworthy, however, that the same treatment in the absence of CD28 signaling resulted in accelerated allograft rejection. In contrast, treatment with PDL1-Ig in CD28 knockout mice induced long-term survival of cardiac allografts.[57,58] Therefore, the PD1–PDL pathway is likely to play an essential role in the regulation of alloimmunity and allograft rejection by modulating the nature of the signals delivered via CD28 on T cells.[57]

Antibodies Against Other Receptors on B and T Cells

Anti-CD2 Antibodies

CD2 is a cell adhesion molecule that is found on the surface of T cells and interacts with lymphocyte function-associated antigen-3 (LFA-3). In 1993, Chavin and colleagues[59] demonstrated that the use of combined anti-CD2 and anti-CD3 receptor mAbs induced donor-specific tolerance in a cardiac transplant model in mice. They showed that while treatment of animals with mAbs against either the CD2 or CD3 receptor prolonged graft survival, indefinite allograft survival was observed when anti-CD2 and anti-CD3 antibodies were used in combination. Interestingly, adoptive transfer of naive recipient type cells into these mice broke tolerance and FACS analysis demonstrated that tolerance was not associated with deletion of T cells. These data imply that the mechanism by which combination treatment with anti-CD2 and anti-CD3 induces tolerance is probably "related to suppression and/or anergy of helper and effector cells at the time of allografting, whereas maintenance of tolerance is associated with anergy in the Th cell compartment."[59] In 1997, Sultan

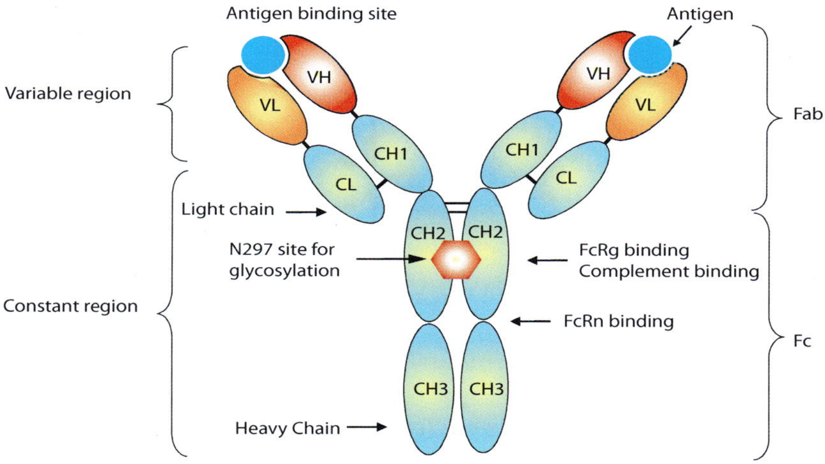

FIGURE 1.1 Typical structure of IgG. IgG is composed of two identical heavy and two identical light chains. Each heavy chain and light chain contains a variable region and a constant region. The antigen-binding sites comprise the variable regions of the heavy and light chains. The heavy constant regions (CH2 and CH3) interact with complement component C1q and Fcγ receptors, and are responsible for the effector functions and pharmacokinetics of the IgG. IgG is also glycosylated via carbohydrate moieties attached at conserved amino acid residue, asparagine at position 297 (N297) in CH2 domain, which is important for Fc interactions with FcRγ. FcRn binds to both CH2 and CH2 domains in the Fc region of the mAb. VH, heavy chain variable domain; VL, light chain variable domain; CH1, CH2, and CH3, heavy chain constant domains; CL, light chain constant domain; Fab, antigen-binding fragment; Fc, constant region fragment; FcRγ, Fcγ receptor; FcRn, neonatal Fc receptor.

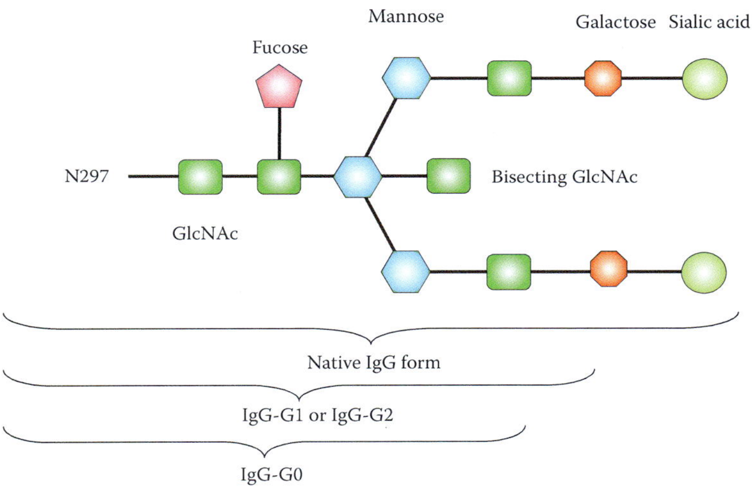

FIGURE 1.2 Schematic representation of glycosylation of IgG_1. IgG carbohydrate moiety attached to N297 of each CH2 domain of IgG_1 in the fully processed (native form) with all potential branching and terminal sugar residues are shown in the figure. When IgG_1 is produced *in vivo*, only a minor fraction of IgG_1 molecules have this carbohydrate side chain configuration. More frequently, partially processed moieties are identified, such as IgG_1 G_0 variants that lack all terminal sialic-acid and galactose residues and G_1 or G_2 glycoforms with one or two galactose residues present in the carbohydrate side chain. GlcNAc, N-acetylglucosamine, N297, asparagine residue at position 297.

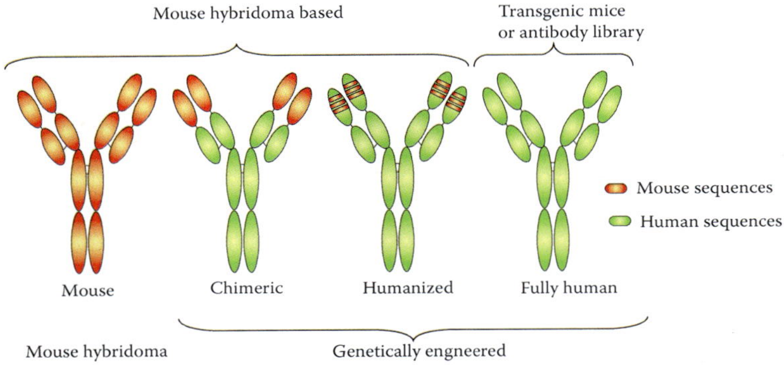

FIGURE 1.3 Different forms of therapeutic mAbs. Mouse hybridoma technology is used to produce mouse mAbs. Genetic engineering techniques are used to develop chimeric and humanized therapeutic mAbs. Chimeric mAbs are constructed by cloning variable region genes derived from a murine source and constant region genes derived from a human source and are produced in transfected cells. Humanized therapeutic mAbs are predominantly derived from a human source except for the CDRs, which are murine and are generated by insertion of murine CDRs to human constant and variable framework regions. Fully human mAbs are generated either by selection of *in vitro* antibody libraries or are produced in transgenic animals by generating human hybridomas.

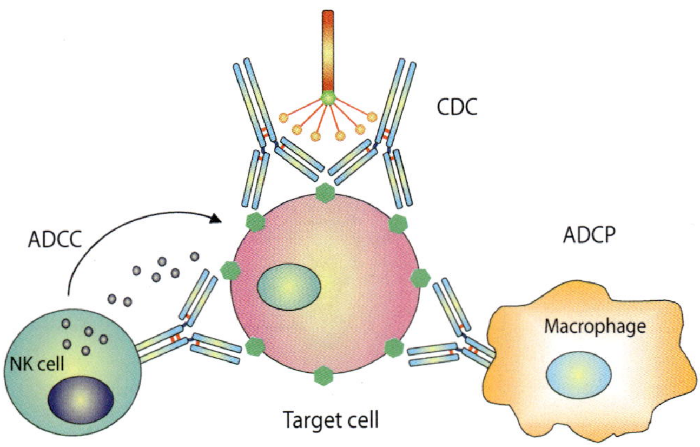

FIGURE 1.4 Typical antibody Fc-dependent effector functions of the mAbs. Following the binding of IgG to a specific target cell, C1q complement factor interacts with the CH2 constant region of the mAb, which leads to the activation of a proteolytic cascade of the classical complement pathway and consequently induces the formation of a membrane-attack complex for the lysis of target cells. This activity is termed complement-dependent cytotoxicity (CDC). Binding of IgG to target cells can also induce antibody-dependent cellular cytotoxicity (ADCC) mediated by immune-effector cells such as macrophages and NK cells which interact with the CH3 domain in the Fc region of the IgG through FcRγIIIA expressed by effector cells. Target cells coated with IgG can also be phagocytosed by macrophages via binding of Fc of IgG to FcRγIIIA expressed on macrophages by an active process called antibody-dependent cellular phagocytosis (ADCP). IgGs can also negatively regulate the cytotoxic response against target cells through FcRIIB, which is expressed on the cell surface of macrophages. IgG_1 and IgG_3 interact with Fc receptors more effectively than IgG_2 or IgG_4, and therefore mediate potent effector functions.

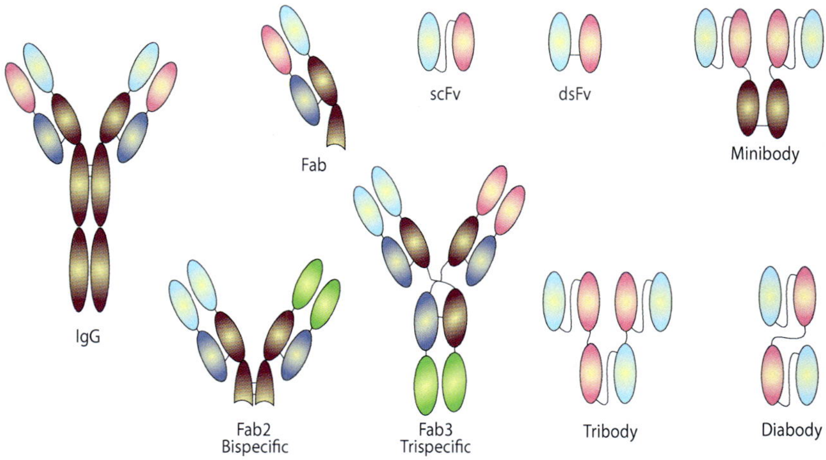

FIGURE 1.5 Different forms of most common antibody fragment formats. Schematic representation of most studied antibody formats including intact IgG molecule along with various fragments are shown in the figure. A variety of antibody fragments, including Fab, scFv, dsFv, and multimeric formats, such as minibodies, diabodies, triabodies, and chemically conjugated Fab multimers, are shown in the figure. Sizes of these fragments vary between approximately 15 to 150 kD.

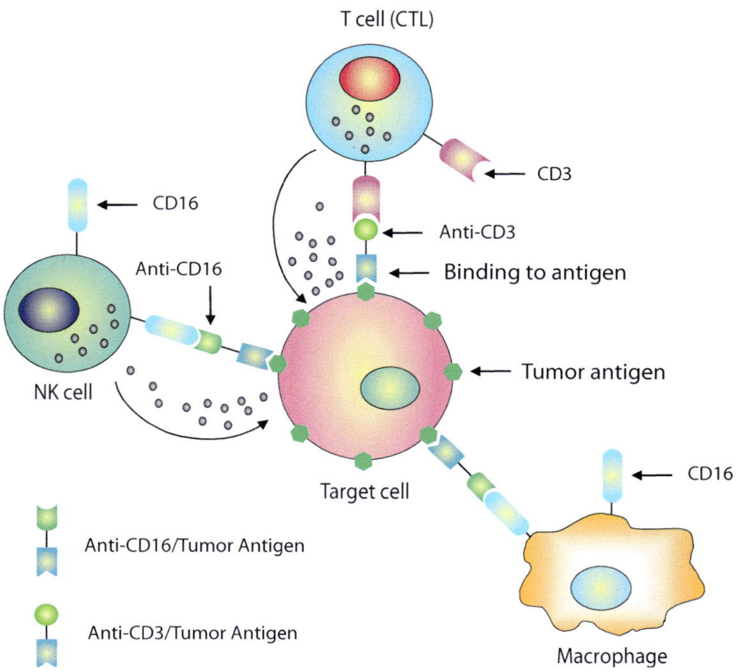

FIGURE 1.6 Mechanism of action of bispecific antibody fusion molecules. Bispecific antibody fragments allow targeting of the target cell by recruiting effector immune cells, such as NK cells and cytotoxic T cells. Bispecifc molecules with binding specificity for the antigen expressed on target cells such as CD19 bind to target cells and at the same time recruit effector cells by binding to receptors expressed on the effector cells. Two bispecific molecules, CD3/CD19 and CD16/CD19, are shown in the figure.

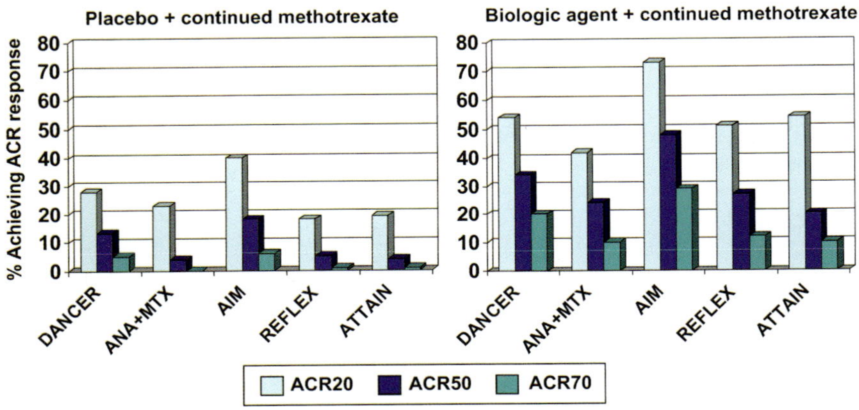

FIGURE 3.3 Frequency and level of ACR response observed with non-TNF antagonist biologic therapies. Patients with active RA despite MTX who continued with this drug and received placebo (left panel) or biologic agent (right panel): rituximab (DANCER and REFLEX),[59,60] anakinra (ANA),[53] and abatacept (AIM and ATTAIN).[66,68] (Adapted from Smolen J, et al. *Lancet* 2007;370:1861–74. With permission from Elsevier.)

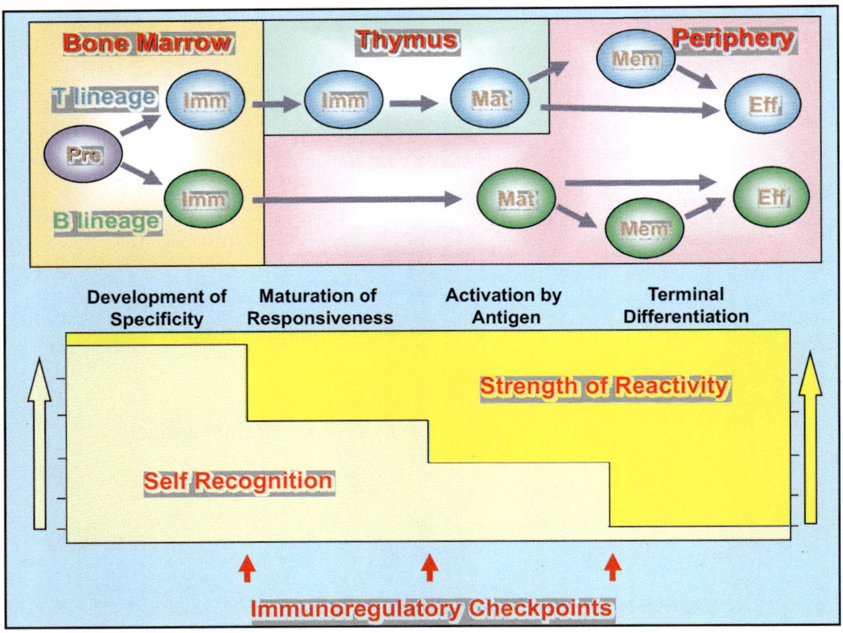

FIGURE 4.1 Schematic representation of the ontogeny of the adaptive immune system and the avoidance of autoreactivity. Initially uncommitted precursor cells develop into specific B cells and T cells in the bone marrow and thymus (T cells only). They mature further to be able to respond with appropriate intracellular signaling to encounter with antigen. After such an encounter, they further differentiate to become effector cells, which carry out the protective functions of the immune system. As they pass through the various stages of ontogeny, they become increasingly reactive to extracellular stimuli, and they are culled of their autoreactive specificities through various checkpoints. Alterations in any of these steps can cause inefficiencies in the tolerization checkpoints, and eventually result in the activation of potentially pathogenic autoreactive effector cells. Pre, precursor cell; Imm, immature cell; Mat, mature cell; Eff, effector cell.

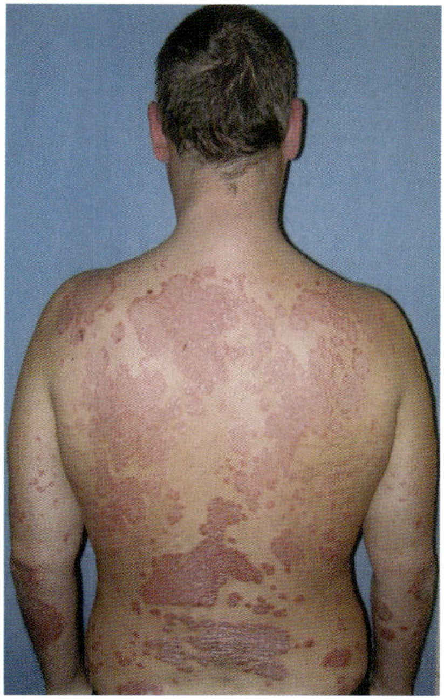

FIGURE 6.1 Plaque-type psoriasis characterized by erythematous, scaly lesions.

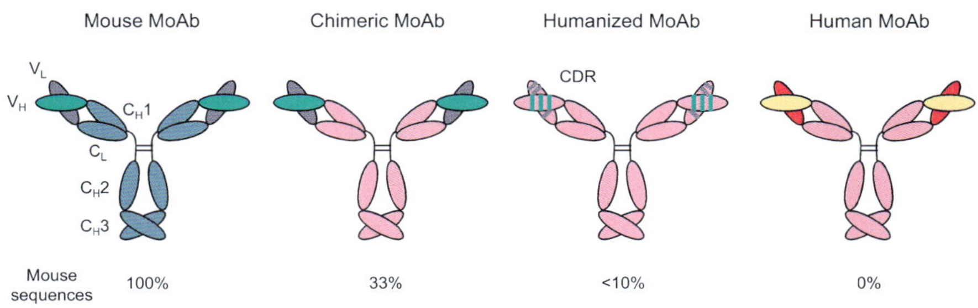

FIGURE 13.1 Structures of therapeutic mAbs. Mouse antibodies are obtained by hybridoma technology following immunization of mice. Mouse/human chimeric antibodies are derived by the variable domains (VH and VL) of the parent mouse antibody and constant domains (CH1, CH2, CH3, and CL) of human antibody (retained 33 percent mouse sequences). Humanized antibodies are constructed by grafting the CDR from the parent mouse antibody to a human antibody (<10 percent mouse sequences). Human antibodies can be produced using transgenic animals or phage display technologies (0 percent mouse sequence).

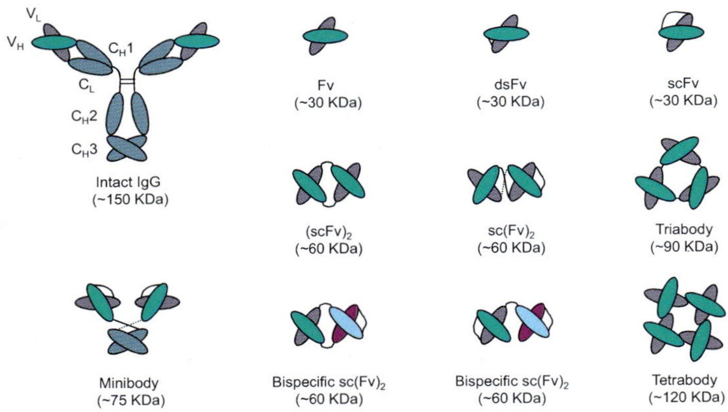

FIGURE 13.2 Schematic structures of engineered antibody fragments. The variable regions of heavy (VH) and light chains (VL) contribute to the antigen binding (Fv fragment). Disulfide-stabilized Fv fragments are molecules in which the VH and VL are stabilized by an interchain disulfide bond. The VH and VL domains can be connected by a peptide linker to form single-chain Fv (scFv). The scFv fragments can form multimers such as $(scFv)_2$ diabody, triabody, and tetrabody depending on the linker length and the V-domain orientation. The bivalent $sc(Fv)_2$ can be generated by connecting two scFvs covalently. Bispecific diabodies can also be engineered by using two different Fv domains.

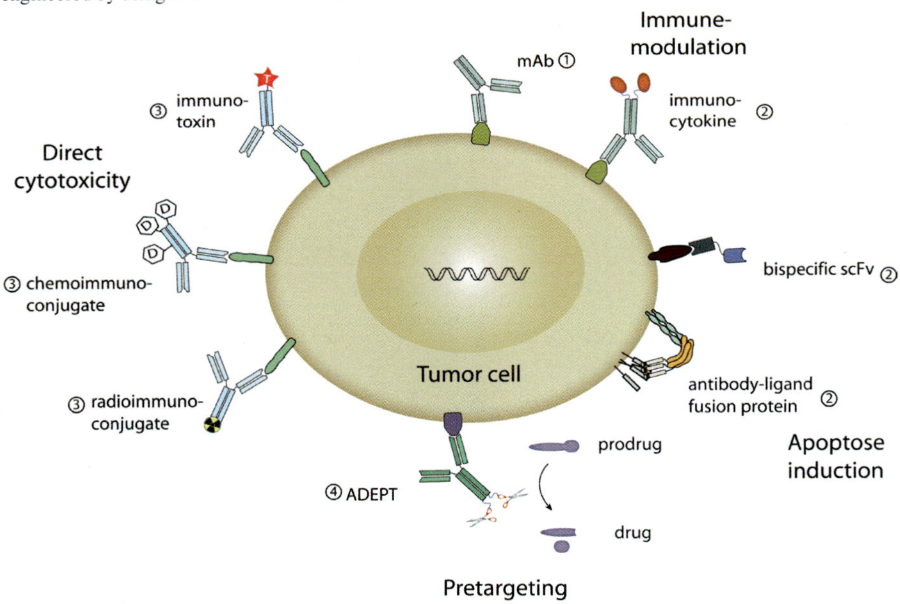

FIGURE 14.1 Concepts of therapeutic antibodies. Binding of naked mAbs to tumor cells may result in the destruction of the tumor cells by ADCC or CDC. Similarly, the immune response can be modulated by targeting cytokines or immunomodulatory molecules either by diabodies or antibody–ligand fusion proteins to the tumor. In addition, antibody–ligand fusion proteins can induce apoptosis to targeted cell as well as bystander cell by, for example, presenting FasL. The targeted cell can be killed directly by conjugation of cytotoxic drugs (D), toxins (T), or radionucleotides to the mAbs. The direct targeting approach might require the homogeneous expression of antigen in the tumor cell population at the most although dependent on the radionucleotides used, radioimmunoconjugates can have a bystander effect. The ADEPT approach specifically aims at this effect by targeting an enzyme to the tumor cell and subsequently delivers a prodrug which can be converted to a chemotherapeutic drug by the targeted enzyme.

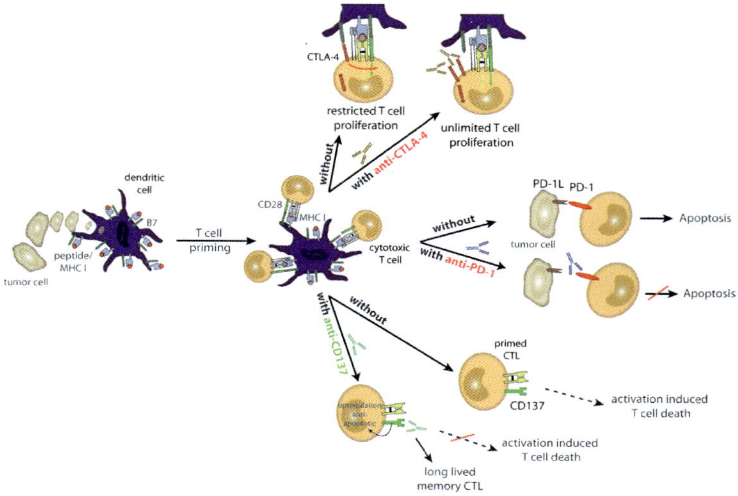

FIGURE 14.2 Immunostimulatory mAb. Dendritic cells pick up tumor cells and present peptide/MHC class I complexes to T cells. Cytotoxic T cells (CTLs) which possess the appropriate T-cell receptor recognizing this peptide/MHC class I complex will be primed. In the normal setting the following T-cell proliferation will be impaired by CTLA-4/CD28 interactions. In the presence of anti-CTLA-4 mAb, however, T-cell proliferation will be unrestricted (upper part). In addition, PD-1/PD-1L interaction between T cell and cancer cell will lead to apoptosis of the T cells. This can be prevented by the administration of anti-PD-1 mAb (middle part). Primed CTLs also express CD137. Ligand binding, which can be substituted by anti-CD137 mAb, prevents activation-induced T-cell death by leading to upregulation of antiapoptotic proteins (lower part). In addition, CD137 is required for the stable differentiation of CTLs and the establishment of long-lived memory CTLs.

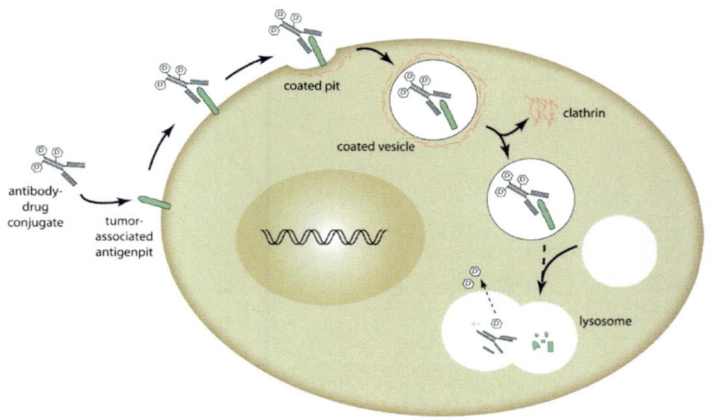

FIGURE 14.3 Internalization of antibody–drug conjugates. To regain their cytotoxic activity, the cytotoxic agent has to be cleaved from the chemoimmunoconjugate. Uptake of antibodies predominantly occurs via the clathrin-mediated endocytosis pathway. After binding the respective antigen associated with coated pits, the antibody–drug conjugates will readily be endocytosed initiating their transit via several stages of transport and endosomal vesicles and finally end up in a lysosome. Subsequently, linkers and antibody will be cleaved releasing the cytotoxic agent which—after exit from the lysosomal compartment—exerts its cytotoxic effect.

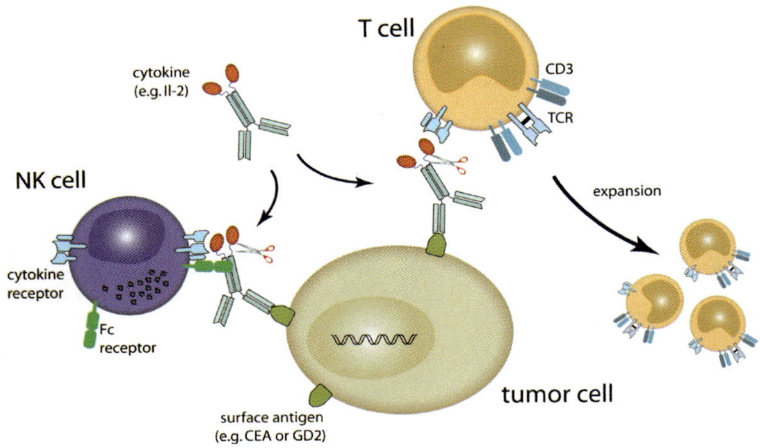

FIGURE 14.4 Working mechanism of immunocytokines exemplified for tumor-targeted IL-2. Cytokines fused to mAb specific for a tumor-associated antigen allows the enrichment of these in the tumor microenvironment. In the case of IL-2 it enhances ADCC mediated by F_c-receptor positive effector cells like NK cells. In addition, tumor-targeted IL-2 stimulates the expansion of T cells. Because the linker contains a plasmin-cutting site, the high concentrations of plasmin at the tumor site enable the cleavage of IL-2 from the fusion protein.

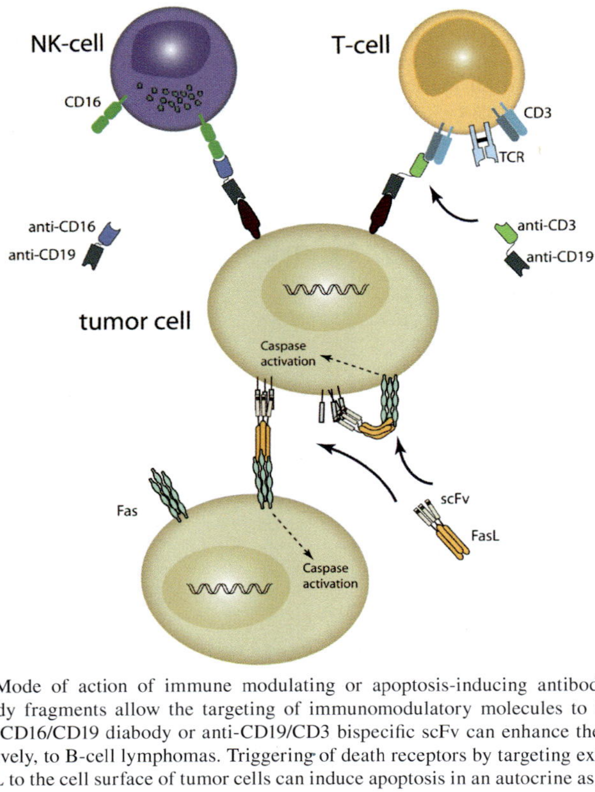

FIGURE 14.5 Mode of action of immune modulating or apoptosis-inducing antibody fusion proteins. Bispecific antibody fragments allow the targeting of immunomodulatory molecules to the tumor cell. As exemplified, anti-CD16/CD19 diabody or anti-CD19/CD3 bispecific scFv can enhance the NK cell or T-cell response, respectively, to B-cell lymphomas. Triggering of death receptors by targeting extrinsic factors such as FasL or TRAIL to the cell surface of tumor cells can induce apoptosis in an autocrine as well as a paracrine way.

and colleagues, using a human skin allograft injury model in immunodeficient mice, showed that blocking CD2 interactions by administration of murine mAb to human LFA-3 protected skin grafts from leukocyte infiltration and injury.[59a]

Anti-CD4 and Anti-CD8 Antibodies

CD4 and CD8 are transmembrane glycoproteins found on the surface of T cells and DCs cells and in the case with T cells, "associate" with the TCR. The CD4/TCR complex recognizes class II MHC molecules while CD8/TCR complexes bind specifically to class I MHC molecules. In 2001, Alters and colleagues depleted $CD4^+$ and $CD8^+$ cells using anti-CD4 mAbs, leading to allograft tolerance. They showed that anti-CD4-mediated tolerance to pancreatic islets transplanted from an A/J (IEk) donor to a diabetic C57B1/6 (B6) (IE-) recipient occurred via clonal anergy. This was partially overcome *in vitro* by the addition of recombinant IL-2.[60] Pearson and colleagues showed similar induction of tolerance to cardiac and skin allografts in mice after a full dose treatment with mAb against CD4.[61,62] Treatment with anti-CD4 and anti-CD8 mAb has also been important in establishing specific tolerance following bone marrow transplantation.[63]

Anti-IL-2 Receptor Antibodies

The IL-2 receptor is a heterotrimeric protein expressed on the surface of lymphocytes. It is made up of α (CD25), β (CD122), and γ (CD132) chains. The α and β chains are involved in IL-2 binding, and the γ chain, along with the β chain, contributes to the signaling following IL-2 binding to the receptor. During activation of T lymphocytes, the IL-2 receptor is upregulated on the cell surface. Resting regulatory T cells also express high levels of CD25 which is often used as a marker along with FoxP3 for regulatory T cells. In 1991, Brown and colleagues used a humanized antibody to the IL-2R α chain (CD25), anti-Tac-H, in cynomolgus monkeys undergoing heterotopic cardiac allografting.[64] This treatment prolonged allograft survival from 9 days in untreated animals to 20 days in animals that received anti-Tac-H antibodies. In 1994, Tinubu and colleagues used a humanized antibody to the IL-2R β chain (CD122), anti-Mik beta 1, and also found prolonged cardiac allograft survival in cynomolgus monkeys.[65] Gassel and colleagues developed a combination protocol using subtherapeutic amounts of CsA and mAbs to CD25, NDS-61, and found prolonged liver allograft survival in 84 percent of the rats treated. Addition of antibodies against the intercellular adhesion molecule-1 (CD54) resulted in prolongation of liver allografts in 100 percent of the animals.[66]

Antibodies Directed to Cytokines

Proinflammatory cytokines play a major role in T-cell activation and differentiation and are known to contribute to allotransplant rejection. The following section describes some of the antibodies used to block these cytokines in experimental transplantation (Table 8.2).

Anti-IL-1 Antibodies

Both IL-1α and IL-1β are produced by macrophages, monocytes, and DCs, and play an important role in the inflammatory response. In 2001, Smith and colleagues

TABLE 8.2
Anticytokine Experimental Treatments, and Transplant Rejection

Surface Molecule	Transplant Model	Effect on Transplant Rejection	Species	Ref.
IL-1	HLA-A2-transgenic tracheal allograft	Reduced obliterative airway disease	Mouse	67
IL-2	Heart	Prolonged allograft survival	Rat	68–70
TNF	Heart, HLA-A2-transgenic tracheal allograft	Prolonged allograft survival; reduced obliterative airway disease	Mouse, rat	71–77
IFNγ	Heart, kidney, islets	Prolonged allograft survival; amelioration of glycemic levels	Mouse, rat, dog	78–80

used anti-IL-1 and anti-TNFα neutralizing antibodies in HLA-A2-transgenic tracheal allograft mouse models.[66a] They observed that allografts in untreated animals exhibited obliterative airway disease (OAD) by day 28 posttransplantation, while anti-TNF-treated recipients showed no evidence of OAD, even at day 60. Anti-IL-1-treated recipients showed airway epithelium changes by day 28 but minimal evidence of OAD by day 60 after transplantation.[67]

Anti-IL-2 Antibodies

IL-2 is upregulated following allotransplantation, resulting in increased T effector cell proliferation as well as an increase in alloreactive cytotoxic T lymphocytes (CTL). A series of papers from the laboratory of Orita and colleagues showed that the use of an anti-IL-2 mAb or antisera increased cardiac allograft survival in rats.[68–70] Rats treated with a control antibody exhibited an average survival of 9 days. Those treated with the anti-IL2 mAb, 8H 10, exhibited an average allograft survival of 25 days with accumulation of the intravenously administered mAb in the grafted heart.[70]

Anti-TNF Antibodies

TNFα is a cytokine involved in systemic inflammation and is part of a group of cytokines that stimulate the acute-phase reaction. In 1990 and 1991, Imagawa and colleagues[71,72] described a series of experiments that utilized anti-TNF antibodies in prolonging cardiac allograft survival in rats. They showed that intraparenteral or intravenous administration of anti-TNF antibodies from day 1 to day 10 following a heart transplant increased the mean survival to 15 to 16 days versus 11 days for nontreated controls. Serum samples showed increased cytotoxic activity in control animals, correlating with an increase in circulating TNF. This activity was partially attenuated in animals receiving anti-TNF antibodies. These data demonstrated that TNF levels are increased during rejection.[71] They also showed that cotreatment of recipients with anti-TNF and antilymphotoxin antibodies[72] or anti-TNF and cyclosporine[73] increased cardiac allograft survival in rats by 21 and greater than 30 days, respectively. Similar results were obtained by Bolling and colleagues using the anti-TNF and cyclosporine combination therapy on rats receiving cardiac allografts.[74] Wei and colleagues further showed that anti-TNF antibody treatment not only neutralized TNF but also suppressed TNF production from isolated spleen cells from treated animals[75] and altered MHC class II expression.[76] In 1995, Gerber and colleagues were able to prolong murine cardiac allograft survival, using microspheres containing anti-TNF neutralizing antibodies.[77]

Anti-IFNγ Antibodies

In 1990, Rosenberg and colleagues studied the effect of *in vivo* administration of anti-interferon-gamma (IFNγ) mAbs on rejection of MHC class II-disparate skin allografts, and compared that to MHC class I-disparate skin allografts placed on the same mouse. They found that anti-IFNγ mAb blocked the rejection of MHC class II-mismatched skin allografts, but had no effect on the rejection of MHC class I-disparate skin allografts. This study helped to support the concept that allotransplant rejection requires IFNγ-induced MHC class II expression on keratinocytes of the allograft.[78] In 1994, Fuller and colleagues showed that administration of anti-IFNγ mAbs (ADI-1) prolonged canine renal allograft survival to 22 to 25 days, compared to 9 to 13 days for animals given an isotype control.[79] In 2007, Satoh and colleagues showed that streptozotocin-induced diabetes was ameliorated in mice receiving anti-IFNγ antibody and 200 islets, resulting in normoglycemia. Results were improved with only 100 islets when anti-TNFα and anti-IL-1β antibodies were administered in conjunction with anti-IFNγ antibody, wild-type diabetic mice.[80]

PROTEIN-BASED THERAPIES IN CLINICAL TRANSPLANTATION

Much of the success that has been achieved by replacing, with healthy allografts, irreversibly diseased human organs has relied on lifelong administration of chemical immunosuppressants. A major limitation of these agents is the toxicity that results from their rather nonselective reactivity with cells other than those involved in the immune response to the allograft. The clinical introduction of biologic agents in the form of ALS or antithymocyte globulin (ATG) in the mid 1960s represented the first effort in humans to target for suppression specific T-cell subsets suspected to participate in the recipient's rejection response.

Polyclonal antisera are designed to selectively deplete the infiltrating mononuclear cells that are observed in allograft biopsies coincident with rejection.[81–84] The addition of these agents to clinical treatment protocols represented a major therapeutic advance; and, in fact, they continue to be utilized in more than one-half of U.S. renal allograft recipients, now more than four decades later.[85] The preparations currently used are typically prepared by immunizing horses or rabbits with human lymphoid or thymic cells. Being polyclonal in derivation, the antisera react with a broad spectrum of molecules involved in T-cell activation.[86] Accordingly, their efficacy has been attributed not only to simple T-cell cytotoxicity, but also to interruption of donor antigen recognition, blocking of adhesion molecules, modulation of T-cell activation molecules, induction of apoptosis, interference with leukocyte endothelial interactions, and even expansion of the regulatory T-cell population.[87] These multiple mechanisms help to explain the clinical observations of not only lymphopenia, but also a decreased incidence of delayed graft function[88] and even apparent tolerance induction in some ALS-treated recipients. Unfortunately, the broad range of antibody specificities in polyclonal preparations also leads to undesired adverse effects including thrombocytopenia, anemia, and viral infections.

The next milestone to be achieved using biological agents for immunosuppression of human allograft recipients was first reported in 1981 following the development of cell hybridization techniques that could yield mAbs to single cell surface determinants.[89] The CD3 receptor is a protein complex that comprises four distinct chains (CD3γ, CD3δ, and two CD3ε chains) in mammals, that associates with the TCR and the ζ-chain to form an activation signal in T cells. Initial clinical trials successfully utilized a PAN T-cell reactive agent, OKT3 mAb, which targets the CD3 complex associated with the TCR expressed on mature T cells. This murine-derived mAb proved to be markedly effective in reversing even steroid-resistant rejection.[90,91] In 1999, Woodle and colleagues used huOKT3[γ1(Ala-Ala), also called Tepizumab, a genetically engineered, humanized derivative of the parental murine OKT3 mAb, whose CH2 region was mutagenized to prevent FcR-binding and thus eliminate T-cell activation properties, in phase I clinical trials for the treatment of acute renal allograft rejection. Rejection has been reversed in five of seven patients treated with huOKT3[γ1(Ala-Ala), five kidney transplants and two kidney-pancreas transplants, with recurrent rejection observed in one patient.[92] This observation of recurrent rejection after successful reversal of earlier episodes has been reported regularly, emphasizing that donor-specific tolerance in humans is unlikely via anti-CD3 antibody approaches. In 2004, Hering and colleagues pretreated six patients with type 1 diabetes prior to receiving single-donor islet transplants with a regimen that included humanized non-FcR-binding CD-3-specific antibodies, hOKT3γ1 (Ala-Ala). Of six patients, four became insulin independent and exhibited no occurrences of hypoglycemia. The four insulin-independent patients exhibited prolonged $CD4^+$ T-cell lymphocytopenia, inverted CD4:CD8 ratios, and increased levels of $CD4^+CD25^+$ regulatory T cells while receiving conventional immunosuppressive therapy.[93] Like the ALS preparations, OKT3 is still utilized clinically, but now only for rarely observed recalcitrant rejection episodes. The remarkable clinical efficacy of this mAb in the initial trials and the capacity to generate mAbs against virtually any cell surface or secreted antigen has subsequently spurred the development and testing of many additional agents. Potential targets include the functionally important molecules depicted in Figure 8.1. To date, mAbs directed against CD2, CD3, CD4, CD11A/CD18 (LFA-1), CD20, CD25 (IL-2R), CD52, CD54 (ICAM), and CD154 (CD40L) have been evaluated clinically (Table 8.1).

A further advance in the production of these biologic agents has been the incorporation of recombinant DNA technology to provide chimeric or humanized murine preparations. With this methodology, novel antibody molecules that are largely human protein, but retain short sequences from the rodent variable region, have been constructed in the effort to limit infusion-related toxicity and prolong the half-life of the administered mAb.[94–96]

The anti CD-25 preparations Simulect and Daclizumab are the most commonly administered mAbs today. As part of multiagent induction regimens, about 30 percent of U.S. renal allograft recipients currently receive these agents.[85] These mAbs are less costly than ATG and have been shown to reduce the incidence of rejection compared to placebo. Nevertheless, the incidence of rejection episodes is typically 1.5 to 2 times greater in anti-CD25-treated patients than in those receiving ATG, making the mAbs less attractive, particularly for higher-risk recipients.

The most recent addition to the T-cell mAb armamentarium is the more broadly reactive agent Alemtuzumab. This agent is a humanized mAb reactive against CD52. Alemtuzumab is used most commonly for induction therapy.[97,98]

Acute antibody-mediated rejection (acute humoral rejection; AHR) of organ allografts historically resulted in allograft loss in most recipients. Circulating donor-specific alloantibodies is diagnostic of AHR in kidney allografts. Removal plus the suppression of further antibody production was impossible until recently. To date, one mAb specifically directed against B cells has become available. Rituximab is a chimeric anti-CD20 antibody that was originally introduced for treatment of refractory B-cell lymphomas.[99] While the optimal protocol to treat AHR has still to be developed, the use of rituximab to deplete B cells and suppress antibody production together with plasmapheresis and IV-Ig has been utilized to alleviate humoral rejection symptoms.[100] In 2004, Raj and colleagues were able to treat autoimmune hemolytic anemia (AIHA), a condition that can occur as a result of hematopoietic stem cell transplantation (HSCT), with monthly injections of rituximab following allogeneic nonmyeloablative bone marrow transplantation in a child with sickle cell disease.[101]

It is now being increasingly utilized in transplant recipients for reversal of preexisting alloimmunity, reversal of antibody-mediated rejection, and treatment of post-transplant lymphoproliferative disease (PTLD).

The most recent attempt to provide increasingly potent and less toxic biologic immunosuppressants has been the production of immunoglobulin fusion proteins. CTLA-4, the homologue of CD28, has been of particular interest because CD28 is one of the most critical costimulatory receptors for T-cell activation. The first recombinant fusion protein developed was CTLA4-Ig (abatacept), designed to competitively inhibit CD28 interactions with its ligands. This fusion protein, though effective in the treatment of rheumatoid arthritis,[102] was less potent in transplant recipients. A high avidity mutant, LEA29Y (belatacept), has subsequently become available.[103] Initial clinical trials with LEA29Y have been extremely encouraging.[104]

Amevive (Alefacept) is a fusion protein that includes the extracellular CD2-binding portion of the human LFA-3. This construct has been shown in patients with psoriasis to inhibit interactions between LFA-3 and CD2 and also to deplete $CD2^+$ T cells, presumably by providing bridging to cytotoxic cells.[105] These patients were remarkably responsive to Amevive treatment. Clinical trials in transplant recipients are now being initiated.

CONCLUSIONS

Rejection by the immune system remains the main obstacle to large-scale transplantation of most allogeneic organs and tissues. While chronic immunosuppression of recipient individuals with calcineurin inhibitors has proved effective in preventing acute rejection of some organ allografts such as kidney and hearts, such treatment is nephrotoxic and associated with susceptibility to infections and neoplasia. In addition, a large proportion of transplants succumb to chronic rejection, a phenomenon induced by $CD4^+$ T cells activated through the indirect allorecognition pathway, which is resistant to current immunosuppressive treatments. These observations underscore the need for selective immunotherapies capable of preventing or suppressing proinflammatory antidonor T and B cells without compromising the entire host's immune system. Over the past 20 years, a number of protein-based therapies have been designed in an attempt to achieve transplantation tolerance. Among those, treatment of recipients with soluble MHC proteins and peptides as well as therapies using antibodies directed to key T and B cell costimulation molecules have been shown to ensure long-term tolerance to allografts in rodent experimental models. Some of these strategies are being tested in transplanted nonhuman primates and patients. However, alloreactive memory T cells present prior to transplantation in primates (but not in rodents) represent a formidable hurdle to successful transplantation tolerance using these therapies in patients. Therefore, blocking donor-reactive memory T and B cells in patients represents an essential challenge in clinical transplantation.

REFERENCES

1. Rosenberg, A.S. and A. Singer, Cellular basis of skin allograft rejection: An *in vivo* model of immune-mediated tissue destruction. *Annu Rev Immunol*, 1992. 10: 333–58.
2. Benichou, G., A. Valujskikh, and P.S. Heeger, Contributions of direct and indirect T cell alloreactivity during allograft rejection in mice. *J Immunol*, 1999. 162(1): 352–58.
3. Benichou, G., Direct and indirect antigen recognition: The pathways to allograft immune rejection. *Front Biosci*, 1999. 4: D476–80.
4. Benichou, G., et al., Limited T cell response to donor MHC peptides during allograft rejection. Implications for selective immune therapy in transplantation. *J Immunol*, 1994. 153(3): 938–45.
5. Liu, Z., et al., Mapping of dominant HLA-DR determinants recognized via the indirect pathway. *Transplant Proc*, 1997. 29(1–2/01): 1014–15.
6. Suciu-Foca, N., et al., Role of indirect allorecognition in chronic rejection of human allografts. *Transplant Proc*, 1996. 28(1): 404–5.
7. Shirwan, H., Chronic allograft rejection. Do the Th2 cells preferentially induced by indirect alloantigen recognition play a dominant role? *Transplantation*, 1999. 68(6): 715–26.
8. Najafian, N., et al., Enzyme-linked immunosorbent spot assay analysis of peripheral blood lymphocyte reactivity to donor HLA-DR peptides: Potential novel assay for prediction of outcomes for renal transplant recipients. *J Am Soc Nephrol*, 2002. 13(1): 252–59.
9. Burlingham, W.J. and F.H. Claas, Soluble MHC in allograft recipients. *Hum Immunol*, 1999. 60(5): 401–2.
10. Ghio, M., et al., Soluble HLA class I and Fas ligand molecules in blood components and their role in the immunomodulatory effects of blood transfusions. *Leuk Lymphoma*, 2000. 39(1–2): 29–36.
11. Puppo, F., et al., Soluble human MHC class I molecules induce soluble Fas ligand secretion and trigger apoptosis in activated CD8(+) Fas (CD95)(+) T lymphocytes. *Int Immunol*, 2000. 12(2): 195–203.
12. Zavazava, N. and D. Kabelitz, Alloreactivity and apoptosis in graft rejection and transplantation tolerance. *J Leukoc Biol*, 2000. 68(2): 167–74.

13. Liu, Z., et al., Limited usage of T cell receptor V beta genes by allopeptide-specific T cells. *J Immunol*, 1993. 150(8 Pt 1): 3180–86.
14. Oluwole, S.F., et al., Induction of specific unresponsiveness to rat cardiac allografts by pretreatment with intrathymic donor major histocompatibility complex class I antigens. *Transplant Proc*, 1993. 25(1 Pt 1): 299–300.
15. Oluwole, S.F., et al., Induction of transplantation tolerance to rat cardiac allografts by intrathymic inoculation of allogeneic soluble peptides. *Transplantation*, 1993. 56(6): 1523–27.
16. Shirwan, H., et al., Intrathymic immune regulation with donor class I allopeptides leads to the development of immunoregulatory cells that maintain tolerance to cardiac allografts. *Transplantation Proc*, 2001. 33(1–2): 80.
17. Shirwan, H., L. Barwari, and N.S. Khan, Immune nonresponsiveness to cardiac allografts by intrathymic inoculation of donor class I allopeptides is associated with high levels of transcripts for Th2 cytokines in the graft. *Transplantation Proc*, 1999. 31(1–2): 123–24.
18. Oluwole, O.O., et al., Indirect allorecognition in acquired thymic tolerance: Induction of donor-specific permanent acceptance of rat islets by adoptive transfer of allopeptide-pulsed host myeloid and thymic dendritic cells. *Diabetes*, 2001. 50(7): 1546–52.
19. Zavazava, N., et al., Oral feeding of an immunodominant MHC donor-derived synthetic class I peptide prolongs graft survival of heterotopic cardiac allografts in a high-responder rat strain combination. *J Leukoc Biol*, 2000. 67(6): 793–800.
20. Akiyama, Y., et al., Intratracheal delivery of a single major histocompatibility complex class I peptide induced prolonged survival of fully allogeneic cardiac grafts and generated regulatory cells. *Hum Immunol*, 2002. 63(10): 888–92.
21. Nisco, S., et al., Induction of allograft tolerance in rats by an HLA class-I-derived peptide and cyclosporine A. *J Immunol*, 1994. 152(8): 3786–92.
22. Clayberger, C. and A.M. Krensky, Immunosuppressive peptides corresponding to MHC class I sequences. *Curr Opin Immunol*, 1995. 7(5): 644–48.
23. Andrassy, J., et al., Tolerance to noninherited maternal MHC antigens in mice. *J Immunol*, 2003. 171(10): 5554–61.
24. Watschinger, B., et al., Mechanisms of allo-recognition. Recognition by in vivo-primed T cells of specific major histocompatibility complex polymorphisms presented as peptides by responder antigen-presenting cells. *Transplantation*, 1994. 57(4): 572–76.
25. Sayegh, M.H., et al., Mechanisms of acquired thymic unresponsiveness to renal allografts. Thymic recognition of immunodominant allo-MHC peptides induces peripheral T cell anergy. *Transplantation*, 1994. 58(2): 125–32.
26. Sayegh, M.H., et al., Induction of immunity and oral tolerance with polymorphic class II major histocompatibility complex allopeptides in the rat. *Proc Natl Acad Sci USA*, 1992. 89(16): 7762–66.
27. Hancock, W.W., et al., Differential effects of oral versus intrathymic administration of polymorphic major histocompatibility complex class II peptides on mononuclear and endothelial cell activation and cytokine expression during a delayed-type hypersensitivity response. *Am J Pathol*, 1994. 144(6): 1149–58.
28. Otto, C., et al., Prolongation of small bowel allograft survival with a sequential therapy consisting of a synthetic MHC class II peptide and temporarily low-dose cyclosporine A. *Hum Immunol*, 2002. 63(10): 880–87.
29. Chicz, R.M., et al., Specificity and promiscuity among naturally processed peptides bound to HLA-DR alleles. *J Exp Med*, 1993. 178(1): 27–47.
30. Benichou, G. and E. Sercarz, The presentation of self-peptides: Tolerance and competition. *Int Rev Immunol*, 1990. 6(1): 75–88.
31. Benichou, G., et al., Immunogenicity and tolerogenicity of self-major histocompatibility complex peptides. *J Exp Med*, 1990. 172(5): 1341–46.

32. Shen, X., et al., Peptides corresponding to CD4-interacting regions of murine MHC class II molecules modulate immune responses of CD4+ T lymphocytes in vitro and in vivo. *J Immunol*, 1996. 157(1): 87–100.
33. Choksi, S., B.A. Jameson, and R. Korngold, A structure-based approach to designing synthetic CD8alpha peptides that can inhibit cytotoxic T-lymphocyte responses. *Nat Med*, 1998. 4(3): 309–14.
34. Zechel, M.A., et al., Modulation of antigen presentation and class II expression by a class II-associated invariant chain peptide. *J Immunol*, 1996. 156(11): 4232–39.
35. Fedoseyeva, E.V., et al., *De novo* autoimmunity to cardiac myosin after heart transplantation and its contribution to the rejection process. *J Immunol*, 1999. 162(11): 6836–42.
36. Mares, D.C., et al., Type V collagen modulates alloantigen-induced pathology and immunology in the lung. *Am J Respir Cell Mol Biol*, 2000. 23(1): 62–70.
37. Benichou, G., et al., Induction of autoimmunity after allotransplantation. *Front Biosci*, 2007. 12: 4362–69.
38. Yasufuku, K., et al., Oral tolerance induction by type V collagen downregulates lung allograft rejection. *Am J Respir Cell Mol Biol*, 2001. 25(1): 26–34.
39. Fedoseyeva, E.V., et al., Modulation of tissue-specific immune response to cardiac myosin can prolong survival of allogeneic heart transplants. *J Immunol*, 2002. 169(3): 1168–74.
40. Lafferty, K.J. and A.J. Cunningham, A new analysis of allogeneic interactions. *Aust J Exp Biol Med Sci*, 1975. 53(1): 27–42.
41. Vadivel, N., S. Trikudanathan, and A. Chandraker, Transplant tolerance through costimulation blockade—Are we there yet? *Front Biosci*, 2007. 12: 2935–46.
42. Linsley, P.S. and J.A. Ledbetter, The role of the CD28 receptor during T cell responses to antigen. *Annu Rev Immunol*, 1993. 11: 191–212.
43. McAdam, A.J., A.N. Schweitzer, and A.H. Sharpe, The role of B7 co-stimulation in activation and differentiation of CD4+ and CD8+ T cells. *Immunol Rev*, 1998. 165: 231–47.
44. Dong, V.M., et al., Mechanisms of targeting CD28 by a signaling monoclonal antibody in acute and chronic allograft rejection. *Transplantation*, 2002. 73(8): 1310–17.
45. Montgomery, S.P., et al., Combination induction therapy with monoclonal antibodies specific for CD80, CD86, and CD154 in nonhuman primate renal transplantation. *Transplantation*, 2002. 74(10): 1365–69.
46. Larsen, C.P., et al., Long-term acceptance of skin and cardiac allografts after blocking CD40 and CD28 pathways. *Nature*, 1996. 381(6581): 434–38.
47. Larsen, C.P. and T.C. Pearson, The CD40 pathway in allograft rejection, acceptance, and tolerance. *Curr Opin Immunol*, 1997. 9(5): 641–47.
48. Grewal, I.S. and R.A. Flavell, The role of CD40 ligand in costimulation and T-cell activation. *Immunol Rev*, 1996. 153: 85–106.
49. Shimizu, K., et al., Host CD40 ligand deficiency induces long-term allograft survival and donor-specific tolerance in mouse cardiac transplantation but does not prevent graft arteriosclerosis. *J Immunol*, 2000. 165(6): 3506–18.
50. Kawai, T., et al., Costimulatory blockade for induction of mixed chimerism and renal allograft tolerance in nonhuman primates. *Transplant Proc*, 2001. 33(1–2): 221–22.
51. Yoshinaga, S.K., et al., T-cell co-stimulation through B7RP-1 and ICOS. *Nature*, 1999. 402(6763): 827–32.
52. Ozkaynak, E., et al., Importance of ICOS-B7RP-1 costimulation in acute and chronic allograft rejection. *Nat Immunol*, 2001. 2(7): 591–96.
53. Calderhead, D.M., et al., Cloning of mouse Ox40: A T cell activation marker that may mediate T–B cell interactions. *J Immunol*, 1993. 151(10): 5261–71.

54. Paterson, D.J., et al., Antigens of activated rat T lymphocytes including a molecule of 50,000 Mr detected only on CD4 positive T blasts. *Mol Immunol*, 1987. 24(12): 1281–90.
55. Vu, M.D., et al., Critical, but conditional, role of OX40 in memory T cell-mediated rejection. *J Immunol*, 2006. 176(3): 1394–401.
56. Agata, Y., et al., Expression of the PD-1 antigen on the surface of stimulated mouse T and B lymphocytes. *Int Immunol*, 1996. 8(5): 765–72.
57. Ito, T., et al., Analysis of the role of negative T cell costimulatory pathways in CD4 and CD8 T cell-mediated alloimmune responses in vivo. *J Immunol*, 2005. 174(11): 6648–56.
58. Ozkaynak, E., et al., Programmed death-1 targeting can promote allograft survival. *J Immunol*, 2002. 169(11): 6546–53.
59. Chavin, K.D., et al., Combined anti-CD2 and anti-CD3 receptor monoclonal antibodies induce donor-specific tolerance in a cardiac transplant model. *J Immunol*, 1993. 151(12): 7249–59.
59a. Sultan, P., Schechner, J.S., McNiff, J.M., Hochman, P.S., Hughes, C.C., Lorber, M.I., Askenase, P.W., and Pober, J.S., Blockade of CD2-LFA-3 interactions protects human skin allografts in immunodeficient mouse/human chimeras. *Nat. Biotechnol.*, 1997. 15(8): 759–62.
60. Alters, S.E., et al., Anti-CD4 mediates clonal anergy during transplantation tolerance induction. *J Exp Med*, 1991. 173(2): 491–94.
61. Pearson, T.C., et al., The assessment of transplantation tolerance induced by anti-CD4 monoclonal antibody in the murine model. *Transplantation*, 1993. 55(2): 361–67.
62. Pearson, T.C., et al., Induction of transplantation tolerance in adults using donor antigen and anti-CD4 monoclonal antibody. *Transplantation*, 1992. 54(3): 475–83.
63. Sharabi, Y. and D.H. Sachs, Mixed chimerism and permanent specific transplantation tolerance induced by a nonlethal preparative regimen. *J Exp Med*, 1989. 169(2): 493–502.
64. Brown, P.S., Jr., et al., Anti-Tac-H, a humanized antibody to the interleukin 2 receptor, prolongs primate cardiac allograft survival. *Proc Natl Acad Sci USA*, 1991. 88(7): 2663–67.
65. Tinubu, S.A., et al., Humanized antibody directed to the IL-2 receptor beta-chain prolongs primate cardiac allograft survival. *J Immunol*, 1994. 153(9): 4330–38.
66. Gassel, H.J., et al., Tolerance of rat liver allografts induced by short-term selective immunosuppression combining monoclonal antibodies directed against CD25 and CD54 with subtherapeutic cyclosporine. *Transplantation*, 2000. 69(6): 1058–67.
66a. Smith, C.R., Jaramillo, A., Lu, K.C., Higuchi, T., Kaleem, Z., and Mohanakumar, T., Prevention of obliterative airway disease in HLA-A2-transgenic tracheal allografts by neutralization of tumor necrosis factor. *Transplantation*, 2001. 72(9): 1512–18.
67. Jones, N.D., et al., Differential susceptibility of heart, skin, and islet allografts to T cell-mediated rejection. *J Immunol*, 2001. 166(4): 2824–30.
68. Ohnishi, T., K. Sakagami, and K. Orita, Treatment with monoclonal anti-IL2 antibody prolongs cardiac allograft survival in rats. *Transplant Proc*, 1987. 19(5): 4304–7.
69. Osaki, T., K. Sakagami, and K. Orita, Prolongation of cardiac allograft survival in rats by treatment with anti-interleukin 2 antiserum. *Acta Med Okayama*, 1988. 42(2): 77–78.
70. Sakagami, K., et al., The effect of anti-interleukin 2 monoclonal antibody treatment on the survival of rat cardiac allograft. *J Surg Res*, 1989. 46(3): 262–66.
71. Imagawa, D.K., et al., Anti-tumor necrosis factor antibody enhances allograft survival in rats. *J Surg Res*, 1990. 48(4): 345–48.
72. Imagawa, D.K., et al., The role of tumor necrosis factor in allograft rejection. II. Evidence that antibody therapy against tumor necrosis factor-alpha and lymphotoxin enhances cardiac allograft survival in rats. *Transplantation*, 1990. 50(2): 189–93.

73. Imagawa, D.K., et al., The role of tumor necrosis factor in allograft rejection. III. Evidence that anti-TNF antibody therapy prolongs allograft survival in rats with acute rejection. *Transplantation*, 1991. 51(1): 57–62.
74. Bolling, S.F., S.L. Kunkel, and H. Lin, Prolongation of cardiac allograft survival in rats by anti-TNF and cyclosporine combination therapy. *Transplantation*, 1992. 53(2): 283–86.
75. Wei et al., Inhibition of tumor necrosis factor production by lymphocytes from anti-TNF antibody-treated, cardiac-allografted rats. *J Surg Res*, 1994. 56(6): 601–5.
76. Wei et al., Anti-TNF antibody modulates cytokine and MHC expression in cardiac allografts. *J Surg Res*, 1999, 81(2): 123–8.
77. Gerber, D.A., et al., Prolongation of murine cardiac allograft survival by microspheres containing TNF alpha and IL1-beta neutralizing antibodies. *J Drug Target*, 1995. 3(4): 311–15.
78. Rosenberg, A.S., et al., Specific prolongation of MHC class II disparate skin allografts by in vivo administration of anti-IFN-gamma monoclonal antibody. *J Immunol*, 1990. 144(12): 4648–50.
79. Fuller, L., et al., Characterization of anti-canine cytokine monoclonal antibodies specific for IFN-gamma: Effect of anti-IFN-gamma on renal transplant rejection. *Tissue Antigens*, 1994. 43(3): 163–69.
80. Satoh, M., et al., Successful islet transplantation to two recipients from a single donor by targeting proinflammatory cytokines in mice. *Transplantation*, 2007. 83(8): 1085–92.
81. Beiras-Fernandez, A., E. Thein, and C. Hammer, Induction of immunosuppression with polyclonal antithymocyte globulins: An overview. *Exp Clin Transplant*, 2003. 1(2): 79–84.
82. Guttmann, R.D. and C. Flemming, Sequential biological immunosuppression. Induction therapy with rabbit antithymocyte globulin. *Clin Transplant*, 1997. 11(3): 185–92.
83. Shield, C.F., 3rd, et al., Use of antithymocyte globulin for reversal of acute allograft rejection. *Transplantation*, 1979. 28(6): 461–64.
84. Starzl, T.E., et al., The use of heterologous antilymphoid agents in canine renal and liver homotransplantation and in human renal homotransplantation. *Surg Gynecol Obstet*, 1967. 124(2): 301–8.
85. Andreoni, K.A., et al., Kidney and pancreas transplantation in the United States, 1996–2005. *Am J Transplant*, 2007. 7(5 Pt 2): 1359–75.
86. Bourdage, J.S. and D.M. Hamlin, Comparative polyclonal antithymocyte globulin and antilymphocyte/antilymphoblast globulin anti-CD antigen analysis by flow cytometry. *Transplantation*, 1995. 59(8): 1194–2000.
87. Lopez, M., et al., A novel mechanism of action for anti-thymocyte globulin: induction of CD4+CD25+Foxp3+ regulatory T cells. *J Am Soc Nephrol*, 2006. 17(10): 2844–53.
88. Goggins, W.C., et al., A prospective, randomized, clinical trial of intraoperative versus postoperative Thymoglobulin in adult cadaveric renal transplant recipients. *Transplantation*, 2003. 76(5): 798–802.
89. Kohler, G. and C. Milstein, Continuous cultures of fused cells secreting antibody of predefined specificity. *Nature*, 1975. 256(5517): 495–97.
90. Cosimi, A.B., et al., Treatment of acute renal allograft rejection with OKT3 monoclonal antibody. *Transplantation*, 1981. 32(6): 535–39.
91. Cosimi, A.B., et al., Use of monoclonal antibodies to T-cell subsets for immunologic monitoring and treatment in recipients of renal allografts. *N Engl J Med*, 1981. 305(6): 308–14.
92. Woodle, E.S., et al., Phase I trial of a humanized, Fc receptor nonbinding OKT3 antibody, huOKT3gamma1(Ala-Ala) in the treatment of acute renal allograft rejection. *Transplantation*, 1999. 68(5): 608–16.

93. Hering, B.J., et al., Transplantation of cultured islets from two-layer preserved pancreases in type 1 diabetes with anti-CD3 antibody. *Am J Transplant*, 2004. 4(3): 390–401.
94. Amlot, P.L., et al., Prolonged action of a chimeric interleukin-2 receptor (CD25) monoclonal antibody used in cadaveric renal transplantation. *Transplantation*, 1995. 60(7): 748–56.
95. Marx, J.L., Antibodies made to order. *Science*, 1985. 229(4712): 455–56.
96. Vincenti, F., et al., Interleukin-2-receptor blockade with daclizumab to prevent acute rejection in renal transplantation. Daclizumab Triple Therapy Study Group. *N Engl J Med*, 1998. 338(3): 161–65.
97. Bal, V., et al., Antigen presentation by keratinocytes induces tolerance in human T cells. *Eur J Immunol*, 1990. 20(9): 1893–97.
98. Watson, C.J., et al., Alemtuzumab (CAMPATH 1H) induction therapy in cadaveric kidney transplantation—Efficacy and safety at five years. *Am J Transplant*, 2005. 5(6): 1347–53.
99. Pescovitz, M.D., Rituximab, an anti-cd20 monoclonal antibody: history and mechanism of action. *Am J Transplant*, 2006. 6(5 Pt 1): 859–66.
100. Venetz, J.P. and M. Pascual, New treatments for acute humoral rejection of kidney allografts. *Expert Opin Invest Drugs*, 2007. 16(5): 625–33.
101. Raj, A., S. Bertolone, and A. Cheerva, Successful treatment of refractory autoimmune hemolytic anemia with monthly rituximab following nonmyeloablative stem cell transplantation for sickle cell disease. *J Pediatr Hematol Oncol*, 2004. 26(5): 312–14.
102. Kremer, J.M., et al., Treatment of rheumatoid arthritis by selective inhibition of T-cell activation with fusion protein CTLA4Ig. *N Engl J Med*, 2003. 349(20): 1907–15.
103. Larsen, C.P., et al., Rational development of LEA29Y (belatacept), a high-affinity variant of CTLA4-Ig with potent immunosuppressive properties. *Am J Transplant*, 2005. 5(3): 443–53.
104. Vincenti, F. and M. Luggen, T cell costimulation: A rational target in the therapeutic armamentarium for autoimmune diseases and transplantation. *Annu Rev Med*, 2007. 58: 347–58.
105. Ellis, C.N. and G.G. Krueger, Treatment of chronic plaque psoriasis by selective targeting of memory effector T lymphocytes. *N Engl J Med*, 2001. 345(4): 248–55.

9 Biologic Therapies for Type 1 Diabetes

Tom L. Van Belle and Matthias G. von Herrath

CONTENTS

ABSTRACT

Type 1 diabetes mellitus (T1D), also called juvenile diabetes, is a chronic autoimmune disorder that precipitates in genetically susceptible individuals by environmental factors.[1] The body's own immune system attacks the β-cells in the islets of Langerhans of the pancreas, destroying or damaging them sufficiently to reduce and eventually eliminate insulin production, leading to hyperglycemia. Since the discovery of "insulin replacement" in the early 1920s and its first life-saving application in a patient with juvenile diabetes, the hope has lingered that a cure for the disease is within reach. Today, however, despite the many remaining challenges in the field of T1D immunotherapy, hope is approaching reality. This chapter summarizes current protein- or cell-based therapeutics that are approved for clinical trials or are still in preclinical research. These therapies include endocrinologist approaches to increase β-cell mass and/or function, (non-)autoantigen-specific immune intervention, or combinations thereof. Last, a peek over the fence of preclinical research will provide insights in what might enter the pipeline of therapy development.

PRECLINICAL IMMUNOLOGICAL EVENTS IN T1D

Type 1 diabetes mellitus (T1D) can occur at any age, but it usually starts in people younger than 30, and can cause serious health complications including ketoacidosis, kidney failure, heart disease, stroke, and blindness. Patients are often diagnosed with T1D when they see a doctor for having clinical signs like excessive thirst, urination, and hunger. These symptoms result from the underlying hyperglycemia that is in turn caused by an insulin deficiency. In the preclinical phase, several events go unnoticed (so far). In addition to autoantibodies, the preclinical disease stage is characterized by the generation of activated, self-reactive lymphocytes that infiltrate the pancreas and selectively destroy the insulin-producing β-cells present in the islets of Langerhans.[2] This persistent, targeted destruction occurs "silently" and may go undetected for many years. By the time the first clinical symptoms become apparent, a majority of the patient's β-cells have been destroyed or rendered dysfunctional, making the individual dependent on insulin injections for survival. The administration of insulin is not a cure—even when used to maintain tight glycemic control, it does not halt the persistent autoimmune response. As patients with T1D require life-long insulin therapy, they still have a high risk of medical complications, and therefore preventative or curative therapies are urgently needed. The remaining β-cell mass at diagnosis creates a window of opportunity for the interventional strategies discussed below, because β-cell recovery and regeneration can lead to re-establishment of normoglycemia following immune-based interventions in animal models.

TO PREVENT OR TO INTERVENE, THAT'S THE QUESTION

The genetic susceptibility component of T1D allows some degree of targeting primary preventive care to family members of patients diagnosed with T1D. However, there is no complete inheritance of the disease, making preventive treatment difficult at best, detrimental at worst. Therefore, the two stages that are more acceptable to start medical treatment are (1) prior to clinical onset but after the appearance of autoantibodies (secondary prevention) and (2) immediately after onset (intervention). Prevention is preferred over intervention, because it has the advantage that the organ damage has not progressed as far as a critical point of "no return" and perhaps the self-regenerative capacity of β-cells is still better preserved. However, prevention depends on an unambiguous prediction/identification of at-risk individuals. In addition, long-term complications from immunosuppressive therapies such as reactivation of endogenous viruses (e.g., Epstein–Barr virus, EBV) and secondary cancers have to be avoided, which raises the bar for suitable drugs rather high.

Identification of Prediabetic Individuals

Pre-T1D describes the period when islet β-cell destruction is ongoing but glucose homeostasis is still maintained. *De facto,* the best definition of a prediabetic individual

TABLE 9.1
Identification of prediabetic individuals

1.	Genetic screening:	Relationship to diabetic individual (usually a first-degree relative) or identified to have high-risk HLA genotypes, like DR3-DQ2/DR4-DQ8[3]
2.	Serologic screening:	Serum autoantibodies associated with islet β-cells are found (ICA: GAD65, ICA512, and insulin)[4]
3.	Metabolic testing:	First phase insulin production is low enough to give ≥ 50% risk for diabetes within the next 5 years and/or impaired fasting glucose or impaired glucose tolerance[5,6]

is an individual who will actually develop diabetes. In T1D, we now have the ability to identify individuals at high risk by the sequence of events summarized in Table 9.1. This approach will enhance the success rate of preventive strategies dramatically and was considered crucial for identifying at-risk individuals to start secondary preventive treatment in a National Institutes of Health (NIH) consensus report on prevention of T1D. Identifying prediabetics is very expensive (estimated at $1,200 per subject, for autoantibody screening and two intravenous glucose tolerance tests), so a cost-effective approach for preventive trials requires extensive and international collaboration.

Preventive Trials

Many preventive trials are based on research in the non-obese diabetic (NOD) mouse model for T1D, which has improved our understanding of disease pathophysiology. A common tenet is that it is easy to prevent T1D in NOD if treatment is initiated early but more difficult later in the disease. Although this is conceivable given the progressive loss of islets to a point of no return, a recent comprehensive review of agents that are used to prevent T1D in NOD mice concluded that the therapeutic outcome of *prevention* protocols is not dependent on treatment onset time.[7] The good news here is that maybe some preventive strategies have a good chance at curing the disease even during an advanced status of β-cell destruction. Less promising is that so far, hardly any candidate preventive or therapeutic agent identified in NOD mice has been successful in preventing/curing human disease, indicating that the bar for reverting human hyperglycemia is set much higher or, conversely, immunosuppression is much better tolerated at higher dosages in animal models.

Current preventive trials using antigen-specific approaches are mainly based on different formulations of insulin. The Diabetes Prevention Trial-1 (DPT-1) assessed the use of oral and parental insulin in a large randomized, controlled, blinded prevention screening of about 100,000 first- and second-degree relatives of patients with T1D for risk at enrollment.[8] At their conclusion, neither the parenteral[9] nor the oral[8] insulin therapies showed evidence of protection from T1D development (Ref. 10: www.diabetestrialnet.org). However, post-hoc analysis suggested that individuals with pre-existing autoimmunity to insulin might benefit from oral insulin treatment. This is now addressed in a follow-up study. Non-antigen-specific preventive approaches include pilots and efficacy studies using ketotifen,[11] cyclosporine,[12] nicotinamide,[13] or combinations thereof.[14–16] Again, no beneficial effects were observed.

Currently, Vitamin D3,[17] hydrolyzed cow's milk (TRIGR[18]), and docosahexaenoic acid (DHA[19]) are in ongoing pilot studies or phase I preventive trials.

FAILURE BETWEEN BENCH AND BEDSIDE

Multiple factors have to be considered when moving from animal models into clinical trials, including, but not limited to the following: inappropriate translation of route, dose, frequency, and timing of antigen administration; late administration of vaccine, leading to exacerbation rather than dampening of the ongoing immune response; no clear knowledge of the regulatory versus pathogenic epitopes on the targeted tissue as disease progresses; limited vaccine efficacy (bystander suppression not effective enough).

ROLE OF THE PATIENT HISTORY IN THE SUCCESS OR FAILURE OF TREATMENTS

There are several different types of patients that enter the clinic. First, the time between onset of disease and actual diagnosis in the clinic can play a major role (early versus late diagnosis). Second is the speed of islets destruction by the immune system (Figure 9.1). As such, little intervention time is left for a patient with fast destruction and late diagnosis compared to an early diagnosed patient with slow islet destruction.

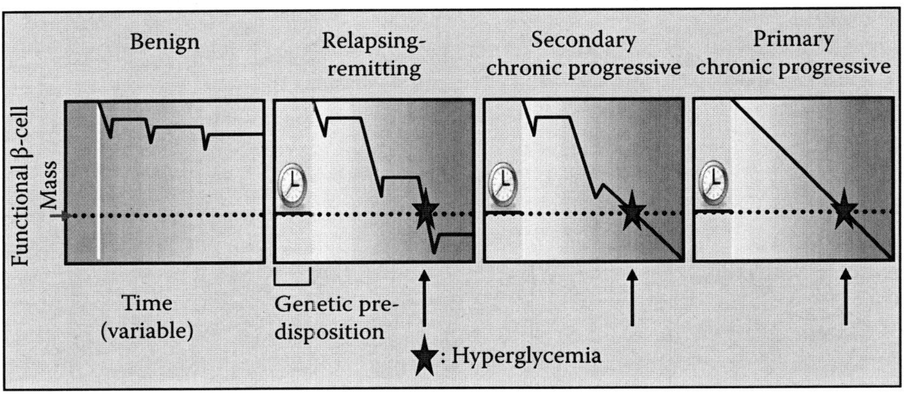

FIGURE 9.1 Multiple models of progression toward onset of T1D. In nonsusceptible individuals, some degree of β-cell loss occurs with age. In patients with autoimmune T1D, diabetes can start precipitating at any age by environmental triggers. The immune system then attacks the β-cells in the pancreas, resulting in hyperglycemia when only about 20 to 35 percent of the β-cells are left. This ultimate stage is preceded by a relatively long phase: the prediabetic phase, between triggering of T1D and hyperglycemia. In this phase autoantibodies can be detected and help with prediabetic diagnosis. Although reliable β-cell quantifying techniques are not yet available, one can envision several models of disease heterogeneity depending on the progression profile and the speed of β-cell destruction. Some individuals with T1D may have a benign form and never progress to overt hyperglycemia. In those individuals that do progress to metabolic dysfunction, immunological attack may be present as a chronically progressive form or display relapsing–remitting kinetics. (Based in part on Reference 40.)

Influence of Viral Infection on Type I Diabetes

Virus infections have been found to both accelerate and abrogate ongoing autoimmune processes. This dualism has certainly hampered attempts to establish a causal relationship between viral infections and T1D. An interesting example is the prevention of T1D by LCMV (Lymphocytic Choriomeningitis Virus) infection of the NOD mouse and the transgenic Rat Insulin Promotor (RIP)-LCMV mouse model.[20,21] Infection of prediabetic mice with ongoing insulitis but not clinically manifested T1D with LCMV results in strong inflammation at sites other than the target site of autoimmune destruction. This was shown to have a significant impact on trafficking of autoaggressive lymphocytes that now recirculate from the islets to peripheral lymphoid sites where stronger inflammatory signals, among them IP-10, are present.[20]

On the other hand, acceleration of T1D onset resulting from molecular mimicry or cross-reactivity between foreign and host components can occur at the T- and B-cell levels.[22,23] Studies in mice showed that mimicry can indeed break tolerance to autoantigens and thereby accelerate disease.[24] However, in this study, diabetes did not develop unless T-cell receptor (TCR) signal transduction was enhanced,[25] so it is deemed unlikely that mimicry alone is sufficient to lead to T1D. Also, Coxsackie B virus can accelerate diabetes when given during a crucial prediabetic phase.[26,27]

So, can viral or microbial infection cause precipitation of autoimmune diabetes in nonpredisposed, naive individuals? Several experimental data argue that this is not a likely scenario. For instance, for autoantigenic immunizations in conjunction with inflammatory signals (e.g., Toll-like receptor ligation) to precipitate disease in naive animals, autoreactive T cells have to be present and already activated.[28–30] Also, in a noninflamed state, β-cells express very few major histocompatibility complex (MHC) class I molecules, which renders them essentially nonrecognizable for killing by cytotoxic T lymphocytes (CTLs). Any viral infection leading to the release of α/β-interferons (α/β-IFNs), and consequently the upregulation of MHC class I, will "unmask" the β-cells to the immune system. This transient event will not lead to β-cell destruction unless a significant number of activated autoaggressive CTLs are able to reach the islets.[31] Therefore, it is believed that an effect of viral infection on autoimmunity is more likely to occur if an autoimmune process is already established, as would be the case in prediabetic individuals.

In conclusion, viruses can accelerate or even abrogate, but are unlikely to initiate T1D onset without other contributing factors. At this time, it is impossible to predict the influence of viral infections on the success or failure of therapies in clinical trials. Caution should be taken in the other direction, the influence of the drug (e.g., nonspecific immune suppressants) on the flare-ups of dormant or chronic infections, as was observed in the European anti-CD3 trial for transient reactivation of latent EBV.[32]

NEW-ONSET T1D TRIALS

At diagnosis, the immune system has dramatically reduced β-cell mass and function, and consequently the insulin-producing capacity of patients with T1D. Current treatments are either substitutive (e.g., the many formulations of insulin) or palliative (long-term use of anti-inflammatory and/or immune suppressive agents). As these types of treatments are often only partly efficacious and potentially toxic, research in the last decade has focused on the regeneration of β-cells and especially on inducing immune tolerance (Table 9.2 and Figure 9.2).

Replacing β-Cell Deficiency in Recent-Onset T1D

Substitutive treatment with many forms of insulin, especially combined with technological advances like continuous blood glucose sensors (phase III) or the closed loop system (phase I), has led to improved metabolic control of the disease. As inadequate endogenous insulin levels result from reduced β-cell mass and/or function, three strategies aim at restoring this: (1) islet replacement by transplantation of β-cells or pancreas from deceased individuals, (2) islet neogenesis from progenitor cells or islet regeneration from existing β-cell mass (EGF [epidermal growth factor] + gastrin, by Transition Therapeutics), (3) stimulation of insulin secretion (exendin-4, also exenatide, marketed as Byetta by Amylin). The latter two strategies could be used in combination for the best result.

Antigen Non-Specific Intervention Trials in T1D

A major culprit in T1D is the destructive action of the immune system that creates a β-cell mass shortage. Therefore, the profitable solutions described above might not be able to expand enough β-cells to permanently maintain normoglycemia, unless the immune response can be dampened concomitantly by, for example, anti-CD3 and autoantigen-specific therapy.

Interventional trials with nicotinamide, bacille Calmette-Guerin (BCG), diazoxide, or PRO-DIAB (oral protease mix) have shown no effect on T1D.[33–36] Several clinical trials have demonstrated successful induction of remission of T1D by systemic immune modulation with cyclosporin, but the unacceptable side effects paired with the required chronic use have forced suspension of this therapy.[37] However, this limited success has opened a gateway for the use of therapeutic biologic compounds that promote immune tolerance, such as monoclonal antibodies (mAbs) and soluble ligands directed to functionally relevant receptors on immune cells. For instance, the antidiabetogenic capacity of antilymphocyte serum (ALS),[38] anti-CD4, and FcR-non-binding anti-CD3 antibodies is in part the result of the transient depletion of (auto)immune cells, freezing the ongoing β-cell destruction.

One of the most promising targeted immunotherapies is the CD3-specific mAb therapy[39] that combines *de facto* autoimmune-specific targeting (autoantigen-specific activated T cells affected) with an immunomodulatory capacity (induction of Treg).[40] In brief, the mode of action is that anti-CD3 treatment modulates the TCR–CD3 complex resulting in the cell becoming "blind" to antigen.[41] In addition, alteration in TCR-mediated signal transduction leads to induction of anergy or apoptosis preferentially in activated T cells (especially Th1),[39] while Tregs seem to be more resistant to this. The apoptosis is partially mediated by CD95-CD95L interactions with neighboring T cells, and might explain why effector T-cell death is most dramatic at

TABLE 9.2
Major Interventional Clinical Trials T1D

Study Name and/or Product Name	Phase	Result or Status	Organizing (Clinical Trial Identifier)
		Antigen Specific	
IFA[a]-Enhanced human insulin B chain	I	Ongoing	ITN[b] (NCT00057499)
Recombinant human GAD65	II, III	Planned	Diamyd; TrialNet (NCT00529399)
Proinsulin-based DNA vaccine	I	Recruiting	BayHill Therapeutics (NCT00453375)
		Antigen Non-Specific	
DiaPep277 (hsp60[c] immunodominant peptide)	II	Ongoing. In phase I: 1-year positive effect (preservation of C-peptide concentration)	DeveloGen, Israel (NCT00058981)
Single course anti-CD3 (hOKT3g1(Ala-Ala))	II	Remission out to 18 months, 36 months positive effects on C-peptide levels	JDRF (NCT00378508)
AbATE (multidose) hOKT3g1(Ala-Ala)	II	Recruiting	ITN (NCT00129259)
Protégé Study (multinational) hOKT3g1(Ala-Ala)	II/III	Recruiting	Macrogenics; JDRF[d] (NCT00385697); US-based
TTEDD (multidose) ChAglyCD3 (TRX4)	II/III	Recruiting. In phase I: reduced insulin requirement out to 18 months	TolerRx; JDRF (NCT00451321); Europe-based
MMF/DZB (Mycophenolate mofetil; Daclizumab)	II	Ongoing	TrialNet (NCT00100178)
Rituximab (antiCD20 mAb)	II	Ongoing	TrialNet (NCT00279305)
START (Antithymocyte globulin, ATG)	II	Recruiting	ITN (NCT00515099)
Campath-1H (Anti-CD52 mAb)	I	Planned	Univ. of Wisconsin (NCT00214214)
		Cell Therapy	
Umbilical cord blood	I/II	Recruiting	Univ. Of Florida; JDRF (NCT00305344)
Autologous DCs	I	Recruiting	Univ. Of Pittsburgh (NCT00445913)

[a] Incomplete Freund's adjuvant (IFA).
[b] Immune Tolerance Network (ITN).
[c] Heat-shock protein 60 (Hsp60).
[d] Juvenile Diabetes Research Foundation (JDRF).

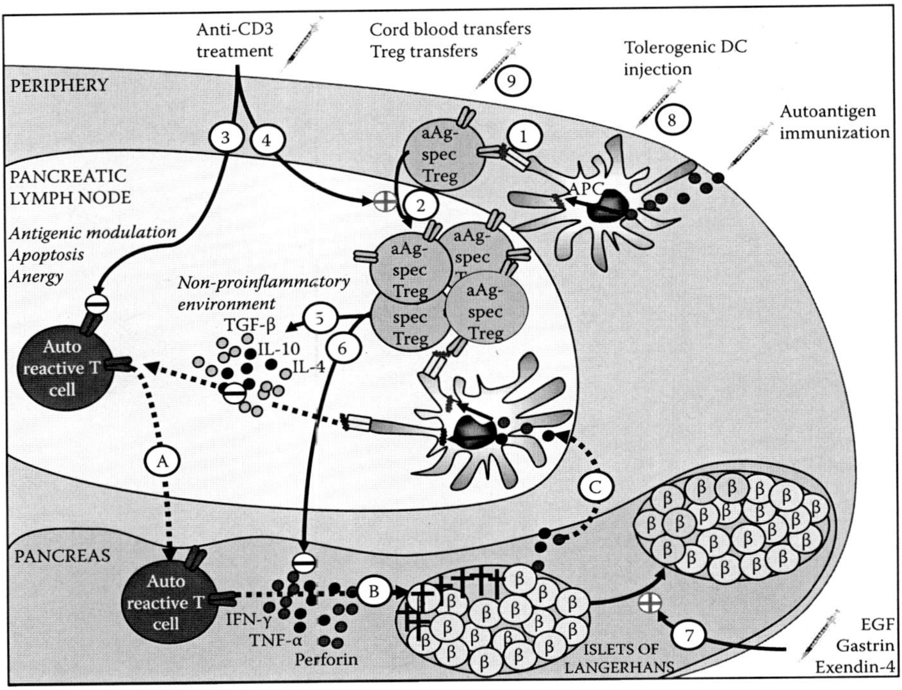

FIGURE 9.2 Therapeutic intervention in the pathogenesis of T1D. (A) Priming of autoreactive effector T cells occurs in susceptible individuals by environmental triggers together with an unbalanced immune response. (B) Autoaggressive T cells migrate to the pancreas and start destroying the β-cells in the islets of Langerhans via cytokine- and perforin-mediated mechanisms. (C) Autoantigens (aAgs: insulin, GAD65, etc.) are presented by antigen-presenting cells (APC) in the lymph nodes draining the pancreas, (D) enabling the expansion of effector T cells and initiating epitope spreading. Intervention strategies: (1) Immunization with aAg via a tolerogenic route (2) induces aAg-specific regulatory T cells (Tregs). (3) Systemic administration of non-FcR-binding anti-CD3 antibodies results in the T cells becoming "blind" to antigen, induces anergy or apoptosis, while also (4) promoting expansion of Treg. (5) The expanded aAg-specific Treg population migrates to the pancreatic lymph node where they create a non-proinflammatory environment that inhibits activation of autoreactive T cells, but also (6) blocks the autoaggressive response via bystander suppressive mechanisms. (7) Administration of EGF or exendin-4 in conjunction with gastrin regenerates β cells and restores the shortage. (8) T-cell stimulation by administered tolerogenic DCs should lead to an aborted immune response. (9) Transfer of Tregs aims at restoring the deficient immune regulation by increasing peripheral Treg numbers.

the site of inflammation, that is, where the T-cell density is highest. On top of this, anti-CD3 treatment might also foster Treg development directly.

This approach has been taken into two clinical trials using two different antibodies specific to human CD3 that are fully humanized immunoglobulin G_1 (IgG_1) with a mutated (hOKT3γ[Ala-Ala], MGA031, teplizumab by Macrogenics; Phase III) or aglycosylated (ChAglyCD3, TRX4 by TolerRx; Phase III) Fc region that prevents them from binding to the Fc receptors.[42,43] In the United States-based phase I and II clinical trials, teplizumab halted progression of recent-onset T1D for more than 1 year in most patients, and 3 years after treatment, patients showed a continued benefit in terms of C-peptide production and use of insulin as compared to control groups.[44,45] In the Europe-based phase II trial, treatment of patients with recent-onset T1D with TRX4 preserved β-cell function very efficiently, maintaining significantly higher levels of endogenous insulin secretion, and a drastic decrease in insulin requirements, even 18 months after single course treatment.[32] The latter study revealed two side-effects: (1) Anti-idiotypic antibodies were detected 2 to 3 weeks after injection of the drug. This could become a problem only when repeated treatment is needed. (2) Transient reactivation of EBV in patients who had already experienced a primary EBV infection. Notably, this subset of patients was/is excluded from participation in the United States-based trials.

Therapy with DiaPep277, derived from the heat-shock protein (hsp) 60, preserved endogenous insulin production in a phase II trial in patients with new-onset T1D. DiaPep277 was initially designed as antigen-specific treatment on the basis that the peptide is a targeted as an autoantigen in T1D as a result of cross-reactivity.[46,47] Since hsp60 activates T cells antigen non-specifically via Toll-like receptor-2,[48] thereby expanding regulatory T cells,[49] Diapep277 treatment might actually result in systemic modulation. So far, no DiaPep277-specific regulatory T cells (Tregs) have been characterized in mice or humans.

Additional intervention trials with systemic agents are currently planned using antithymocyte globulin (ATG, reengineered as Thymoglobulin by Genzyme) that is efficacious in animal models and has been used successfully in clinical transplantation. Preclinical studies have shown that ALS treatment of NOD mice or BB rats with recent-onset T1D can induce remission, supporting such an approach in humans in the phase II Study of Thymoglobulin to Arrest Type 1 Diabetes (START). It is postulated that the broader specificity of the polyclonal drug Thymoglobulin will prove more efficacious at inducing tolerance than current anti-CD3 approaches. An early limited human study with equine ATG with new-onset T1D suggested efficacy in prolonging the honeymoon phase.

Another reagent transmitted from the transplantation world is the B-cell-depleting CD20-specific mAb rituximab (Rituxan; Genentech and Biogen Idec). Given that B cells are the key autoantigen presenting cells in T1D,[50] the rationale for this treatment approach comes from the hypothesis that B-cell depletion using rituximab would eliminate a potentially potent antigen-presenting population and inflammatory islet autoantibodies, without affecting long-lived antibody-producing plasma cells.

Antigen-Specific Intervention Trials in T1D

Nonspecific immune suppression is partially effective in dampening autoreactive responses, but comes with several unwanted side effects limiting their long-term use. Researchers are now trying to take advantage of autoantigens, such as insulin or GAD65, to create therapeutic vaccines to tolerize antigen-specific responses. Although this approach was effective in experimental settings,[51–53] initial antigen-specific trials have shown limited promise in recent-onset T1D.[9,54,55] Current trials include the administration of insulin B chain with incomplete Freund's adjuvant (IFA), which has been highly effective in preventing diabetes in animal models.[56] In another trial, no protection was observed with an altered peptide ligand based on putative major autoantigenic sites in the insulin B chain, which had induced strong T helper 2 (Th2) responses in animal models.[57,58]

Two recent strategies (GAD65- or Proinsulin-based) look promising. First, a proinsulin-based DNA vaccine developed by Bayhill Therapeutics, combining DNA codons for an immunodominant peptide of insulin[59] and immunomodulatory CpG oligonucleotides,[60] has shown striking efficacy in recent-onset diabetes in the NOD mouse, and patients are being recruited in a phase I clinical trial. Second, Diamyd's alum-formulated GAD-65 protein has shown promising results in subjects with late-onset autoimmune diabetes of adulthood (LADA)[61] and positive results in phase II clinical trials, resulting in current preparation for a phase III program. The patients in these trials were selected on the basis of GAD65 autoantibodies, and it is too early to judge whether these studies signify a trend in which GAD65 emerges as a primary autoantigen in human T1D. Also, some investigations suggest that future antigen-based therapies should target what is called ignored determinants of β-cell antigens, as these were proved to inhibit late-stage autoimmune disease more effectively in diabetes-prone mice.[62]

Combination Intervention Trials in T1D

A single effective immunotherapy has not been identified thus far. It is possible that, as in cancer therapy, the complementary effects of combination immunotherapies synergize to the most effective therapy to reverse autoimmunity and to establish tolerance. This might allow for a decrease in the dosage of each drug used—and thereby reduce the side-effects. We propose that (combo)therapy is validated to achieve two major goals: (1) the "freezing" of the active immune system and dampening of any autoreactive response without strong side effects, and (2) the regeneration of a critical β-cell mass in order to maintain euglycemia without repetitive insulin injections.[63]

Most of the combination therapies are still in the preclinical stage. Ongoing phase II trials include the MMF/DZB Phase II Trial to test the combination of mycophenolate (Mofetil, CellCept by Roche) and an anti-CD25 mAb (daclizumab, Zenapax by Roche). The former is an adjuvant drug that selectively inhibits T- and B-lymphocyte proliferation by suppressing the *de novo* purine synthesis; the latter blocks the interleukin 2 (IL-2) signaling pathway, but does not interfere with the generation of regulatory T cells.[64] Both are systemic immune suppressants which have proved efficacy in other autoimmune diseases and, in combination, in preventing acute graft rejection.[65]

Very recently, a phase I/II trial was started on prevention of rejection of transplanted pancreatic islets from deceased donors in T1D recipients by the anti-CD3 mAb teplizumab in conjunction with Sirolimus (rapamycin, inhibiting response to IL-2) and low-dose Tacrolimus (cyclosporin, a calcineurin inhibitor blocking IL-2 production). Of note, earlier studies in mice showed that induction of tolerance, but not immune suppression, by anti-CD3 was completely blocked when tacrolimus was coadministered.[66]

Cell-Based Tolerogenic Therapy

In healthy individuals, regulatory or tolerogenic mechanisms or cells control the autoreactive T cells that escape thymic negative selection and reach the periphery. The cell-based tolerogenic therapy compensates for an assumed insufficiency by transferring cell types with immunomodulatory capacity, like regulatory T cells, umbilical cord blood, or tolerogenic dendritic cells (DCs). Propelled by evidence from animal models, cell-based tolerogenic immunotherapy has gained momentum in recent years (Table 9.2).

A pilot study is approved to test the safety and efficacy of transfer of autologous cord blood to recently diagnosed patients with T1D (ages 1 and up). This approach takes advantage of the capacity to initiate pancreatic regeneration of human umbilical cord blood cells, which also contain a large number of regulatory T cells.

Another approved safety study will evaluate the success achieved with immunoregulatory dendritic cells (iDC) in NOD mice.[67,68] Autologous monocyte-derived DC will be treated *ex vivo* with antisense phosphorothioate-modified oligonucleotides targeting the primary transcripts of the CD40, CD80, and CD86 costimulatory molecules, and then transferred to patients with recent-onset T1D. Several concerns should be raised with this approach: (1) instability of antisense knockdown can lead to stimulatory DC when the costimulatory molecules are sufficiently reexpressed, and (2) the expression of IL-7 by iDC, promoted in the therapy rationale as survival factor for Treg,[67] is also important for naïve $CD4^+$ T cells[69] and memory T cells[70,71] in general, and consequently also for autoreactive T cells.

Cellular immunotherapy with Tr1 cells or expanded Treg cells is currently being tested with the aim of preventing acute graft-versus-host disease (GVHD) in bone marrow transplantation, but could be expanded to prevent graft rejection of transplanted pancreatic islets. The numerous successful adoptive transfer experiments with autoantigen-specific Treg or Tr1 to NOD mice to reestablish immune tolerance[72–74] have to date not resulted in pilot studies in humans.

Experts in the field acknowledge the numerous technical problems that are likely to be encountered: a bona fide set of markers for human regulatory T cells, the number of cells to be transferred, the frequency of transfers, *in vitro* expansion methods, the survival of these cells *in vivo*, correct homing to the target tissue, or the inability to eliminate the transferred cells if there were untoward effects. Therefore, the field is divided into believers and nonbelievers, with some seeing cell-based tolerogenic therapy as a viable, routine clinical approach, while others prefer the use of small molecules and mAbs to steer cells with immunoregulatory capacity *in vivo*.

PROMISING BENCH-SIDE THERAPEUTICS

The preclinical research phase forms an indispensable basis for a therapy development pipeline by providing insights into what to do and what not to do in treatment of patients with T1D. Still, translation to the clinic is often far from smooth for reasons explained above. Preventive or interventional treatments that may (or may not) make it to the clinic are discussed below.

Future combination therapies include (1) anti-CD3 and intranasal insulin, (2) anti-CD3 and exenatide, (3) IL-2 and rapamycin, and (4) exenatide and gastrin. Initiation of phase II trials is to be expected soon, as each individual compound has passed extensive safety and efficacy studies. Suboptimal doses of the FcR-nonbinding anti-CD3 $F(ab')_2$ in conjunction with intranasal administration of proinsulin peptide has been proved to reverse diabetes in two mouse models of diabetes.[75] Noteworthy, insulin-specific $CD25^+FoxP3^+$ Treg, secreting immunomodulatory cytokines, like transforming growth factor beta (TGFβ) and IL-10, were induced and could confer dominant tolerance upon transfer to immune competent recent-onset diabetic recipients. Broader applicability of this combination approach was proved by using anti-CD3 therapy in conjunction with a GAD65-expressing plasmid. Strong synergy was evidenced in the RIP-LMCV-GP (C57Bl/6) model of T1D, but not in NOD mice (Damien Bresson and M.v.H., unpublished results). Interestingly, these data also reveal an important influence of the genetic background to mount a curative response by expanding antigen-specific regulatory T cells *in vivo*.

Although intuitively a bizarre combination, preclinical studies with IL-2 and rapamycin (inhibiting the response to IL-2) have shown impressive results in preventing spontaneous T1D onset in NOD. The rationale behind these experiments is that IL-2 will promote the expansion of $CD4^+CD25^+$ Treg in preference to activating ($CD25^+$) effector T cells if the expansion of the latter can be simultaneously blocked with rapamycin.[76] The mechanism of protection involves a shift from Th1- to Th2- and Th3-type cytokine-producing cells, possibly due to deletion of autoreactive Th1 cells.

Anti-CD3 in conjunction with exenatide combines with the immune tolerance induction by FcR-nonbinding anti-CD3 mAb to the stimulatory action of exenatide on insulin secretion. Therefore, this approach has higher chances of success than the gastrin and exenatide trials, which lack the immune modulatory arm required to stop autoimmune attack of the pancreatic β-cells.

Some groups have found that restoration of DC defects in NOD mice protected against T1D. As such, Flt3L-induced expansion of DC *in vivo* restored the deficiency in $CD4^+CD8^-$ subtype DCs,[77] and granulocyte colony-stimulating factor (G-CSF) particularly recruited immature $CD11c^{low}$ $B220^+$ plasmacytoid DCs (also considered tolerogenic DCs).[78] However, our data show no interventional potential for Flt3-L treatment (T.V.B. and M.v.H., unpublished results). Neonatal treatment of NOD mice with tumor necrosis factor alpha (TNFα) accelerated the onset of T1D by stimulating the maturation of dendritic cells, whereas TNFα blockade in the same time period resulted in a complete absence of diabetes.[79,80] So, a key would be to specifically stimulate the expansion or recruitment of immature DCs.

As B cells are considered to be important autoantigen-presenting cells, new drugs have been developed to specifically target B cells. A T1D phase I trial is planned with Campath-1H (Alemtuzumab, Genzyme), a fully humanized panlymphocyte antigen CD52 specific mAb, that is widely used in the treatment of B-cell chronic lymphocytic leukemia and has good results as preconditioning treatment in transplantation. Other B-cell-directed drugs target pathways that regulate cell survival, such as B-cell activating factor (BAFF, also BLyS). Both BR3-Fc (Genentech, Inc.) and atacicept (TACI-Ig, Zymogenetics) are currently in a phase II trial for application in rheumatoid arthritis (RA), and belimumab (Lymphostat-B; Human Genome Sciences) is currently in a phase II trial for use in systemic lupus erythematosus (SLE). So far, no clinical trials are requested or approved to test any of these BAFF-targeting drugs in T1D.

Current data of cytokine-based therapies in animal models reveal the complexity of using cytokines in immune intervention. Besides a broad spectrum of efficacy as well as side effects, the majority of cytokine administrations revealed a dual effect depending on the timing, dose, and route of administered. For instance, IL-18 accelerated or inhibited T1D in NOD when administrated early or late, respectively.[81,82] Therefore, cytokine immune therapy for T1D requires further research on ideal intervention schemes. However, blockade of cytokines with neutralizing mAbs or scavenging soluble receptors is applied successfully in other autoimmune diseases. Therefore, TNF antagonists, IL-1 receptor antagonists, TLR antagonists, and other anti-inflammatory agents may also be useful in T1D, for example, as adjunctive therapies to augment and/or sustain anti-CD3 therapy. These drugs work by blocking cell migration to inflammatory sites and the enhancement of regulatory T cells, as well as reduction of cytokinemia. Novel in this respect are the data from a genetic scanning effort that suggest a contribution of both IL-21 and IL-21R to genetic susceptibility to T1D.[83] NOD mouse studies demonstrate an essential role for IL-21 in T1D, suggesting IL-21 can be an interesting target for blockade in human T1D.[84,84a,84b]

The role of natural killer (NK) and NK-T cells has been discussed for many years. NK-T cell defects have been described in NOD mice and humans with diabetes.[85,86] In NOD mice, activation of immunoregulatory iNKT cells by the glycolipid ligand alpha-galactosylceramide (alpha-GalCer, Kirin compound 7000) was shown to protect against T1D via an IL-4-dependent mechanism and required Treg activity.[87,88] A preclinical trial is under way to explore the potential of NK-T cell activation by KRN7000 in humans with T1D.

Expression of the NKG2D ligand Rae-1 on activated NK cells in NOD mice, but not C57Bl/6 mice, suggested differential regulation of NKG2D ligand in NOD mice.[89] Treatment with a nondepleting anti-NKG2D mAb during the prediabetic stage completely prevented disease.[89,90] This preventive treatment represents a double-armed approach as it targets both the innate immune system via NK cells and the adaptive immune system by blocking (autoreactive) $CD8^+$ T cells. Currently, no data of interventional anti-NKG2D treatments are published.

Very new avenues for therapeutic strategies might come from a neurologic angle. For a long time, Jamaican healers have treated diabetes mellitus with *Capsicum frutescens* (chili peppers) that contains capsaicin as the active glucoregulatory ingredient.[91] The capsaicin receptor, vanilloid type 1 transient receptor potential receptor (TRPV1), is present on TRPV1$^+$ pancreatic sensory neurons. Deletion of these neurons by capsaicin treatment or genetic ablation of TRPV1 showed that autoreactive CD8$^+$ T cell pools, while still present, did not attack islets, rendering the NOD mice resistant to T1D.[92] This identified TRPV1$^+$ sensory neurons as important elements of diabetes pathoetiology by controlling islet inflammation, raising the possibility that sensory nerve dysfunction may contribute to prediabetes initiation and progression in diabetes-prone humans.

CONCLUSIONS

A global preventive treatment of T1D might prove unrealistic and more individualized therapies might be needed. Ideally, treatment starts when β-cell destruction is ongoing but hyperglycemia is absent. However, treatment at onset will be the most common in the near future. The ideal therapy includes a nonspecific T-cell suppressant such as anti-CD3 in conjunction with antigen-specific induction of Tregs (e.g., Proinsulin, GAD65) and a suitable compound that increases β-cell mass or function. This would have the advantages that the systemic immune suppressant could be used at a lower dose, reducing their side effects, but still creating a gateway for better antigen-specific induction of Tregs. For patients with C-peptide levels > 0.5 (nmol/l), preclinical research suggests that this two-compound therapy might be sufficient. Patients with C-peptide levels ranging from 0.2 to 0.5 might benefit from the addition of pancreatic β-cell regenerative compounds (e.g., gastrin + exenatide), whereas C-peptide levels below 0.2 indicate a need for supplementary pancreatic islet transplantation. The unfortunate reality is that combination therapies using one or more nonapproved drugs are difficult to license in the end. Considerable efforts will need to be made to change our current thinking that each compound needs to be efficacious and licensed on its own first, in order to move forward more expeditiously.

REFERENCES

1. Atkinson, M.A. and G.S. Eisenbarth, Type 1 diabetes: New perspectives on disease pathogenesis and treatment. *Lancet*, 2001. 358(9277): 221–29.
2. Bottazzo, G.F., et al., *In situ* characterization of autoimmune phenomena and expression of HLA molecules in the pancreas in diabetic insulitis. *N Engl J Med*, 1985. 313(6): 353–60.
3. Ludvigsson, J. and B. Lindblom, Human lymphocyte antigen DR types in relation to early clinical manifestations in diabetic children. *Pediatr Res*, 1984. 18(12): 1239–41.
4. Karjalainen, J.K., Islet cell antibodies as predictive markers for IDDM in children with high background incidence of disease. *Diabetes*, 1990. 39(9): 1144–50.
5. Srikanta, S., et al., Pre-type I diabetes. Linear loss of beta cell response to intravenous glucose. *Diabetes*, 1984. 33(8): 717–20.

6. Keskinen, P., et al., First-phase insulin response in young healthy children at genetic and immunological risk for Type I diabetes. *Diabetologia*, 2002. 45(12): 1639–48.
7. Shoda, L.K., et al., A comprehensive review of interventions in the NOD mouse and implications for translation. *Immunity*, 2005. 23(2): 115–26.
8. Skyler, J.S., et al., Effects of oral insulin in relatives of patients with type 1 diabetes: The Diabetes Prevention Trial—Type 1. *Diabetes Care*, 2005. 28(5): 1068–76.
9. DPT, Type I Diabetes Study Group. Effects of insulin in relatives of patients with type 1 diabetes mellitus. *N Engl J Med*, 2002. 346(22): 1685–91.
10. Staeva-Vieira, T., M. Peakman, and M. von Herrath, Translational mini-review series on type 1 diabetes: Immune-based therapeutic approaches for type 1 diabetes. *Clin Exp Immunol*, 2007. 148(1): 17–31.
11. Bohmer, K.P., et al., Linear loss of insulin secretory capacity during the last six months preceding IDDM. No effect of antiedematous therapy with ketotifen. *Diabetes Care*, 1994. 17(2): 138–41.
12. Carel, J.C., et al., Cyclosporine delays but does not prevent clinical onset in glucose intolerant pre-type 1 diabetic children. *J Autoimmun*, 1996. 9(6): 739–45.
13. Lampeter, E.F., et al., The Deutsche Nicotinamide Intervention Study: An attempt to prevent type 1 diabetes. DENIS Group. *Diabetes*, 1998. 47(6): 980–84.
14. Crino, A., et al., A two year observational study of nicotinamide and intensive insulin therapy in patients with recent onset type 1 diabetes mellitus. *J Pediatr Endocrinol Metab*, 2005. 18(8): 749–54.
15. Crino, A., et al., A randomized trial of nicotinamide and vitamin E in children with recent onset type 1 diabetes (IMDIAB IX). *Eur J Endocrinol*, 2004. 150(5): 719–24.
16. Pozzilli, P., et al., Randomized trial comparing nicotinamide and nicotinamide plus cyclosporin in recent onset insulin-dependent diabetes (IMDIAB 1). The IMDIAB Study Group. *Diabet Med*, 1994. 11(1): 98–104.
17. www.clinicaltrials.gov, identifier: NCT00141986.
18. www.trigr.org.
19. www.diabetestrialnet.org.
20. Christen, U., et al., Cure of prediabetic mice by viral infections involves lymphocyte recruitment along an IP-10 gradient. *J Clin Invest*, 2004. 113(1): 74–84.
21. Oldstone, M.B., Prevention of type I diabetes in nonobese diabetic mice by virus infection. *Science*, 1988. 239(4839): 500–2.
22. Srinivasappa, J., et al., Molecular mimicry: frequency of reactivity of monoclonal antiviral antibodies with normal tissues. *J Virol*, 1986. 57(1): 397–401.
23. Wucherpfennig, K.W. and J.L. Strominger, Molecular mimicry in T cell-mediated autoimmunity: Viral peptides activate human T cell clones specific for myelin basic protein. *Cell*, 1995. 80(5): 695–705.
24. Christen, U., et al., A viral epitope that mimics a self antigen can accelerate but not initiate autoimmune diabetes. *J Clin Invest*, 2004. 114(9): 1290–98.
25. Gronski, M.A., et al., TCR affinity and negative regulation limit autoimmunity. *Nat Med*, 2004. 10(11): 1234–39.
26. Horwitz, M.S., et al., Diabetes induced by Coxsackie virus: Initiation by bystander damage and not molecular mimicry. *Nat Med*, 1998. 4(7): 781–85.
27. Serreze, D.V., et al., Acceleration of type 1 diabetes by a coxsackievirus infection requires a preexisting critical mass of autoreactive T-cells in pancreatic islets. *Diabetes*, 2000. 49(5): 708–11.
28. Lang, K.S., et al., Toll-like receptor engagement converts T-cell autoreactivity into overt autoimmune disease. *Nat Med*, 2005. 11(2): 138–45.
29. Hamilton-Williams, E.E., et al., Cutting edge: TLR ligands are not sufficient to break cross-tolerance to self-antigens. *J Immunol*, 2005. 174(3): 1159–63.

30. Zipris, D., et al., TLR activation synergizes with Kilham rat virus infection to induce diabetes in BBDR rats. *J Immunol*, 2005. 174(1): 131–42.
31. Seewaldt, S., et al., Virus-induced autoimmune diabetes: most beta-cells die through inflammatory cytokines and not perforin from autoreactive (anti-viral) cytotoxic T-lymphocytes. *Diabetes*, 2000. 49(11): 1801–9.
32. Keymeulen, B., et al., Insulin needs after CD3-antibody therapy in new-onset type 1 diabetes. *N Engl J Med*, 2005. 352(25): 2598–608.
33. Elliott, R.B. and H.P. Chase, Prevention or delay of type 1 (insulin-dependent) diabetes mellitus in children using nicotinamide. *Diabetologia*, 1991. 34(5): 362–65.
34. Allen, H.F., et al., Effect of Bacillus Calmette-Guerin vaccination on new-onset type 1 diabetes. A randomized clinical study. *Diabetes Care*, 1999. 22(10): 1703–7.
35. Elliott, J.F., K.L. Marlin, and R.M. Couch, Effect of bacille Calmette-Guerin vaccination on C-peptide secretion in children newly diagnosed with IDDM. *Diabetes Care*, 1998. 21(10): 1691–93.
36. Ortqvist, E., et al., Temporary preservation of beta-cell function by diazoxide treatment in childhood type 1 diabetes. *Diabetes Care*, 2004. 27(9): 2191–97.
37. Jenner, M., et al., Cyclosporin A treatment of young children with newly-diagnosed type 1 (insulin-dependent) diabetes mellitus. London Diabetes Study Group. *Diabetologia*, 1992. 35(9): 884–88.
38. Ogawa, N., et al., Short administration of polyclonal anti-T cell antibody (ALS) in NOD mice with extensive insulitis prevents subsequent development of autoimmune diabetes. *J Autoimmun*, 2006. 26(4): 225–31.
39. Chatenoud, L., et al., Anti-CD3 antibody induces long-term remission of overt autoimmunity in nonobese diabetic mice. *Proc Natl Acad Sci USA*, 1994. 91(1): 123–27.
40. Chatenoud, L. and J.A. Bluestone, CD3-specific antibodies: A portal to the treatment of autoimmunity. *Nat Rev Immunol*, 2007. 7(8): 622–32.
41. Chatenoud, L., et al., Human in vivo antigenic modulation induced by the anti-T cell OKT3 monoclonal antibody. *Eur J Immunol*, 1982. 12(11): 979–82.
42. Xu, D., et al., In vitro characterization of five humanized OKT3 effector function variant antibodies. *Cell Immunol*, 2000. 200(1): 16–26.
43. Bolt, S., et al., The generation of a humanized, non-mitogenic CD3 monoclonal antibody which retains in vitro immunosuppressive properties. *Eur J Immunol*, 1993. 23(2): 403–11.
44. Herold, K.C., et al., Anti-CD3 monoclonal antibody in new-onset type 1 diabetes mellitus. *N Engl J Med*, 2002. 346(22): 1692–98.
45. Herold, K.C., et al., A single course of anti-CD3 monoclonal antibody hOKT3gamma1 (Ala-Ala) results in improvement in C-peptide responses and clinical parameters for at least 2 years after onset of type 1 diabetes. *Diabetes*, 2005. 54(6): 1763–69.
46. Raz, I., et al., Beta-cell function in new-onset type 1 diabetes and immunomodulation with a heat-shock protein peptide (DiaPep277): A randomised, double-blind, phase II trial. *Lancet*, 2001. 358(9295): 1749–53.
47. Elias, D., et al., Vaccination against autoimmune mouse diabetes with a T-cell epitope of the human 65-kDa heat shock protein. *Proc Natl Acad Sci USA*, 1991. 88(8): 3088–91.
48. Zanin-Zhorov, A., et al., T cells respond to heat shock protein 60 via TLR2: Activation of adhesion and inhibition of chemokine receptors. *FASEB J*, 2003. 17(11): 1567–69.
49. Sutmuller, R.P., et al., Toll-like receptor 2 controls expansion and function of regulatory T cells. *J Clin Invest*, 2006. 116(2): 485–94.
50. Serreze, D.V. and P.A. Silveira, The role of B lymphocytes as key antigen-presenting cells in the development of T cell-mediated autoimmune type 1 diabetes. *Curr Dir Autoimmun*, 2003. 6: 212–27.

51. Kaufman, D.L., et al., Spontaneous loss of T-cell tolerance to glutamic acid decarboxylase in murine insulin-dependent diabetes. *Nature*, 1993. 366(6450): 69–72.
52. Tisch, R., et al., Immune response to glutamic acid decarboxylase correlates with insulitis in non-obese diabetic mice. *Nature*, 1993. 366(6450): 72–75.
53. Elias, D. and I.R. Cohen, Peptide therapy for diabetes in NOD mice. *Lancet*, 1994. 343(8899): 704–6.
54. Chaillous, L., et al., Oral insulin administration and residual beta-cell function in recent-onset type 1 diabetes: A multicentre randomised controlled trial. Diabete Insuline Orale group. *Lancet*, 2000. 356(9229): 545–49.
55. Ergun-Longmire, B., et al., Oral insulin therapy to prevent progression of immune-mediated (type 1) diabetes. *Ann N Y Acad Sci*, 2004. 1029: 260–77.
56. Muir, A., et al., Insulin immunization of nonobese diabetic mice induces a protective insulitis characterized by diminished intraislet interferon-gamma transcription. *J Clin Invest*, 1995. 95(2): 628–34.
57. Alleva, D.G., et al., Immunological characterization and therapeutic activity of an altered-peptide ligand, NBI-6024, based on the immunodominant type 1 diabetes autoantigen insulin B-chain (9-23) peptide. *Diabetes*, 2002. 51(7): 2126–34.
58. Alleva, D.G., et al., Immunomodulation in type 1 diabetes by NBI-6024, an altered peptide ligand of the insulin B epitope. *Scand J Immunol*, 2006. 63(1): 59–69.
59. Urbanek-Ruiz, I., et al., Immunization with DNA encoding an immunodominant peptide of insulin prevents diabetes in NOD mice. *Clin Immunol*, 2001. 100(2): 164–71.
60. Ho, P.P., et al., An immunomodulatory GpG oligonucleotide for the treatment of autoimmunity via the innate and adaptive immune systems. *J Immunol*, 2003. 171(9): 4920–26.
61. Agardh, C.D., et al., Clinical evidence for the safety of GAD65 immunomodulation in adult-onset autoimmune diabetes. *J Diabetes Complications*, 2005. 19(4): 238–46.
62. Olcott, A.P., et al., Antigen-based therapies using ignored determinants of beta cell antigens can more effectively inhibit late-stage autoimmune disease in diabetes-prone mice. *J Immunol*, 2005. 175(3): 1991–99.
63. Bresson, D. and M. von Herrath, Moving towards efficient therapies in type 1 diabetes: to combine or not to combine? *Autoimmun Rev*, 2007. 6(5): 315–22.
64. Vlad, G., et al., Anti-CD25 treatment and FOXP3-positive regulatory T cells in heart transplantation. *Transpl Immunol*, 2007. 18(1): 13–21.
65. Mastrobuoni, S., et al., Two-dose daclizumab, tacrolimus, mycophenolate mofetil, and steroid-free regimen in de novo cardiac transplant recipients: Early experience. *Transplant Proc*, 2007. 39(7): 2163–66.
66. Chatenoud, L., J. Primo, and J.F. Bach, CD3 antibody-induced dominant self tolerance in overtly diabetic NOD mice. *J Immunol*, 1997. 158(6): 2947–54.
67. Harnaha, J., et al., Interleukin-7 is a survival factor for $CD4^+$ $CD25^+$ T-cells and is expressed by diabetes-suppressive dendritic cells. *Diabetes*, 2006. 55(1): 158–70.
68. Machen, J., et al., Antisense oligonucleotides down-regulating costimulation confer diabetes-preventive properties to nonobese diabetic mouse dendritic cells. *J Immunol*, 2004. 173(7): 4331–41.
69. Masse, G.X., et al., Gamma(c) cytokines provide multiple homeostatic signals to naive CD4(+) T cells. *Eur J Immunol*, 2007. 37(9): 2606–16.
70. Surh, C.D., et al., Homeostasis of memory T cells. *Immunol Rev*, 2006. 211: 154–63.
71. Dooms, H., et al., Interleukin-2 enhances $CD4^+$ T cell memory by promoting the generation of IL-7R alpha-expressing cells. *J Exp Med*, 2007. 204(3): 547–57.

72. Tang, Q., et al., In vitro-expanded antigen-specific regulatory T cells suppress autoimmune diabetes. *J Exp Med*. 2004. 199(11): 1455–65.
73. Tarbell, K.V., et al., $CD25^+$ $CD4^+$ T cells, expanded with dendritic cells presenting a single autoantigenic peptide, suppress autoimmune diabetes. *J Exp Med*, 2004. 199(11): 1467–77.
74. You, S., et al., Presence of diabetes-inhibiting, glutamic acid decarboxylase-specific, IL-10-dependent, regulatory T cells in naive nonobese diabetic mice. *J Immunol*, 2004. 173(11): 6777–85.
75. Bresson, D., et al., Anti-CD3 and nasal proinsulin combination therapy enhances remission from recent-onset autoimmune diabetes by inducing Tregs. *J Clin Invest*, 2006. 116(5): 1371–81.
76. Rabinovitch, A., et al., Combination therapy with sirolimus and interleukin-2 prevents spontaneous and recurrent autoimmune diabetes in NOD mice. *Diabetes*, 2002. 51(3): 638–45.
77. O'Keeffe, M., et al., Fms-like tyrosine kinase 3 ligand administration overcomes a genetically determined dendritic cell deficiency in NOD mice and protects against diabetes development. *Int Immunol*, 2005. 17(3): 307–14.
78. Kared, H., et al., Treatment with granulocyte colony-stimulating factor prevents diabetes in NOD mice by recruiting plasmacytoid dendritic cells and functional CD4(+) CD25(+) regulatory T-cells. *Diabetes*, 2005. 54(1): 78–84.
79. Jacob, C.O., et al., Prevention of diabetes in nonobese diabetic mice by tumor necrosis factor (TNF): Similarities between TNF-alpha and interleukin 1. *Proc Natl Acad Sci USA*, 1990. 87(3): 968–72.
80. Lee, L.F., et al., The role of TNF-alpha in the pathogenesis of type 1 diabetes in the nonobese diabetic mouse: Analysis of dendritic cell maturation. *Proc Natl Acad Sci USA*, 2005. 102(44): 15995–6000.
81. Oikawa, Y., et al., Systemic administration of IL-18 promotes diabetes development in young nonobese diabetic mice. *J Immunol*, 2003. 171(11): 5865–75.
82. Rothe, H., et al., IL-18 inhibits diabetes development in nonobese diabetic mice by counterregulation of Th1-dependent destructive insulitis. *J Immunol*, 1999. 163(3): 1230–36.
83. Asano, K., et al., Molecular scanning of interleukin-21 gene and genetic susceptibility to type 1 diabetes. *Hum Immunol*, 2007. 68(5): 384–91.
84. Sutherland, A.P., et al., IL-21 is required for the development of type 1 diabetes in NOD mice. *Diabetes*, 2009.
84a. Spolski, R., et al., IL-21 signaling is critical for the development of type 1 diabetes in the NOD mouse. *Proc Natl Acad Sci USA*, 2008. 105(37): 14028–33.
84b. Datta, S. and N. E. Sarvetnick, IL-21 limits peripheral lymphocyte numbers through T-cell homeostatic mechanisms. *PLoS ONE*, 2008. 3(9): e3118.
85. Kukreja, A., et al., NKT cell defects in NOD mice suggest therapeutic opportunities. *J Autoimmun*, 2002. 19(3): 117–28.
86. Kukreja, A., et al., Multiple immuno-regulatory defects in type-1 diabetes. *J Clin Invest*, 2002. 109(1): 131–40.
87. Ly, D., et al., Protection from type 1 diabetes by invariant NK T cells requires the activity of $CD4^+CD25^+$ regulatory T cells. *J Immunol*, 2006. 177(6): 3695–704.
88. Hussain, S., et al., Role of regulatory invariant CD1d-restricted natural killer T-cells in protection against type 1 diabetes. *Immunol Res*, 2005. 31(3): 177–88.
89. Ogasawara, K., et al., Impairment of NK cell function by NKG2D modulation in NOD mice. *Immunity*, 2003. 18(1): 41–51.
90. Ogasawara, K., et al., NKG2D blockade prevents autoimmune diabetes in NOD mice. *Immunity*, 2004. 20(6): 757–67.

91. Tolan, I., D. Ragoobirsingh, and E.Y. Morrison, Isolation and purification of the hypoglycaemic principle present in *Capsicum frutescens*. *Phytother Res*, 2004. 18(1): 95–96.
92. Razavi, R., et al., TRPV1$^+$ sensory neurons control beta cell stress and islet inflammation in autoimmune diabetes. *Cell*, 2006. 127(6): 1123–35.

Section III

Biologics for Hematological Malignancies

10 Biologic Therapies for Acute Myeloid Leukemia/ Myelodysplastic Syndrome

Javier Pinilla-Ibarz and Celeste Bello

CONTENTS

KEY WORDS: monoclonal antibodies, lintuzumab, acute myeloid leukemia, gentuzumab ozogamicin

ABSTRACT

Monoclonal antibodies (mAbs) are a relatively new class of biologic agents that target specific receptors on cancer cells. These types of drugs have shown significant improvement in the treatment of certain hematologic disorders. Leukemias represent an ideal target for the development of such therapies as malignant cells are easily accessible in the blood and bone marrow of these patients. Antibodies against CD33, a cell-surface glycoprotein, are the best example. The humanized anti-CD33 lintuzumab has modest activity on acute myeloid leukemia (AML), but it can eliminate minimal residual disease in acute promyelocytic leukemia. Gentuzumab ozogamicin (GO), an anti-CD33–calicheamicin conjugate has activity in patients with relapsed

AML and appears promising in combination with chemotherapy for newly diagnosed patients.

INTRODUCTION

Acute myeloid leukemia (AML), with the exception of the good prognostic karyotypes t(15;17), inv16, and t(8;22), has a dismal prognosis in adults. Despite the advances in supportive care of these patients, the treatment has not significantly changed during recent years. Induction therapy with cytarabine and an anthracycline achieved complete remission (CR) in 50 to 70 percent of patients with 20 to 40 percent long-term survivors.[1]

Monoclonal antibodies (mAbs) are a relatively new class of biologic agents that target specific receptors on cancer cells. These types of drugs have shown significant improvement in the treatment of certain hematologic disorders. The most well known example is the improvement in outcomes with the use of rituxan in the treatment of lymphomas. The success of rituxan has fueled the development of other mAbs for the treatment of leukemias because of the easy access of malignant cells in the blood and bone marrow. Furthermore, the well-characterized immunophenotype of leukemias make this disease an ideal target for the identification of antigenic proteins. Most of the studies have focused on CD33, a cell-surface glycoprotein found in myeloid leukemias. Naked and conjugated antibodies have been studied extensively.[2] Other targets such as CD45, CD66, CD44, and CD123 are also under investigation. The following section discusses the use of mAbs in the treatment of AML.

ANTI-CD33 MAB, LINTUZUMAB

The sialic acid binding receptor CD33 is expressed on myeloid progenitor cells, monocytes, and macrophages. In addition, approximately 90 percent of AML cells express CD33. The exact cellular function of CD33 is unknown but binding of this receptor by the antibody has been shown to result in antibody-dependent cellular cytotoxicity (ADCC).[3] Recent data suggest that ligation of CD33 could block *in vitro* proliferation and induce apoptosis in AML cells, likely through activation of the protein tyrosine phosphatases SHP-1 and SHP-2 and recruitment of Syk protein tyrosine kinase.[4,5] Initial studies using a murine anti-CD33 mAb, M195, showed antileukemic effects in humans.[6] The use of this drug was, however, limited by the development of human antimouse antibodies (HAMA) and the inability to recruit human immune effector functions *in vivo*.[6] It did, however, demonstrate specificity and rapid internalization by CD33-positive cells making it a promising treatment and drug delivery option in AML.

HuM195 (lintuzumab, SNG-33) is a humanized immunoglobulin G_1 (IgG_1) mAb against CD33 made by grafting complementarity-determining regions of murine M195 into a human IgG_1 framework and backbone. This humanized anti-CD33 antibody was shown to bind the CD33 antigen on human cells with an eightfold greater avidity than its mouse counterpart, to mediate leukemia cell killing *in vitro* by recruiting human immune effector functions thought to be necessary in ADCC, and to fix human complement.[3] In an initial phase I trial, 13 patients with relapsed or refractory AML received six doses of lintuzumab (0.5 to 10 mg/m^2) every 72 hours. Pharmacology and leukemia targeting was similar to the murine mAb, but no significant development of human antimouse antibodies was seen.[7]

A following study investigating the role for high-dose lintuzumab included 10 patients with relapsed or refractory AML. Patients were treated with two courses of lintuzumab (12 to 36 mg/m^2 daily for 4 days) every 2 weeks. An increase of the plasma half-life of the antibody of 7 days was seen compared with 38 hours in the previous trial. Activity was seen in one patient with refractory AML (with less than 10 percent blast in bone marrow) achieving a long-term CR and three patients with a reduction of the percentage of bone marrow blast.[8] An additional 50 patients with relapsed or refractory AML were treated in a phase I/II trial and reported by Feldman et al.[9] These patients were once again randomized to receive HuM195 at a dose of 12 or 36 mg/m^2. Of the 49 evaluable patients, 2 CRs and 1 CR except for full platelet recovery were observed. All three responses were in patients receiving the lower dose (12 mg/m^2) of HuM195. This study showed that the antileukemic effects of lintuzumab as a single agent were modest at best, and prompted the development of a trial using lintuzumab combined with chemotherapy.

In 2005, Feldman et al.[10] published the results of a phase III trial that followed 192 patients with relapsed or refractory AML. These patients were randomized to receive induction chemotherapy with MEC regimen (mitoxantrone, etoposide, and cytarabine) with or without lintuzumab. No differences in adverse events or treatment-related mortality between the two groups were seen. The group treated with induction chemotherapy and lintuzumab had a slight increase in CR (23.4 versus 20.6 percent) and a CR$^+$CR with incomplete platelet recovery (36 versus 28 percent). However, these results were not statistically significant. The addition of lintuzumab produced the greatest benefit in patients whose first remission lasted 7 to 12 months.

Review of the results from these earlier trials did show an interesting finding. Objective responses were seen only in those patients with a low leukemic burden. The earlier phase I/II trial found that all three responders had a bone marrow blast count of less than 30 percent.[9] As a result of this finding, a new phase I, single-arm dose-escalation trial exploring escalating doses of HuM195 (SGN-33) (1.5 to 8 mg/kg) in patients with high risk myelodysplastic syndrome (MDS) and low leukemic burden AML (<60 percent blasts on bone marrow biopsy) has been reported. Entry criteria included CD33 expression of more than 50 percent of bone marrow as determined by flow cytometry. Patients received weekly doses for 5 weeks. If clinical benefit was demonstrated, patients were able to receive further every-other-week infusions. Dose-limiting toxicity was not observed at 8 mg/kg per week. Among 18 patients with AML, 4 achieved CR and 1 CR with incomplete platelet recovery. Complete responses were observed in older patients with AML who were not candidates for intensive therapies. Based on these data, the drug is currently being investigated in additional further clinical trials as single agent or in combination with lenalidomide.[11]

Acute promyelocytic leukemia (APL) is a subtype of AML characterized by the translocation between chromosome 15 (PML gene) and 17 (retinoic acid receptor alpha, RARα). Reverse transcription polymerase chain reaction (RT-PCR) monitoring of the fusion mRNA is useful to assess the postremission therapy in patients in clinical remission with high risk of relapse.[12] The presence of CD33 in APL stem cells made the use of lintuzumab very attractive for this disease. This led to a study by Jurcic et al.[13] that evaluated lintuzumab in patients with APL. Their study included 31 patients with APL who obtained a clinical remission with all-*trans* retinoic acid (ATRA). These patients then received lintuzumab twice weekly for 3 weeks followed by three cycles of idarubicin and cytarabine as consolidation and six monthly courses of lintuzumab as maintenance. Of 24 evaluable patients for conversion of positive RT-PCR assays, 12 became RT-PCR negative after lintuzumab without additional therapy. The 5-year disease-free survival rate was 93 percent compared with 73 percent among historical controls receiving ATRA followed by chemotherapy. A subsequent postremission study with lintuzumab followed by arsenic trioxide in 15 patients with newly diagnosed APL using a risk adopted approach based on RT-PCR monitoring has been reported.[14] Four of seven patients with a positive RT-PCR after induction became negative after lintuzumab and all patients achieved molecular remission after arsenic trioxide. Of the 15 patients, 14 were alive with a median follow-up of 50 months. These studies support the investigation of lintuzumab in the setting of minimal residual disease in other types of myeloid leukemias.

ANTI-CD33 CONJUGATES, GENTUZUMAB OZOGAMICIN

The fact that CD33 is internalized into leukemic cells under binding with anti-CD33 paved the way for development of conjugates to increase the efficacy of mAb. Gemtuzumab ozogamicin (GO; Mylotarg; Wyeth-Ayerst Pharmaceuticals, Radnor, PA) is an anti-CD33 mAb conjugated with calicheamicin. Calicheamicin is an antitumor antibiotic that breaks double-stranded DNA. In the phase I trial evaluating GO, leukemic cells were eliminated from the peripheral blood and bone marrow of 8 of the 40 patients (including three CR), and the drug was reasonably well tolerated.[15] This led to the phase II trial by Siever et al.,[16] which evaluated 142 patients with CD33-positive AML in first relapse. These patients all received GO as a 2-hour intravenous infusion, at a dose of 9 mg/m^2, at 2-week intervals for two doses. Thirty percent of these patients obtained a remission, which was defined as a marrow blast count less than 5 percent, recovery of neutrophils to at least 1,500/uL, and red blood cell (RBC) and platelet transfusion independence. Subsequently, an analysis of a larger cohort of 277 patients with AML in first relapse treated with the same schedule in three different clinical trials including the previously described has been reported.[17] CR was achieved in 35 patients (13 percent) and 36 patients (13 percent) had a CRp. Response was independent of age, but the median duration of response was brief (6.4 months for CR and 4.5 months for CRp). The treatment-related side effects observed with GO included hypotension in 4 percent of the patients several hours after administration, grade 3 to 4 elevations of hepatic enzymes in 18 percent, and hyperbilirubinemia in 29 percent (likely because of dissociation of calicheamicin within the liver), and grade 4 neutropenia and thrombocytopenia in virtually all patients. No cardiac or cerebellar toxicity was associated with GO. Hepatic veno-occlusive disease (VOD) was seen in 5 percent of patients but in the setting of stem cell transplant, as 0.9 percent of patients who did not undergo this procedure had this complication. Additional studies have been reported on the increased risk of VOD (48 to 64 percent) in patients after undergoing stem cell transplant.

A frontline study by the EORTC and GIMEMA explored the use of GO in patients older than 60 years with previously untreated AML. The response rate was 33 percent for patients aged 61 to 75 years but only 5 percent in older patients, and the median duration of response was 6.1 months for patients with CR and 2.2 months for patients with CRp.[18] This and other studies led to the Food and Drug Administration (FDA) approval of GO on May 17, 2000, for the treatment of patients with CD33-positive AML in first relapse who are 60 years of age or older.

GO has also been tested in combination with induction chemotherapy. Because of this toxicity profile the dose has been reduced. In a study in younger patients by the MRC, the maximum tolerated dose was 3 mg/m^2.[19] Another study explored a 6 mg/m^2 dose in combination with standard induction for newly diagnosed AML in a younger population; 36 of 43 evaluable patients (84 percent) achieved a CR.[20]

A recent study by Fianchi et al.[21] evaluated 53 patients over the age of 60 with untreated or primary refractory/relapsed AML. These patients were treated with two cycles of cytarabine, GO, and granulocyte colony-stimulating factor (G-CSF). The overall response rate was 57 percent with a median disease-free survival and overall survival of 8 and 9 months, respectively.

Currently, there are several ongoing trials to assess the effectiveness of GO in different clinical settings. One of these trials is the MRC AML15 trial which is designed to evaluate the use of GO in first-line induction and consolidation regimens for patients with AML who are under the age of 60.[21] Patients are randomized to receive induction with one of the following: Daunorubicin/Ara-C, Ara-C/Daunorubicin/Etoposide, or Fludarabine/Ara-C/Idarubicin/G-CSF with or without GO. Patients who obtain a CR will then be randomized to undergo consolidation with or without GO. Preliminary results evaluating 1,115 patients after induction therapy showed that the addition of GO to induction chemotherapy, regardless of the regimen used, had an improvement in the 3-year relapse rate (37 versus 52 percent) without adding significant extra toxicity and improved 3-year disease-free survival in the GO arm (51 versus 40 percent). Data from the consolidation groups have not yet been reported.

Another study is the SWOG106 trial, which is a prospective trial designed to evaluate GO in induction therapy and as maintenance therapy. In this study patients who have untreated *de novo* AML and who are under the age of 61 are randomized to receive standard induction therapy alone or induction therapy combined with GO. Patients who obtain a CR will then receive consolidation and then be randomized to receive maintenance therapy with GO or not. This trial is still currently accruing patients and no data have been reported to date.

GO has also been tested in patients with relapsed APL. Estey et al.[22] treated 19 *de novo* APL patients with a combination ATRA and GO. Eight courses of ATRA and GO were given as consolidation; 16 patients achieved a clinical CR and 12 of 12 evaluable patients were RT-PCR negative within 2 to 4 months of clinical remission with a median duration of 5 months. In a trial by Lo-Coco et al.,[23] 16 patients with molecular relapsed APL were treated with two doses of GO. Patients who obtained a molecular remission after the second dose then received a third dose of GO. Overall molecular remission (MR) was obtained in 81 percent. Of the 14 responders, 7 (50 percent) remained in sustained MR for a median of 15 months (range, 7 to 31 months). These data indicate that GO is effective as single-agent treatment for patients with molecularly relapsed APL.

In addition to conjugating anti-CD33 antibodies with a toxin, studies using radiolabeled mAbs against CD33 have been promising. These studies are discussed later in this chapter in the section on radiolabeled antibodies.

ANTI-CD44

CD44 is a transmembrane glycoprotein expressed on the surface of normal and leukemic myeloid cells.[24] It mediates the adhesion of these cells to hyaluronan, a component of the extracellular matrix.[25] It is felt that CD44 plays an important role in tumor growth and metastatic potential, and also has some bearing on prognosis. Studies to evaluate CD44 as a cell marker or in serum have shown a correlation with higher levels and poorer outcomes in non-Hodgkin's lymphoma, colon cancer, and head and neck cancer.

The role of CD44 in leukemias is still being evaluated. Studies have shown that antibodies to CD44 alter *in vitro* myelopoiesis of bone marrow cultures. Two monoclonal anti-CD44 antibodies, H90 and A3D8, were found to induce terminal differentiation and inhibit proliferation of leukemic cell lines.[26] The proposed mechanism of action is thought to involve kinase inhibition. A study by Gadhoum et al.[27] showed that CD44 ligation stabilizes the cyclin-dependent kinase inhibitor $p27^{Kip1}$ (p27) protein, resulting in increased inhibition of kinase activity with resultant cell growth arrest. To date, there have been no clinical trials involving humans using these mAbs, but research in this area appears promising.

ANTI-CD123

AML leukemic stem cells have a high expression of the interleukin-3 (IL-3) receptor alpha chain, also called CD123. Lock et al.[28] tested an mAb against CD123 (7G3) in mouse models and human cell lines. This mAb was shown to inhibit leukemic stem cell proliferation and also engraftment of these cells when injected into the bone marrow of mice. Although only in early development, these results demonstrated that an mAb against CD123 could block homing and proliferation of leukemic stem cells in the bone marrow microenvironment.

RADIOLABELED ANTIBODIES

The use of mAbs conjugated with radioisotopes has recently shown promise in the treatment of AML. Radioactive isotopes emit charged particles that cross the cell membrane and cause damage to DNA and other cellular components. This damage results in cell death by apoptosis or necrosis. Through this activity, the radioisotopes improve the cytotoxicity of the mBAs.[29] Studies involving anti-CD33, anti-CD45, and anti-CD66 labeled with radioisotopes, such as iodine-131 (131I), yttrium-90(90Y), rhenium-188 (188Re), and alpha particles, such as bismuth-213 (213Bi) or actinium-225 (225Ac), have demonstrated the ability to deliver significant doses of radiation to target organs with acceptable toxicity.[30]

Radiolabeled Anti-CD33

Several different compounds using anti-CD33 mAbs conjugated with a radioisotope have been investigated. One of the first studies used M195, a murine anti-CD33 mAb, bound to iodine-131. Schwartz et al.[31] performed a phase I trial involving 24 patients with relapsed or refractory AML treated with escalating doses of ^{131}I-M195. In the study, 22 patients had reductions in the percentage of bone marrow blasts, and three achieved CRs. However, profound myelosuppression occurred and resulted in 8 patients requiring bone marrow transplantation. Based on these results, ^{131}I-M195 and the humanized product ^{131}I-HuM195 were tested as part of a conditioning regimen for allogeneic stem cell transplant.[32] In this study, 31 patients with relapsed or refractory AML, accelerated or blastic phase chronic myelogenous leukemia (CML), or advanced MDS were treated with ^{131}I-M195 or ^{131}I-HuM195 followed by busulfan, cyclophosphamide, and allogeneic stem cell infusion. Toxicities were no greater than those seen with the busulfan/cyclophosphamide conditioning regimen alone. Of the 13 evaluable patients, 8 had decreases in bone marrow blasts immediately after treatment with ^{131}I-labeled anti-CD33 antibodies, and three patients remained in CR for 5 to 8 years following transplant. The results of this study are not better than those using the conditioning regimen of busulfan and cyclophosphamide alone, but it does support a novel way to deliver radiation to the bone marrow with very few extramedullary side effects.

Another radioconjugated anti-CD33 mAb compound is ^{90}Y-HuM195. 90Y is a radioactive element that emits β emissions and does not emit any γ radiation. Therefore, patients do not need to be isolated as in those treated with ^{131}I compounds, and 90Y can safely be given in the outpatient setting. Like ^{131}I-HuM195, the dose-limiting toxicity of ^{90}Y-HuM195 in phase I trials was myelosuppression.[33] Currently, this drug is being evaluated as part of a reduced intensity regimen for allogeneic stem cell transplant in CD33-positive leukemias.

A third radiolabeled anti-CD33 conjugate that has shown some promise in the treatment of leukemias is ^{213}Bi-HuM195. This compound uses the α emitting radioisotope bismuth-213. This substance also has the advantage of emitting only short range but high energy radiation; thus, the patients do not have to be isolated and there is less extramedullary toxicity. A phase I study done by Jucic et al.[34] evaluated 17 patients treated with ^{213}Bi-HuM195. All 17 patients developed myelosuppression with a median time to recovery of 22 days. No significant extramedullary toxicity was seen. A phase I/II study using ^{213}Bi-HuM195 after patients were cytoreduced with cytarabine demonstrated a complete response in 2 of the 19 patients treated.[35] Unfortunately, both of these relapsed at 9 and 12 months, respectively. This study demonstrates the antileukemic effect of this drug, but clearly the ideal regimen has not been established. Actinium-225-HuM195, another alpha particle with a longer half-life, is currently under clinical investigation.

Radiolabeled Anti-CD45

The leukocyte common antigen, CD45, has been another active area of monoclonal drug development for the treatment of AML. It is expressed on almost all leukocytes and approximately 90 percent of AML clones. Like CD33, use of monoclonal anti-CD45 alone showed only modest results. As a result, the majority of studies have evaluated anti-CD45 conjugated to radioisotopes. A phase I/II study using ^{131}I-CD45 as part of a conditioning regimen for allogeneic stem cell transplant showed a trend toward lower treatment-related mortality and improved disease-free survival.[36] However, as mentioned earlier, the beta/gamma particle emitting ^{131}-I has limitations in that it requires patient isolation due to significant radioactivity and the dose is limited by extramedullary toxicity. Therefore, the non-gamma-producing isotope ^{90}Y has been studied as an alternative to ^{131}I. A study using nonhuman models showed that ^{90}Y-anti-CD45 antibody was effective in delivering targeted radiation to hematopoietic tissue.[37] Human studies have not been reported to date.

Radiolabeled Anti-CD66

Another area of drug development is using mAbs against CD66 conjugated to radioactive isotopes. The CD66 antigen is expressed on normal myeloid cells but not on AML blasts.[38] Conjugating anti-CD66 antibodies with the radioactive isotope rhenium-188 (^{188}Rh) has been studied as a conditioning regimen in allogeneic stem cell transplant. The rationale behind using ^{188}Rh is that it emits β particles which are necessary to achieve a "crossfire killing" effect since the CD66 mAb does not actually bind to blast cells. However, one of the benefits of ^{188}Rh over ^{131}I is that it emits very few γ particles thereby reducing the amount of patient isolation required during treatment.[39] A phase I/II study using ^{188}Rh anti-CD66 as a stem cell conditioning regimen showed similar results to those using ^{131}I bound to anti-CD45 and anti-CD33.[40] It also demonstrated an effective safety profile. However, like the other radioisotope mAb conjugates, no phase III trials have been done to prove its efficacy compared to standard conditioning regimens.

CONCLUSIONS

Over the past three decades there has been little improvement in the treatment of AML. Targeted immunotherapy with mAbs, either naked or conjugated with toxins or radioisotopes, offers a novel way to improve outcomes in the treatment of AML. Through several mechanisms such as triggering apoptosis, ADDC, complement mediated cytotoxicity, and the generation of reactive oxygen species, mAbs have been shown to be effective in the treatment of certain hematologic malignancies. The potential benefits of such targeted therapy include enhanced cellular cytotoxicity and less systemic toxicity. Leukemia appears to be especially well suited to treatment with mAbs due to the accessibility of the leukemic cells in the blood, bone marrow, spleen, and lymph nodes. In addition, the immunophenotypes of the different types of leukemia are well defined, allowing for the identification of antigen targets. The further development of unconjugated and conjugated mAbs shows great promise in the treatment of acute leukemias.

REFERENCES

1. Cassileth PA, Harrington DP, Appelbaum FR et al. Chemotherapy compared with autologous or allogenic bone marrow transplantation in the management of acute myeloid leukemia in first remission. *N Engl J Med.* 1998;339:1649–56.
2. Jucic JG. Ab therapy of AML: Native and anti-CD33 Ab and drug conjugates. *Cytotherapy.* 2008;10:7–12.
3. Caron PC, Co MS, Bull MK, Avdalovic NM, Scheinberg DA. Biological and immunological features of humanized M195 (anti-CD33) monoclonal antibodies. *Cancer Res.* 1992;52:6761–67.
4. Vitale C, Romagnani C, Puccetti A et al. Surface expression and function of p75/AIRM-1 o CD33 in acute myeloid leukemias: engagement of CD33 induces apoptosis of leukemic cells. *Proc Natl Acad Sci USA*. 2001;98:5764–69.
5. Balaian L, Zhong RK, Ball E et al. The inhibitory effect of anti-CD33 monoclonal antibodies of AML cell growth correlates with Syk and/or ZAP-70 expression. *Exp Hematology*. 2003;5:363–71.
6. Scheinberg DA et al. A Phase I trial of monoclonal antibody M 195 in acute myelogenous leukemia: Specific bone marrow targeting and internalization of radionuclide. *J Clin Oncol*. 1991;9:478–90.
7. Caron PC, Jurcic JG, Scott A et al. A phase IB trial of humanized monoclonal antibody M195 (anti-CD33) in myeloid leukemia: Specific targeting without immunogenicity. *Blood* 1994;83:160–8.
8. Caron PC, Dumont L, Scheinberg DA. Supersaturating infusional humanized anti-CD33 monoclonal antibody HuM195 in myelogenous leukemia. *Clin Cancer Res.* 1998;4:1421–28.
9. Feldman EJ et al. Treatment of relapsed/refractory acute myeloid leukemia (AML) with humanized anti-CD33 monoclonal antibody HuM195. *Leukemia.* 2003;17:314–18.
10. Feldman EJ, Brandwein J, Stone R et al. Phase III randomized multicenter study of a humanized anti-CD33 monoclonal antibody, lintuzumab, in combination with chemotherapy, versus chemotherapy alone in patients with refractory or first-relapsed acute myeloid leukemia. *J Clin Oncol.* 2005;23:4110–16.
11. Raza A, Jurcic JG, Roboz GJ et al. Complete remissions observed in acute myeloid leukemia following prolonged exposure to SNG-33 (lintuzumab), a monoclonal antibody targeting CD33. *Blood.* 2007;110(11) (abstract 159).
12. Jurcic JG, Nimer SD, Scheinberg DA et al. Prognostic significance of minimal residual disease detection and PML/RAR alpha isoform type: Long-term follow up in acute promyelocytic leukemia. *Blood.* 2001;98:2651–56.
13. Jurcic JG, DeBlasio T, Dumont L et al. Molecular remission induction with retinoic acid and anti-CD33 monoclonal antibody HuM195 in acute promyelocytic leukemia. *Clin Cancer Res.* 2000;6:372–80.
14. Mulford DA, Maslak PG, Weiss MA et al. Reducing standard postremission chemotherapy in acute promyelocytic leukemia with risk-adapted therapy. *Blood.* 2003;102 (abstract 2287):619a.
15. Sievers EL, Appelbaum FR, Spielberger RT et al: Selective ablation of acute myeloid leukemia using antibody-targeted chemotherapy: A phase I study of an anti-CD33 calicheamicin immunoconjugate. *Blood.* 1999;93:3678–84.
16. Sievers EL et al. Efficacy and safety of gemtuzumab ozogamicin in patients with CD33-positive acute myeloid leukemia in first relapse. *J Clin Oncol.* 2001;19:3244–54.

17. Larson RA, Sievers EL, Satdtmauer EA et al. Final report of the efficacy and safety of gemtuzumab ozogamicin (GO) in patients with CD33 positive acute myeloid leukemias in first recurrence. *Cancer* 2005;104:1442–52.
18. Amadori S, Suciu S, Stasi R et al. Gentuzumab ozogamicin as single agent treatment for the frail 61 years of age and older with acute myeloid leukemia. *Leukemia.* 2005;19:1768–73.
19. Kell WJ, Burnett AK, Chopra R et al. A feasibility study of simultaneous administration of gemtuzumab ozogamicin with intensive chemotherapy in induction and consolidation in younger patients with acute myeloid leukemia. *Blood.* 2003;102:4277–83.
20. Fianchi L et al. Gemtuzumab ozogamicin, citosine arabinoside, G-CSF combination (G-AraMy) in the treatment of elderly patients with poor-prognosis acute myeloid leukemia. *Ann. Oncol.* 2008;19:128–34.
21. Burnett AK, Kell WJ, Goldstone AH et al. The addition of gemtuzumab ozogamicin to induction chemotherapy for AML improves disease free survival without extra toxicity: Preliminary analysis of 1115 patients in the MRC AML15 Trial [abstract]. *Blood.* 2006;108:8a.
22. Estey E, Giles FJ, Beran M et al. Experience with gemtuzumab ozogamicin and all-*trans* retinoic acid in untreated promyelocytic leukemia. *Blood.* 2002;99:4222–24.
23. LoCoco F et al. Gemtuzumab ozogamicin (GO) as a single agent for molecularly relapsed acute promyelocytic leukemia. *Blood.* 2004;104(7):1995–99.
24. Kansas GS, Miurhead MJ, Dailey MO. Expression of the CD11/CD18, leukocyte adhesion molecule, and CD44 adhesion molecules during normal myeloid and erythroid differentiation in humans. *Blood.* 1990;76:2483–92.
25. Smadja-Joffe F, Legras S, Girard N et al. CD44 and hyaluronan binding by human myeloid cells. *Leuk Lymphoma.* 1996;21:407–20.
26. Charrad RS, Gadhoum Z, Qi J et al. Effects of anti-CD44 monoclonal antibodies on differentiation and apoptosis of human myeloid leukemia cell lines. *Blood.* 2002;99:290–99.
27. Gadhoum Z et al. CD44: A new means to inhibit acute myeloid leukemia cell proliferation via $p27^{Kip1}$. *Blood.* 2004;103:1059–68.
28. Lock RB et al. CD123 (IL-3 receptor α chain) neutralization by a monoclonal antibody selectively eliminates human acute myeloid leukemic stem cells. *Blood* (ASH Annual Meeting Abstracts). 2007;110:161.
29. Jonathan EC, Bernhard EJ, McKenna WG. How does radiation kill cells? *Curr Opin Chem Biol.* 1999;3:77–83.
30. Burke JM and Jurcic JG. Radioimmunotherapy of leukemia. *Adv Pharmacol.* 2004;51:185–208.
31. Schwartz MA et al. Dose escalation trial of M195 labeled with iodine 131 for cytoreduction and marrow ablation in relapsed or refractory myeloid leukemias. *J Clin Oncol.* 1993;11:294–303.
32. Burke JM et al. Cytoreduction with iodine-131-anti-CD33 antibodies before bone marrow transplantation for advanced myeloid leukemias. *Bone Marrow Transplant.* 2003;32:549–56.
33. Jurcic JG et al. Potential for myeloablation with yttrium-90-HuM195 in myeloid leukemia [abstract]. *Proc ASCO.* 2000;19:8a.
34. Jurcic JG et al. Targeted alpha particle immunotherapy for myeloid leukemia. *Blood.* 2002;100:1233–39.
35. Mulford D et al. Sequential therapy with cytarabine and bismuth-213 labeled HuM195 for acute myeloid leukemia. *Blood* (ASH Annual Meeting Abstracts). 2004;104 (abstract 1790).
36. Pagel JM et al. 131I–anti-CD45 antibody plus busulfan and cyclophosphamide before allogeneic hematopoietic cell transplantation for treatment of acute myeloid leukemia in first remission. *Blood.* 2006;107(5):2184–91.

37. Nemecek ER, Hamlin DK, Fisher DR et al. Biodistribution of 90Y-labeled anti-CD45 antibody in a nonhuman primate model. *Clin Cancer Res*. 2005;11:787–94.
38. Bordessoule D, Jones M, Gatter K, Mason D. Immunohistological patterns of myeloid antigens: Tissue distribution of CD13, CD14, CD16, CD31, CD36, CD65, CD66 and CD67. *Br J Haematol*. 1993;83:370–83.
39. Kotzerke J, Glatting G, Seitz U et al. Radioimmunotherapy for the intensification of conditioning prior to stem cell transplantation: Differences in dosimetry and biokinetics of Re-188 and Tc-99m-labeled monoclonal anti NCA-95 antibodies. *J Nucl Med*. 2000;41:531–37.
40. Bunjes D et al. Rhenium 188-labeled anti-CD66 (a, b, c, e) monoclonal antibody to intensify the conditioning regimen prior to stem cell transplantation for patients with high-risk acute myeloid leukemia or myelodysplastic syndrome: results of a phase I-II study. *Blood*. 2001;98(3):565–72.

11 Biologic Therapies for Non-Hodgkin Lymphoma and Hodgkin Disease

Stephen M. Ansell

CONTENTS

KEY WORDS: monoclonal antibody, non-Hodgkin lymphoma, Hodgkin's disease

ABSTRACT

Non-Hodgkin lymphoma (NHL) and Hodgkin lymphoma (HL) are B-cell malignancies that together account for an estimated 75,000 new cases in the United States annually. Many patients with these diseases are successfully treated with cytotoxic therapy; however, the addition of biologic therapy in the form of the anti-CD20 antibody rituximab has significantly improved the outcome of patients, particularly those with B-cell NHL. The success of this agent has led to the development of multiple other monoclonal antibodies directed at targets on lymphoma cells and these therapies are now in clinical trials to determine their utility in the treatment of patients with NHL and HL.

INTRODUCTION

Non-Hodgkin lymphoma (NHL) is the fifth most common type of cancer in the United States.[1] An estimated 60,000 to 65,000 new cases of NHL are diagnosed in the United States annually.[1] Data from the National Cancer Institute have shown the incidence of NHL increased by 3 percent annually in the United States in the last three decades, but seems to have stabilized in the 1990s with some variation in the rates according to age, ethnicity, and gender.[2–4] Currently, approximately half of all patients with aggressive lymphomas may be cured with cytotoxic therapy; however, most indolent lymphomas remain incurable with current therapy.[5] New therapies are needed to more effectively treat patients with NHL.

Approximately 8,000 to 9,000 new cases of Hodgkin lymphoma (HL) are diagnosed annually in the United States.[1] The incidence of this disease has a bimodal pattern with the highest incidence seen in young adults as well as in elderly patients.[6] The selection of the appropriate therapy is based on accurately assessing the stage of disease.[7] Patients with early stage disease are treated with combined modality strategies utilizing abbreviated courses of combination chemotherapy followed by involved-field radiation therapy, while those with advanced stage disease receive a longer course of chemotherapy without radiation therapy. Currently, more than 80 percent of all newly diagnosed patients are expected to be long-term survivors. Although many patients respond well to initial therapies and have durable long-term remissions, a subset of patients have resistant disease and commonly relapse even after aggressive high-dose therapy. Therapeutic options for patients who have relapsed following high-dose therapy and autologous stem cell transplantation are limited and new therapies are clearly needed for these patients.

Significant advances in technology and in the manufacture of monoclonal antibodies have allowed these agents to be developed for clinical use in patients with NHL and HL. Research over the last 20 years has focused on the development of antibodies directed at targets on lymphoma cells and has established that monoclonal antibody therapy of NHL and HL is safe and effective. This chapter highlights the monoclonal antibodies currently in development and clinical use for patients with NHL and HL, and discusses the mechanisms of action of these agents as well as the results of clinical trials using these antibodies in patients with lymphoma. The initial focus is on the anti-CD20 antibody rituximab, with a subsequent discussion of other antibodies directed at targets on the malignant B cell.

RITUXIMAB

Rituximab is a chimeric murine/human monoclonal antibody that binds specifically to the antigen CD20 located on pre-B and mature B lymphocytes.[8,9] It is composed of human immunoglobulin G_1 (IgG_1) and kappa constant regions and variable antigen recognition sites from the murine CD20 monoclonal antibody IDEC-2B8.[10] CD20 is expressed on >90 percent of B-cell NHL[11] and is not found on hematopoietic stem cells, normal plasma cells, or other normal tissues.[12] Studies suggest that CD20 acts as a calcium channel and plays a role in B-cell activation, proliferation, and differentiation. Furthermore, CD20 is a good target for monoclonal antibody therapy as it is not internalized. The Fab domain of rituximab binds to the CD20 antigen on B lymphocytes and the Fc domain recruits immune effector functions to mediate lysis of the B cell.[10]

Mechanisms of Action

The major reason for selecting CD20 as an appropriate target in NHL was the fact that the antigen is expressed at high levels on malignant cells and is not downregulated after antibody binding. *In vitro*, rituximab induces apoptosis,[13,14] complement-dependent cytotoxicity,[15,16] and antibody-dependent cellular cytotoxicity (ADCC)[17,18] in malignant B-cell lines. Most data suggest that the predominant effector mechanism is antibody-dependent cell-mediated cytotoxicity, with a minor role for complement. The efficacy of rituximab has been shown to require mononuclear cells as effector cells for ADCC.[10,15] Recent data have suggested that monocytes are the dominant effector cells for B-cell depletion,[19] but neutrophils have also been shown to function as effector cells in ADCC induced by rituximab.[17–20]

In some but not all lymphoma cell lines evaluated, rituximab binding to CD20 can lead to apoptosis. Currently, there is no consensus regarding the precise pathways that are most important, nor the degree to which this mechanism contributes to the clinical response of lymphoma to rituximab.[13,14] While the exact role of the transmembrane CD20 molecule is not known, it is involved in B-cell differentiation and activation.[12,21] CD20 can act as a calcium channel,[22] either directly or by binding to or activating a calcium channel. Binding by rituximab, especially if cross-linked, initiates a cascade of intracellular signals including the activation of Jun N-terminal (JNK), extracellular signal-regulated (ERK), and p38 mitogen-activated protein kinases (p38MAPK).[23] These signals may play a role, at least in part, in rituximab-mediated cell killing. Rituximab-mediated apoptosis is thought to be a consequence of caspase-3 activation,[24] whereas the FAS ligand/FAS death pathway does not appear to be necessary. The role of bcl-2-dependent pathways remains unclear. One study found that rituximab binding led to a decrease in bcl-2 expression and sensitization to chemotherapy acting via mitochondrial pathways.[25] This was shown to be due to downregulation of interleukin 10 (IL-10), resulting in decreased activation of STAT3 through an IL-10/IL-10R autocrine loop. In contrast, others have shown that bcl-2-expressing as well as bcl-2-nonexpressing cell lines were equally sensitive to anti-CD20 apoptosis,[13] with no significant alterations in bcl-2 levels seen after anti-CD20 treatment.

Intracellular signal transduction may be dependent on the density of CD20, as well as the clustering of CD20, and the amount of antibody present.[14] After CD20 is bound, it rapidly redistributes into a cell membrane fraction called a lipid raft,[26,27] which may aid in signal activation through the src-family tyrosine kinases. These membrane changes may be long lived, and constitute one of several potential explanations for late responses to antibody therapy.

A variety of effector cell populations including monocytes/macrophages, natural killer (NK) cells, and activated granulocytes mediate rituximab-directed cell lysis. Depletion of select effector cell populations including monocytes, NK cells, and granulocytes in animal models decreases the efficacy of rituximab therapy.[28] Also, ADCC is regulated by the selective use of specific Fcγ receptors (FcγR). The antitumor effect is enhanced in mice lacking inhibitory Fcγ RII2B receptors while the antitumor effect is totally absent in mice lacking the FcγR subunit responsible for sending the activation signal to the immune effector cell.[29] A clinical correlative analysis of low affinity FcγRIII CD16 genetic polymorphisms in rituximab-treated patients demonstrated that those who were homozygous for the higher affinity allele of CD16 had significantly higher clinical response rates to rituximab than did subjects who were heterozygotes or homozygous for the lower affinity allele.[30] These data suggest that ADCC is a significant mechanism of action for monoclonal antibodies such as rituximab.

There has been substantial but conflicting evidence regarding the role of complement-mediated cytotoxicity in the antitumor effect of rituximab. *In vitro*, the expression of the complement inhibitory proteins on the target cell correlated with complement-mediated lysis.[15,16] This effect was important for the action of rituximab but was not shown to be important for the action of B1, another anti-CD20 antibody.[31] Furthermore, in an animal model, rituximab was shown to have antitumor effects in wild-type mice, but not in syngeneic knockout mice lacking C1q, a critical component of the complement pathway, suggesting that an intact complement pathway is necessary for antitumor activity of rituximab.[32] Conversely, the expression of inhibitory molecules on cells from patient biopsy specimens does not reliably predict clinical responses to rituximab.[33] At this time, the role of complement in the antitumor activity of rituximab remains unclear.

Clinical Use of Rituximab

Rituximab has become an established component of therapy for patients with B-cell NHL due to clinically significant antitumor effects as well as its ease of administration. In the initial phase I study of rituximab, patients with previously treated B-cell NHL received weekly infusions for 4 weeks. The treatment was well tolerated and approximately half of the patients responded to therapy, the majority of which were partial responses.[34,35] No immunological responses to rituximab, such as human antichimeric antibody responses, were noted. These observations were confirmed in a phase II trial of patients with low grade or follicular NHL.[36] Patients treated in this study had a 48 percent overall response rate (ORR) with a median duration of response of 11.6 months. The main toxicity was infusion-related reactions, with few other toxicities noted. Depletion of B cells from the peripheral blood was not associated with significant hypogammaglobulinemia or infections.

In the phase I trial, no true maximum tolerated dose was reached due to lack of toxicity, so the dose and schedule were selected somewhat empirically and based on the amount of available antibody for use in the trial. Thus, higher doses, more frequent administration, additional doses, and administration earlier in the disease course are issues currently being investigated. Owing to its efficacy and tolerability in relapsed NHL, trials of rituximab as induction therapy have been performed and initial results suggest higher responses than in relapsed patients.[37,38] There have been two studies that enrolled previously untreated patients with advanced stage follicular NHL and treated them with a standard course of rituximab without maintenance. Colombat et al. found an ORR of 73 percent with 27 percent of patients achieving a complete response (CR) or unconfirmed complete response (Cru).[37] Witzig et al. treated 36 patients with the same schedule and found a 72 percent ORR with 36 percent CR.[39] In a similar study by Hainsworth,[40] responding patients or those with stable disease at 6-month intervals received subsequent courses of rituximab for up to 2 years. The eventual ORR was 65 percent, with 27 percent CRs, and the 1 and 2 year progression-free survival rates were 69 and 67 percent, respectively.[38,40] The high response rate in chemotherapy-naive patients observed in these studies should be interpreted cautiously because potential issues such as patient selection and the long-term implications on time to progression and time to next chemotherapy are not yet known. However, these results strongly supported the inclusion of rituximab as upfront therapy in a comparative randomized trial. This randomized study in previously untreated patients that compares re-treatment with rituximab at the time of disease progression to a rituximab maintenance strategy is nearing completion.

Other studies have evaluated the duration of rituximab therapy. In a single arm phase II study utilizing eight weekly doses of rituximab, an ORR of 57 percent was observed.[41] The median duration of response in this study was 13.4+ months. In the above-mentioned phase II study by Hainsworth et al.,[40] subjects who had not progressed after an initial 4-week course of rituximab received additional 4-week courses of rituximab at 6 month intervals for up to 2 years. As outlined above, the results from this trial are also promising. It appears therefore that additional doses or prolonged administration of rituximab results in higher response rates and longer durations of response than those seen with the weekly-for-4-weeks dosing schedule. This is supported by the results of a randomized study comparing no further treatment after 4 weekly doses of rituximab to prolonged rituximab administration (375 mg/m^2 every 2 months four times). At a median follow-up of 35 months, the median event-free survival (EFS) was 12 months in the no further treatment arm versus 23 months in the prolonged treatment arm ($p = 0.02$).[42]

Synergy between rituximab and chemotherapy drugs *in vitro* has led to clinical trials exploring rituximab combined with cytotoxic agents in both indolent and aggressive non-Hodgkin lymphomas. Czuczman et al. treated patients with relapsed or refractory low grade or follicular lymphoma with rituximab and CHOP (cyclophosphamide, doxorubicin, vincristine, prednisone), and found an ORR of 95 percent with CR rate of 55 percent.[43] Toxicity was similar to that seen with CHOP alone. The median progression-free survival was not reached with a median follow-up of 65 months. The combination of CHOP and rituximab has also been examined in trials involving patients with previously untreated aggressive NHL. In a phase II study, patients with previously untreated diffuse large cell and follicular large cell lymphoma were treated with six cycles of rituximab in combination with CHOP chemotherapy.[44] The ORR was 94 and 93 percent of patients who achieved partial response (PR) or CR remained in remission with a median follow-up of 26 months.

Initially, the role of rituximab in combination with chemotherapy was studied in elderly patients with diffuse large B-cell lymphoma. Two phase III studies have been conducted comparing CHOP plus rituximab (R-CHOP) to CHOP alone in elderly patients with aggressive B-cell NHL. In a French multicenter (GELA) study, R-CHOP was superior to CHOP alone with respect to CR rate (75 vs. 63 percent), progression during treatment (9 vs. 22 percent), and 2 year overall survival (70 vs. 57 percent).[45] A similar North American Inter-group study comparing R-CHOP to CHOP was recently reported.[46] Although the administration of rituximab differed from the GELA study, the results from the Inter-group study supported the findings of the GELA study. Similar results were found in a randomized study treating younger patients with chemotherapy with or without rituximab (MinT trial).[47]

In view of the fact that ADCC plays a significant role in the activity of rituximab, agents that can activate immune effector cells are being given to improve the efficacy of rituximab therapy. In a phase II study of rituximab and interferon alpha 2a, the response rate to rituximab plus interferon was approximately the same (45 percent) as other studies of rituximab alone, but the duration of response was longer (22.3 months).[48] A second study using this combination showed similar results with an ORR of 70 percent and a median duration of remission of 19 months.[49]

Agents such as IL-2 and IL-12 have been given in combination with rituximab. IL-2 in combination with rituximab has been shown to be well tolerated, and resulted in increased numbers of effector cells.[50] A phase II clinical trial of this combination in rituximab-refractory indolent lymphoma was subsequently done. The study enrolled 57 patients with 54 evaluable for response; however, only five responses (one complete and four partial) were observed.[51]

The combination of rituximab and IL-12 upregulated gamma interferon and inducible protein-10 (IP-10, CXCL10) expression and increased NK cell lytic activity.[52] A randomized phase II study was done to determine the efficacy of IL-12 in combination with rituximab in patients with B-cell NHL.[53] In the study, 58 patients with relapsed B-cell NHL were randomized to receive concurrent treatment with rituximab and IL-12 (arm A) or rituximab with subsequent treatment with IL-12 after documented nonresponse or progression after rituximab (arm B). The overall response rate was 37 percent (11 of 30) in arm A and 52 percent (13 of 25) in arm B. It was concluded that the concomitant use of IL-12 and rituximab had modest disease activity, but the sequential administration of IL-12 after rituximab did not result in additional clinical responses.

Studies evaluating immunostimulatory CpG oligodeoxynucleotides (ODN) showed that they activate monocytes and NK cells, and induce production of both type 1 interferon and IL-12.[54] CpG ODN also induce upregulation of CD20 by primary malignant B cells,[55] making the combination of this agent and rituximab a promising approach.

Multiple other agents are currently being combined with rituximab for a variety of B-cell diseases. It is becoming clear that rituximab can improve the response rate to therapy and leads to improved long-term clinical outcomes in patients with B-cell lymphoma.

OTHER MONOCLONAL ANTIBODIES IN THE THERAPY OF LYMPHOMA

Rituximab has served as the prototype in the development of monoclonal antibody therapy of NHL. However, other monoclonal antibodies directed at different antigens expressed on lymphoma cells are currently being evaluated in clinical trials (Table 11.1).

OTHER ANTI-CD20 ANTIBODIES

IMMU-106 (hA20)

A new humanized anti-CD20 monoclonal antibody, IMMU-106, has been evaluated to elucidate its action as a therapy for lymphoma. The mechanisms of cytotoxicity of IMMU-106 were found to be similar to rituximab, and included direct apoptosis, antibody-dependent cellular cytotoxicity, and complement-mediated cytotoxicity.[56] IMMU-106 was also found to be very similar to rituximab in terms of antigen-binding specificity, binding avidity, and dissociation constant. In an early phase clinical trial, 68 patients received IMMU-106 at 750 (n = 3), 375 (n = 27), 200 (n = 11), 120 (n = 21), or 80 mg/m^2 (n = 6).[57] IMMU-106 was generally well tolerated, with shorter infusion times at lower doses. In the study, 32 patients with follicular lymphoma were treated and 15 (47 percent) responded with 7 (22 percent) CR/CRus. Another 17 patients with histologies other than follicular lymphoma had 6 (38 percent) responses,

TABLE 11.1
Select Antibodies in Clinical Trials in NHL

Name	Target	Biologic Role of Target
Ofatumumab	CD20	B-cell differentiation/activation
Epratuzumab	CD22	B-cell adhesion/homing/activation
Galiximab	CD80	T–B interaction, costimulation
Siplizumab	CD2	T-cell activation/adhesion
Mapatumumab	TRAIL-R1	Apoptosis
SGN-40/CP-870,893	CD40	T–B interaction, costimulation, survival
MDX-060/SGN-30	CD30	Proliferation/apotosis/tolerance
Atacicept	BLyS	B-cell growth and survival
Bevacizumab/VEGF-Trap	VEGF	Angiogenesis

with 1 CRu in a marginal zone NHL patient. At a median follow-up of 11 months, 9 of 21 patients continue to remain in remission.

Ofatumumab (HuMax-CD20)

Two novel human anti-CD20 antibodies (2F2 and 7D8) were characterized by Teeling et al. and preclinical studies demonstrated greater complement-dependent cytotoxicity with these antibodies than with rituximab.[58] Additionally the fully human structure offered potentially favorable attributes with respect to immunogenicity and pharmacokinetics. From this work, ofatumumab (the HuMax-CD20 human IgG_1 monoclonal antibody 2F2) is undergoing evaluation in phase I/II clinical trials (specifically in follicular lymphoma and chronic lymphocytic leukemia [CLL]). Hagenbeek and colleagues recently reported on 40 patients with follicular lymphoma who were treated with this agent.[59] Subjects were treated with 4 weekly infusions of 300 to 1,000 mg and no dose-limiting toxicity was seen. Of 24 subjects, 12 demonstrated objective responses, including 4 CR and 1 Cru. These data suggest that this antibody is well tolerated and has significant clinical activity.

^{90}Y-Ibritumomab Tiuxetan and ^{131}I Tositumomab

Radioimmunotherapy (RIT) involves the linking of a radionuclide to an antibody to form a radioimmunoconjugate (RIC). In RIT, the goal is to attach a radionuclide with high energy but with a short path length to the antibody. This will focus the radiation on the tumor while sparing the effects of radiotherapy on the nearby normal tissue. Although there are many radionuclides that can be used for RIT, iodine-131 (^{131}I) and yttrium-90 (^{90}Y) have been the ones most commonly used to date. The target antigen has been CD20 in most cases. Over the last 15 years phase I, II, and III trials have been performed with single agent ^{131}I tositumomab or ^{90}Y-ibritumomab tiuxetan. These trials enrolled patients with relapsed disease and adequate bone marrow function with less than 25 percent marrow involvement with NHL. The results showed a high ORR with an acceptable toxicity profile and ^{90}Y-ibritumomab tiuxetan was approved by the Food and Drug Administration (FDA) in 2002 and ^{131}I tositumomab in 2003. Both agents are now commercially available in the United States.

Initial studies involving use of myeloablative and nonmyeloablative doses of ^{131}I tositumomab, an anti-CD20 antibody conjugated with ^{131}I, were done and showed that this agent could be safely administered.[60,61] Trials evaluating ibritumomab tiuxetan, a murine anti-CD20 antibody conjugated with the radioisotope ^{90}Y,[62,63] have demonstrated an ORR and CR rate that was higher than would be expected with rituximab alone, but with durations of response that were similar to those seen with the unlabeled antibody. There has been no normal organ toxicity other than reversible myelosuppression that occurs approximately 6 weeks from the dose of the RIC and lasts several weeks. In a randomized trial comparing ^{90}Y ibritumomab tiuxetan to rituximab alone, the ORR was 80 percent with ^{90}Y ibritumomab tiuxetan and 56 percent with rituximab, with 30 and 16 percent CR rates, respectively.[64] A comparative study of ^{131}I tositumomab and ^{90}Y ibritumomab tiuxetan is being considered.

The current studies with RIT are evaluating the agents in sequence with chemotherapy and stem cell transplantation. There are several phase I/II trials using Bexxar and Zevalin with high-dose chemotherapy and stem cell support that are ongoing. Other studies evaluating RIT in combination with chemotherapy are being performed. A trial of previously untreated, advanced stage follicular lymphoma treated with CHOP followed 4 to 8 weeks later by ^{131}I tositumomab reported an ORR of 90 percent with 67 percent CRs and a 2-year progression-free survival estimate of 81 percent.[65] Due to the significant results seen with ^{90}Y ibritumomab tiuxetan and ^{131}I tositumomab in clinical trials, radiolabeled monoclonal antibodies targeting molecules other than CD20 are also in development.[66–68] Future studies will need to focus on the safety and efficacy of using these agents early in the disease course, and also evaluate the utility of administering multiple doses of RIT.

Myelosuppression associated with these agents has raised concerns about the feasibility and toxicity associated with subsequent therapy administered to patients with progressive disease. Recent publications, however, have suggested that response rates and toxicity associated with subsequent treatments appear no different from that seen after other chemotherapy regimens.[69,70] It should be noted that these were retrospective analyses and this issue will require further prospective study.

Antibodies Targeting Other Cell Surface Molecules

Epratuzumab

This humanized monoclonal antibody binds to CD22, a surface antigen that is B-cell restricted and is expressed by greater than 80 percent of B-cell malignancies.[71] CD22 is rapidly internalized when bound, and can be reexpressed on the cell membrane after modulation. In an initial phase I/II clinical trial, epratuzumab was safely administered weekly for 4 weeks to subjects with CD22-expressing B-cell malignancies.[72] In the trial, 18 percent of patients treated achieved an objective response, including three CRs. All responses were in patients with follicular NHL. A preliminary evaluation of the combination of epratuzumab and rituximab in B-cell NHL has yielded encouraging findings.[73] The combination therapy was well tolerated with infusion-related toxicities comparable to those seen with antibody monotherapy. Objective responses were demonstrated in the majority of patients (66 percent of follicular patients), and the number of patients achieving a CR or CRu (60 percent in follicular patients) appeared to be greater than that expected with rituximab alone.[74] Furthermore, a recent pilot study has also shown that epratuzumab can be safely combined with rituximab and CHOP chemotherapy and the combination (ER-CHOP) appeared effective.[75] However, whether ER-CHOP is superior to R-CHOP will require further comparative clinical trials.

CMC-544

CMC-544 is a CD22-targeted immunoconjugate of calicheamicin and exerts a potent cytotoxic effect against CD22^{+} B-cell lymphoma. In murine models, CMC-544 was efficacious when given either during the early initiation phase as well as during the late established phase of the disease.[76] *In vitro*, the growth-inhibitory activity of CMC-544 when combined with rituximab was additive.[77] In an initial phase I study, 34 previously treated patients with B-cell NHL were enrolled.[78] Significant adverse events included thrombocytopenia, asthenia, and neutropenia. The nadir of the thrombocytopenia was 9 ± 2 days postinfusion and the platelet counts recovered spontaneously. Complete and/or partial responses were observed in the 0.8 mg/m^2 (1 of 3 patients), 1.34 mg/m^2 (3 of 9), 1.8 mg/m^2 (2 of 5), and 2.4 mg/m^2 (2 of 5) cohorts, and in the first 6 patients in the expanded maximum tolerated dose cohort (4 of 6 patients). A phase II study of CMC-544 in combination with rituximab in follicular or diffuse large B-cell NHL patients is in progress.

Apolizumab

Hu1D10 (apolizumab) is a humanized HLA-DR specific antibody that targets the 1D10 antigen on NHL cells. The expression of this class II antigen is not limited to B cells and expression of 1D10 on malignant B cells is present in approximately 50 percent of lymphoma cases. Binding of class II molecules by apolizumab can induce rapid apoptosis mediated by a pathway that appears to be different from that mediated by other antibodies.[79] In a phase I dose-escalation study, increased toxicity was seen at higher dose levels, and pharmacokinetic studies demonstrated marked intersubject variability in the clearance of apolizumab. Interestingly, responses in patients with rituximab-refractory follicular lymphoma were seen. Also, the median time to response was greater than 100 days which is longer than that seen typically with rituximab.[80] Serum from one responding patient contained autologous antilymphoma IgG, suggesting this patient may have developed an active humoral antilymphoma immune response.[81] These studies suggest a possible unique mechanism of action for this antibody. However, the subsequent phase II study failed to confirm the responses seen in the phase I trial[82] and further studies will be necessary to confirm the efficacy of this antibody.

Alemtuzumab

The utility of this monoclonal antibody targeting CD52 on lymphoma cells has been limited due to decreased penetration into nodal sites of disease, low response rates, and infectious complications. In an initial study, 14 percent of patients with B-cell lymphomas achieved a partial remission. Opportunistic infections were diagnosed in 14 percent of patients and 18 percent of patients had bacterial septicemia. Death related to infectious complications occurred in 6 percent of patients treated.[83] In a second study, 2 of the 16 patients with non-bulky NHL achieved a CR and 1 patient achieved a PR. One of two patients with minimal residual NHL achieved a molecular CR. Due to excessive infectious complications observed in the study, the trial was terminated early.[84]

Two recent studies have evaluated alemtuzumab in combination with CHOP chemotherapy for patients with peripheral T-cell lymphoma (PTCL). In the first study, 24 patients with PTCL received CHOP and alemtuzumab was added at 30 mg subcutaneously on day –1.[85] Complete remissions were seen in 17 (71 percent) patients, 1 patient had a partial remission, and 6 had stable/progressive disease. At a median follow-up of 16 months (range, 5 to 42 months), 14 patients were alive, 9 had died from progressive disease, and 1 had died from pneumonia while in CR. The most frequent side effects were grade 4 neutropenia and cytomegalovirus (CMV) reactivation. Major infections were Jacob-Creutzfeldt (JC) virus reactivation, pulmonary invasive aspergillosis, *Staphylococcus* sepsis, and pneumonia. In a second study, 20 patients with newly diagnosed PTCLs were enrolled.[86] The treatment consisted of CHOP chemotherapy plus alemtuzumab (10 mg i.v. on day 1 and 20 mg i.v. on day 2 in the first cycle, then 30 mg i.v. on day 1 in the subsequent cycles) every 3 weeks. In all, 13 complete responses (65 percent) and 3 partial responses (15 percent) were confirmed, giving an overall response rate of 80 percent. The estimated event-free survival at 1 year was 43.3 percent. The most severe hematologic adverse event was grade 4 neutropenia in 18 patients (90 percent). Also, febrile neutropenia was observed in 11 patients (55 percent); 5 patients (25 percent) experienced CMV reactivation, while 3 patients developed CMV diseases, such as pneumonitis or retinitis. There were 2 treatment-related deaths. Based on the high incidence of the adverse infectious and hematologic events, the study was closed after 20 of the planned 43 patients had been enrolled.

Galiximab

CD80 has also recently been identified as a target for monoclonal antibody therapy. In addition to direct apoptosis, ADCC, and complement activation, galiximab may also alter the interaction between malignant B cell and host T cells by blocking the interaction between CD80 and its ligands, CD28 and CTLA4. The preliminary clinical evaluation of this monoclonal antibody in patients with lymphoma suggested that it can be safely administered, and has clinical activity.[87] In an initial clinical trial, 37 patients received galiximab treatment with an overall response rate of 11 percent (two CRs and two PRs).[88] However, nearly half of all patients (49 percent) had a decrease in indicator lesions. A subsequent phase I/II study was conducted to evaluate galiximab in combination with a standard course of rituximab.[89] Patients with follicular NHL who had relapsed or failed primary therapy were enrolled. The overall response rate at the recommended phase II dose of galiximab (500 mg/m^2) was 66 percent: 19 percent CR, 14 percent Cru, and 33 percent PR. The median progression-free survival was 12.1 months. A phase III, randomized study evaluating clinical benefit of rituximab versus the combination has been initiated.

Siplizumab (MEDI-507)

Siplizumab is a humanized monoclonal antibody directed against CD2 that has preclinical efficacy in T-cell leukemia models.[90] The results of two early phase studies of siplizumab in patients with CD2-positive lymphoproliferative disease have been reported.[91,92] The first study of 9 patients included 6 with acute T-cell leukemia (ATL), 2 with large granular lymphocyte leukemia (LGL), and 1 with cutaneous T-cell lymphoma (CTCL).[91] In this study, 1 patient with ATL achieved a PR but treatment was discontinued for CMV antigenemia. In a second study, bi-weekly infusions of MEDI-507 over 3 consecutive days were given at total doses of 0.7, 3.4, or 4.8 mg/kg.[92]. The study enrolled 16 patients: 9 with PTCL, 6 with CTCL, and 1 with NK-LGL. Two responses were observed, one PR in an NK-LGL patient at 3.4 mg/kg and one CR in a PTCL patient at 3.4 mg/kg. MEDI-507 was well tolerated and showed antitumor activity.

Denileukin Diftitox

Denileukin diftitox is a fusion protein combining diphtheria toxin and IL-2 that targets tumor cells expressing the IL-2 receptor. It has efficacy in both T-cell and B-cell NHL and also in both CD25$^+$ and CD25$^-$ lymphomas. Initial studies were done in patients with stage Ib to IVa CTCL. In a phase III study,[93] patients with biopsy-proven CTCL that expressed CD25 on ≥ 20 percent of lymphocytes were assigned to 9 or 18 μg/kg/day of denileukin diftitox administered for 5 consecutive days every 3 weeks for up to eight cycles. Overall, 30 percent of the 71 patients with CTCL treated with denileukin diftitox had an objective response (20 percent PR; 10 percent CR). To also determine the efficacy of denileukin diftitox in PTCL, a further phase II study was done.[94] Patients with CTCL were excluded from this study. Patients received denileukin diftitox 18 μg/kg/day × 5 days every 3 weeks for up to eight cycles. Objective responses (6 CRs, 7 PRs) were achieved in 13 patients (48.1 percent). An objective response was achieved in 8 of 13 patients (61.5 percent) with $CD25^+$ tumors (4 CR/4 PR) and 5 of 11 patients (45.5 percent) with $CD25^-$ tumors (2 CR/3 PR). The median progression-free survival was 6 months (range, 1 to 38+ months).

Similar studies have also been done in patients with B-cell NHL.[95] In patients with relapsed or refractory B-cell NHL, denileukin diftitox was again administered intravenously at a dose of 18 μg/kg once daily for 5 days every 3 weeks, for up to eight cycles. Of the 45 patients assessable for response, 3 CRs (6.7 percent) and eight PRs (17.8 percent) were observed, for an overall response rate of 24.5 percent. For responding patients, the median time to treatment failure was 7 months, with a median follow-up in survivors of 18 months (range, 9 to 28 months), and the projected progression-free survival at 20 months was 24 percent (95 percent CI, 0 to 60 percent). Denileukin diftitox was subsequently combined with rituximab in relapsed and refractory B-cell NHL patients.[96] Of the 38 evaluable patients, 30 (80 percent) were rituximab-refractory. The ORR was 32 percent, with 6 CRs and 6 PRs. The median time to progression for responders was 8 months (range: 2 to 36+ months). The ORR was 55 percent (4 CRs, 2 PRs) in 11/14 patients with rituximab-refractory follicular lymphoma, and 100 percent in the three patients with rituximab-sensitive disease.

Antibodies Targeting Tumor Necrosis Factor (TNF) Family Molecules

Anti-CD30 Antibodies

Two anti-CD30 antibodies are currently being developed for use in anaplastic large cell lymphoma and other CD30-positive lymphomas such as HL. MDX-060, a fully human monoclonal antibody, and SGN-30, a chimeric antibody, have been shown to be well tolerated, and clinical responses to both antibodies were seen primarily in patients with anaplastic large cell lymphoma (ALCL).[97,98] A phase I/II trial of MDX-060 in HL and ALCL has been completed.[97] Among the 72 patients treated, clinical responses were observed in 6; 25 patients had stable disease, including 5 who remained free from progression 1 year after treatment. Similarly, a multicenter phase II study of SGN-30 was conducted treating patients with refractory or recurrent CD30-positive ALCL.[98] The study enrolled 39 patients and 2 CRs and 6 PRs were seen. Both MDX-060 and SGN-30 have demonstrated good tolerability, and clinically meaningful antitumor responses, especially in patients with ALCL patients, have been seen.

Anti-TRAIL Antibodies

TNR-related apoptosis-inducing ligand (TRAIL/Apo2L) is a death protein that preferentially kills tumor cells while sparing normal cells. TRAIL has four exclusive receptors, two of which (TRAIL-R1, TRAIL-R2) are death receptors. Agonistic antibodies to TRAIL-R1 (mapatumumab, HGS-ETR1) and TRAIL-R2 (lexatumumab, HGS-ETR2) are currently being explored for cancer therapy. Both monoclonal antibodies activate caspase-8 and induce cell death in human lymphoma cell lines and primary lymphoma specimens.[99] Mapatumumab is currently in clinical trials and is well tolerated with minimal toxicity. Mapatumumab has been safely administered intravenously every 21 days at doses up to 10 mg/kg, and has clinical activity in patients with relapsed or refractory NHL. In a phase II trial of 40 patients, clinical responses (1 CR and 2 PRs) were observed in 3 patients (8 percent), all of whom had been diagnosed with follicular lymphomas.[100] In addition, stable disease was observed in 30 percent (12/40) of the patients.

Anti-CD40 Antibodies

SGN-40 is a humanized monoclonal antibody that targets the CD40 antigen, which is expressed on most B-cell lineage hematologic malignancies. SGN-40 potently inhibits proliferation and also induces apoptosis in B-cell lymphoma cell lines.[101] Furthermore, it mediates antibody-dependent cytotoxicity in both disseminated and subcutaneous $CD40^+$ xenograft models. In these xenograft models, the antitumor activity of SGN-40 was comparable with that of rituximab. In initial clinical studies,[102] 9 patients with NHL received SGN-40 using an intrapatient dose-loading schedule over 5 weeks. SGN-40 was well tolerated up to 8 mg/kg/week with adverse events generally occurring during the dose-loading segment of therapy rather than at the maximum dose evaluated. In the study, 5 patients achieved an objective antitumor response; 4 patients had PRs, including 2 at 3 mg/kg, 1 at 6 mg/kg, and 1 at 8 mg/kg. In the 4 mg/kg cohort, 1 patient had a CR following one cycle of SGN-40 treatment lasting 20 weeks. Of the 5 objective responses, 4 were in patients with aggressive subtypes of NHL, including 3 with diffuse large B-cell lymphoma and 1 with mantle cell lymphoma. Also, CP-870,893, a fully human and selective CD40 agonist monoclonal antibody, has been tested for safety in a phase I dose-escalation study in solid tumors[103] and may be used in patients with NHL in the future.

Antibodies Targeting BLyS and APRIL

BLyS (B lymphocyte stimulator) and APRIL (a proliferation-inducing ligand) regulate B-cell homeostasis and immunoglobulin production and are overexpressed in B-cell malignancies. These molecules support malignant cell growth and antibodies that bind or block their action are currently in development. Atacicept (TACI-Ig), a recombinant fusion protein that inhibits both BLyS and APRIL,[104] has been administered subcutaneously weekly for 5 weeks to sequential patient cohorts at doses of 2, 4, 7, or 10 mg/kg and was well tolerated at all doses.[105] Treatment with atacicept resulted in dose-dependent decreases in immunoglobulin concentrations. Two patients had stable disease at 8 weeks but all other patients had progressive disease.

An antibody that binds to the BLyS receptor BAFF-R (also called BR3) is also in development,[106] as is a decoy receptor BR3-Fc. The anti-BR3 monoclonal antibody combines the depleting capacity of a therapeutic monoclonal antibody with the blockade of BLyS-BAFF-R signaling that results in B-cell survival. In mice, treatment with the antagonistic anti-BR3 antibodies results in quantitatively greater reduction in some B-cell subsets and qualitatively different effects on bone marrow plasma cells compared with BR3-Fc BAFF blockade or with anti-CD20 treatment. BR3-Fc and the anti-BR3 monoclonal antibody are just beginning clinical testing.

Antibodies Targeting Vascular Endothelial Growth Factor

Bevacizumab

Bevacizumab is a humanized monoclonal antibody directed against vascular endothelial growth factor (VEGF-A). Patients with NHL with high serum VEGF levels have an inferior survival compared to patients with low VEGF levels. Bevacizumab has therefore been combined with R-CHOP in patients with diffuse large cell lymphoma. In a phase II clinical trial,[107] bevacizumab was administered on day 1 followed by rituximab (R) and CHOP on day 2 for cycle 1 and day 1 for 8 cycles 2. The responses seen in this study included 5 CRs and 6 PRs, with an overall response rate of 85 percent and a complete response rate of 38 percent. The 12-month progression-free survival (PFS) was 77 percent and a median follow-up of 16.9 months for the surviving patients. The RA-CHOP combination was felt to be generally well tolerated and safe.

VEGF Trap

VEGF Trap is a fusion protein consisting of portions of the human VEGF receptor VEGFR1 (flt-1) and VEGFR2 (KDR) extracellular domains fused to the Fc portion of human IgG_1. It acts by binding and inactivating VEGF in the circulation and in tissues. The VEGF Trap is designed to have an affinity for VEGF similar to the high-affinity receptor VEGFR1 (1 to 5 pM). In preclinical studies, subcutaneously administered VEGF Trap substantially inhibited the growth of a variety of tumors implanted in mice. In an open-label, dose-escalation phase I study,[108] a single dose of VEGF Trap was given to patients with relapsed or refractory solid tumors or lymphoma followed 4 weeks later by 6 weekly doses of the drug. In all, 14 patients were treated at four dose levels: 25, 50, 100, and 200 μg/kg. No Grade 3 or 4 drug-related toxicity was seen and no anti-VEGF Trap antibodies were detected. Grade 1 and 2 toxicities have included reversible proteinuria, fatigue, and constipation. Stable disease has been noted in 2 patients with solid tumors but the patients with lymphoma did not appear to have clinical benefit from the therapy.

CONCLUSION

As discussed above, many new antibodies hold promise for patients with lymphoma. The challenge will be to determine the place of each antibody in the treatment of NHL. Future studies will need to focus on using the various monoclonal antibodies in combination with each other and with chemotherapy to take advantage of their unique targets and possible different mechanisms of action. We will also need to develop a rational approach to the use of monoclonal antibodies based on their efficacy in patients with lymphoma, potential long-term toxicities, as well as the financial constraints of the heath care system.

REFERENCES

1. Jemal A, Murray T, Ward E, Murray T, Xu J, Thun MJ. Cancer statistics, 2007. *CA Cancer J Clin.* 2007; 57: 43–66.
2. Howe HL, Wingo PA, Thun MJ, Ries LA, Rosenberg HM, Feigal EG, Edwards BK. Annual report to the nation on the status of cancer (1973 through 1998), featuring cancers with recent increasing trends. 2001. *J Natl Canc Inst.* Jun 6: 824–842.
3. Clarke CA, Glaser SL. Changing incidence of non-Hodgkin lymphomas in the United States. 2002. *Cancer.* Apr 1: 2015–2023.
4. Groves FD, Linet MS, Travis LB, Devesa SS. Cancer surveillance series: Non-Hodgkin's lymphoma incidence by histologic subtype in the United States from 1978 through 1995. 2000. *J Natl Canc Inst.* Aug; 2: 1240–1251.
5. Ansell SM, Armitage J. Non-Hodgkin lymphoma: Diagnosis and treatment. *Mayo Clin Proc.* 2005 Aug; 80(8): 1087–1097.
6. Glaser SL, Jarrett RF. The epidemiology of Hodgkin's disease. *Baillieres Clin Haematol* 1996; 9: 401–416.
7. Ansell SM, Armitage JO. Management of Hodgkin lymphoma. *Mayo Clin Proc.* 2006 Mar; 81(3): 419–426.
8. Valentine M A, Meier K E, Rossie S, Clark EA. Phosphorylation of the CD20 phosphoprotein in resting B lymphocytes. *J Biol Chem.* 1989; 264: 11282–11287.
9. Einfeld DA, Brown JP, Valentine MA, Clark EA, Ledbetter JA. Molecular cloning of the human B cell CD20 receptor predicts a hydrophobic protein with multiple transmembrane domains. *EMBO J.* 1988; 7: 711–717.
10. Reff ME, Carner K, Chambers KS, Chinn PC, Leonard JE, Raab R, Newman RA, Hanna N, Anderson DR. Depletion of B cells in vivo by a chimeric mouse human monoclonal antibody to CD20. *Blood.* 1994; 83: 435–445.
11. Anderson K C, Bates M P, Slaughenhoupt B, Schlossman SF, Nadler LM. Expression of human B cell-associated antigens on leukemias and lymphomas: A model of human B cell differentiation. *Blood.* 1984; 63: 1424–1433.
12. Tedder TF, Boyd AW, Freedman AS, Nadler LM, Schlossman SFl. The B cell surface molecule B1 is functionally linked with B cell activation and differentiation. *J Immunol.* 1985; 135: 973–979.
13. Shan D, Ledbetter JA, Press OW. Signaling events involved in anti-CD20-induced apoptosis of malignant human B cells. *Cancer Immunol Immunother.* 2000; 48: 673–683.
14. Hofmeister JK, Cooney D, Coggeshall KM. Clustered CD20 induced apoptosis: Src-family kinase, the proximal regulator of tyrosine phosphorylation calcium influx and caspase 3-dependent apoptosis. *Blood Cells Mol Dis.* 2000; 26: 133–143.
15. Golay J, Zaffaroni L, Vaccari T, Lazzari M, Borleri GM, Bernasconi S, Tedesco F, Rambaldi A, Introna M. Biologic response of B lymphoma cells to anti-CD20 monoclonal antibody rituximab in vitro: CD55 and CD59 regulate complement-mediated cell lysis. *Blood.* 2000; 95: 3900–3908.
16. Harjunpaa A, Junnikkala S, Meri S. Rituximab (anti-CD20) therapy of B cell lymphomas: Direct complement killing is superior to cellular effector mechanisms. *Scand J Immunol.* 2000; 51: 634–641.

17. Wurflein D, Dechant M, Stockmeyer B, Tutt AL, Hu P, Repp R, Kalden JR, van de Winkel JG, Epstein AL, Valerius T, Glennie M, Gramatzki M. Evaluating antibodies for their capacity to induce cell-mediated lysis of malignant B cells. *Cancer Res.* 1998; 58: 3051–3058.
18. Elsasser D, Valerius T, Repp R, Weiner GJ, Deo Y, Kalden JR, van de Winkel JG, Stevenson GT, Glennie MJ, Gramatzki M. HLA class II as potential target antigen on malignant B cells for therapy with bispecific antibodies in combination with granulocyte colony-stimulating factor. *Blood.* 1996; 87: 3803–3812.
19. Uchida J, Hamaguchi Y, Oliver JA, Ravetch JV, Poe JC, Haas KM, Tedder TF. The innate mononuclear phagocyte network depletes B lymphocytes through Fc receptor-dependent mechanisms during anti-CD20 antibody immunotherapy. *J Exp Med.* 2004; 199(12): 1659–1669.
20. Hernandez-Ilizaliturri FJ, Jupudy V, Ostberg J, Oflazoglu E, Huberman A, Repasky E, Czuczman MS. Neutrophils contribute to the biological antitumor activity of rituximab in a non-Hodgkin's lymphoma severe combined immunodeficiency mouse model. *Clin Cancer Res.* 2003 Dec 1; 9(16 Pt 1): 5866–5873.
21. Golay JT, Clark EA, Beverley PC. The CD20 (Bp35) antigen is involved in activation of B cells from the G0 to the G1 phase of the cell cycle. *J Immunol.* 1985 Dec; 135(6): 3795–3801.
22. Bubien JK, Zhou LJ, Bell PD, Frizzell RA, Tedder TF. Transfection of the CD20 cell surface molecule into ectopic cell types generates a Ca^{2+} conductance found constitutively in B lymphocytes. *J Cell Biol.* 1993 Jun; 121(5): 1121–1132.
23. Pedersen IM, Buhl AM, Klausen P, Geisler CH, Jurlander J. The chimeric anti-CD20 antibody rituximab induces apoptosis in B-cell chronic lymphocytic leukemia cells through a p38 mitogen activated protein-kinase-dependent mechanism. *Blood.* 2002 Feb 15; 99(4): 1314–1319.
24. Byrd JC, Kitada S, Flinn IW, Aron JL, Pearson M, Lucas D, Reed JC. The mechanism of tumor cell clearance by rituximab in vivo in patients with B-cell chronic lymphocytic leukemia: evidence of caspase activation and apoptosis induction. *Blood.* 2002 Feb 1; 99(3): 1038–1043.
25. Alas S, Ng CP, Bonavida B. Rituximab modifies the cisplatin-mitochondrial signaling pathway, resulting in apoptosis in cisplatin-resistant non-Hodgkin's lymphoma. *Clin Cancer Res.* 2002 Mar; 8(3): 836–845.
26. Deans JP, Li H, Polyak MJ. CD20-mediated apoptosis: signalling through lipid rafts. *Immunology.* 2002 Oct; 107(2): 176–182.
27. Petrie RJ, Deans JP. Colocalization of the B cell receptor and CD20 followed by activation-dependent dissociation in distinct lipid rafts. *J Immunol.* 2002 Sep 15; 169(6): 2886–2891.
28. van Ojik HH, Bevaart L, Dahle CE, Bakker A, Jansen MJ, van Vugt MJ, van de Winkel JG, Weiner GJ. CpG-A and B oligodeoxynucleotides enhance the efficacy of antibody therapy by activating different effector cell populations. *Cancer Res.* 2003; 63: 5595–5600.
29. Clynes RA, Towers TL, Presta LG, Ravetch JV. Inhibitory Fc receptors modulate in vivo cytoxicity against tumor targets. *Nat Med.* 2000; 6: 443–446.
30. Cartron G, Dacheux L, Salles G, Solal-Celigny P, Bardos P, Colombat P, Watier H. Therapeutic activity of humanized anti-CD20 monoclonal antibody and polymorphism in IgG Fc receptor FcgammaRIIIa gene. *Blood.* 2002; 99: 754–758.
31. Cragg MS, Morgan SM, Chan HT, Morgan BP, Filatov AV, Johnson PW, French RR, Glennie MJ. Complement-mediated lysis by anti-CD20 mAb correlates with segregation into lipid rafts. *Blood.* 2003 Feb 1; 101(3): 1045–1052.

32. Di Gaetano N, Cittera E, Nota R, Vecchi A, Grieco V, Scanziani E, Botto M, Introna M, Golay J. Complement activation determines the therapeutic activity of rituximab in vivo. *J Immunol.* 2003; 171: 1581–1587.
33. Weng WK, Levy R. Expression of complement inhibitors CD46, CD55, and CD59 on tumor cells does not predict clinical outcome after rituximab treatment in follicular non-Hodgkin lymphoma. *Blood.* 2001; 98: 1352–1357.
34. Maloney DG, Grillo-Lopez AJ, White CA, Bodkin D, Schilder RJ, Neidhart JA, Janakiraman N, Foon KA, Liles TM, Dallaire BK, Wey K, Royston I, Davis T, Levy R. IDEC-C2B8 (rituximab) anti-CD20 monoclonal antibody therapy patients with relapsed low-grade non-Hodgkins lymphoma. *Blood.* 1997; 90: 2188–2195.
35. Maloney DG, Grillo-Lopez AJ, Bodkin DJ, White CA, Liles TM, Royston I, Varns C, Rosenberg J, Levy R IDEC-C2B8: Results of a phase I multiple-dose trial in patients with relapsed non-Hodgkin's lymphoma. *J Clin Oncol.* 1997; 15: 3266–3274.
36. McLaughlin P, Grillo-Lopez AJ, Link BK, Levy R, Czuczman MS, Williams ME, Heyman MR, Bence-Bruckler I, White CA, Cabanillas F, Jain V, Ho AD, Lister J, Wey K, Shen D, Dallaire BK. Rituximab chimeric anti-CD20 monoclonal antibody therapy for relapsed indolent lymphoma: Half of patients respond to a four-dose treatment program. *J Clin Oncol.* 1998; 16: 2825–2833.
37. Colombat P, Salles G, Brousse N, Eftekhari P, Soubeyran P, Delwail V, Deconinck E, Haioun C, Foussard C, Sebban C, Stamatoullas A, Milpied N, Boue F, Taillan B, Lederlin P, Najman A, Thieblemont C, Montestruc F, Mathieu-Boue A, Benzohra A, Solal-Celigny P. Rituximab (anti-CD20 monoclonal antibody) as single first-line therapy for patients with follicular lymphoma with a low tumor burden: Clinical and molecular evaluation. *Blood.* 2001; 97: 101–106.
38. Hainsworth JD, Litchy S, Burris HA 3rd, Scullin DC Jr, Corso SW, Yardley DA, Morrissey L, Greco FA. Rituximab as first-line and maintenance therapy for patients with indolent non-Hodgkin's lymphoma. *J Clin Oncol.* 2002; 20(20): 4261–4267.
39. Witzig TE, Vukov AM, Habermann TM, Geyer S, Kurtin PJ, Friedenberg WR, White WL, Chalchal HI, Flynn PJ, Fitch TR, Welker DA. Rituximab therapy for patients with newly diagnosed, advanced-stage follicular grade I non-Hodgkin lymphoma: A phase II trial in the North Central Cancer Treatment Group. *J Clin Oncol.* 2005 Feb 20; 23(6): 1103–1108.
40. Hainsworth JD. First-line and maintenance treatment with rituximab for patients with indolent non-Hodgkin's lymphoma. *Semin Oncol.* 2003 Feb; 30(1 Suppl 2): 9–15.
41. Piro LD, White CA, Grillo-Lopez AJ, Janakiraman N, Saven A, Beck TM, Varns C, Shuey S, Czuczman M, Lynch JW, Kolitz JE, Jain V. Extended Rituximab (anti-CD20 monoclonal antibody) therapy for relapsed or refractory low-grade or follicular non-Hodgkin's lymphoma. *Ann Oncol.* 1999; 10(6): 655–661.
42. Ghielmini M, Schmitz SF, Cogliatti SB, Pichert G, Hummerjohann J, Waltzer U, Fey MF, Betticher DC, Martinelli G, Peccatori F, Hess U, Zucca E, Stupp R, Kovacsovics T, Helg C, Lohri A, Bargetzi M, Vorobiof D, Cerny T. Prolonged treatment with rituximab in patients with follicular lymphoma significantly increases event-free survival and response duration compared with the standard weekly x 4 schedule. *Blood.* 2004; 103(12): 4416–4423.
43. Czuczman MS, Grillo-Lopez AJ, White CA, Saleh M, Gordon L, LoBuglio AF, Jonas C, Klippenstein D, Dallaire B, Varns C. Treatment of patients with low-grade B-cell lymphoma with the combination of chimeric anti-CD20 monoclonal antibody and CHOP chemotherapy. *J Clin Oncol.* 1999; 17(1): 268–276.
44. Vose JM, Link BK, Grossbard ML, Czuczman M, Grillo-Lopez A, Gilman P, Lowe A, Kunkel LA, Fisher RI. Phase II study of rituximab in combination with CHOP chemotherapy in patients with previously untreated, aggressive non-Hodgkin's lymphoma. *J Clin Oncol.* 2001; 19(2): 389–397.

45. Coiffier B, Lepage E, Briere J, Herbrecht R, Tilly H, Bouabdallah R, Morel P, Van Den Neste E, Salles G, Gaulard P, Reyes F, Lederlin P, Gisselbrecht C. CHOP chemotherapy plus rituximab compared with CHOP alone in elderly patients with diffuse large-B-cell lymphoma. *N Engl J Med.* 2002 Jan 24; 346(4): 235–242.
46. Habermann TM, Weller EA, Morrison VA, Gascoyne RD, Cassileth PA, Cohn JB, Dakhil SR, Woda B, Fisher RI, Peterson BA, Horning SJ. Rituximab-CHOP versus CHOP alone or with maintenance rituximab in older patients with diffuse large B-cell lymphoma. *J Clin Oncol.* 2006 Jul 1; 24(19): 3121–3127.
47. Pfreundschuh M, Trumper L, Osterborg A, Pettengell R, Trneny M, Imrie K et al. CHOP-like chemotherapy plus rituximab versus CHOP-like chemotherapy alone in young patients with good-prognosis diffuse large-B-cell lymphoma: A randomised controlled trial by the MabThera International Trial (MInT) Group. *Lancet* Oncol. 2006; 7: 379–391.
48. Davis TA, Maloney DG, Czerwinski DK, Liles TM, Levy R. Anti-idiotype antibodies can induce long-term complete remissions in non-Hodgkin's lymphoma without eradicating the malignant clone. *Blood.* 1998; 92: 1184–1190.
49. Sacchi S, Federico M, Vitolo U, Boccomini C, Vallisa D, Baldini L, Petrini M, Rupoli S, Di Raimondo F, Merli F, Liso V, Tabilio A, Saglio G, Vinci G, Brugiatelli M, Dastoli G; GISL. Clinical activity and safety of combination immunotherapy with IFN-alpha 2a and rituximab in patients with relapsed low grade non-Hodgkin's lymphoma. *Haematologica.* 2001; 86(9): 951–958.
50. Friedberg JW, Neuberg D, Gribben JG, Fisher DC, Canning C, Koval M, Poor CM, Green LM, Daley J, Soiffer R, Ritz J, Freedman AS. Combination immunotherapy with rituximab and interleukin 2 in patients with relapsed or refractory follicular non-Hodgkin's lymphoma. *Br J Haematol.* 2002; 117(4): 828–834.
51. Khan KD, Emmanouilides C, Benson DM Jr, Hurst D, Garcia P, Michelson G, Milan S, Ferketich AK, Piro L, Leonard JP, Porcu P, Eisenbeis CF, Banks AL, Chen L, Byrd JC, Caligiuri MA. A phase 2 study of rituximab in combination with recombinant interleukin-2 for rituximab-refractory indolent non-Hodgkin's lymphoma. *Clin Cancer Res.* 2006 Dec 1; 12(23): 7046–7053.
52. Ansell SM, Witzig TE, Kurtin PJ, Sloan JA, Jelinek DF, Howell KG, Markovic SN, Habermann TM, Klee GG, Atherton PJ, Erlichman C. Phase 1 study of interleukin-12 in combination with rituximab in patients with B-cell non-Hodgkin lymphoma. *Blood.* 2002; 99(1): 67–74.
53. Ansell SM, Geyer SM, Maurer MJ, Kurtin PJ, Micallef IN, Stella P, Etzell P, Novak AJ, Erlichman C, Witzig TE. Randomized phase II study of interleukin-12 in combination with rituximab in previously treated non-Hodgkin's lymphoma patients. *Clin Cancer Res.* 2006 Oct 15; 12(20 Pt 1): 6056–6063.
54. Jahrsdorfer B, Weiner GJ. Immunostimulatory CpG oligodeoxynucleotides and antibody therapy of cancer. *Semin Oncol.* 2003; 30(4): 476–482.
55. Jahrsdorfer B, Hartmann G, Racila E, Jackson W, Muhlenhoff L, Meinhardt G, Endres S, Link BK, Krieg AM, Weiner GJ. CpG DNA increases primary malignant B cell expression of costimulatory molecules and target antigens. *J Leukoc Biol.* 2001; 69(1): 81–88.
56. Stein R, Qu Z, Chen S, Rosario A, Shi V, Hayes M, Horak ID, Hansen HJ, Goldenberg DM. Characterization of a new humanized anti-CD20 monoclonal antibody, IMMU-106, and its use in combination with the humanized anti-CD22 antibody, epratuzumab, for the therapy of non-Hodgkin's lymphoma. *Clin Cancer Res.* 2004 Apr 15; 10(8): 2868–2878.

57. Morschhauser F, Leonard JP, Fayad L, Coiffier B, Petillon M, Coleman M, Horne H, Teoh N, Wegener WA, Goldenberg DM. Low doses of humanized anti-CD20 antibody, IMMU-106 (hA20), in refractory or recurrent NHL: Phase I/II results. 2007 ASCO Annual Meeting Proceedings Part I. *J Clin Oncol*. 2007; 25(18S Jun 20 Suppl): 8032.
58. Teeling JL, French RR, Cragg MS, et al. Characterization of new human CD20 monoclonal antibodies with potent cytolytic activity against non-Hodgkin lymphomas. *Blood*. 2004; 104(6): 1793–1800.
59. Hagenbeek A, Plesner T, Walewski, J et al. A novel fully human anti-CD20 monoclonal antibody, first clinical results from an ongoing phase I/II trial in patients with follicular non-Hodgkin's lymphoma [abstract]. *Ann Oncol*. 2005; 16(S5): abstr 114.
60. Kaminski MS, Zasadny KR, Francis IR, Fenner MC, Ross CW, Milik AW, Estes J, Tuck M, Regan D, Fisher S, Glenn SD, Wahl RL. Iodine-131-anti-B1 radioimmunotherapy for B-cell lymphoma. *J Clin Oncol*. 1996; 14(7): 1974–1981.
61. Liu SY, Eary JF, Petersdorf SH, Martin PJ, Maloney DG, Appelbaum FR, Matthews DC, Bush SA, Durack LD, Fisher DR, Gooley TA, Bernstein ID, Press OW. Follow-up of relapsed B-cell lymphoma patients treated with iodine-131-labeled anti-CD20 antibody and autologous stem-cell rescue. *J Clin Oncol*. 1998; 16(10): 3270–3278.
62. Witzig TE, White CA, Wiseman GA, Gordon LI, Emmanouilides C, Raubitschek A, Janakiraman N, Gutheil J, Schilder RJ, Spies S, Silverman DH, Parker E, Grillo-Lopez AJ. Phase I/II trial of IDEC-Y2B8 radioimmunotherapy for treatment of relapsed or refractory CD20(+) B-cell non-Hodgkin's lymphoma. *J Clin Oncol*. 1999; 17(12): 3793–3803.
63. Witzig TE, White CA, Gordon LI, Wiseman GA, Emmanouilides C, Murray JL, Lister J, Multani PS. Safety of yttrium-90 ibritumomab tiuxetan radioimmunotherapy for relapsed low-grade, follicular, or transformed non-Hodgkin's lymphoma. *J Clin Oncol*. 2003; 21(7): 1263–1270.
64. Witzig TE, Gordon LI, Cabanillas F, Czuczman MS, Emmanouilides C, Joyce R, Pohlman BL, Bartlett NL, Wiseman GA, Padre N, Grillo-Lopez AJ, Multani P, White CA. Randomized controlled trial of yttrium-90-labeled ibritumomab tiuxetan radioimmunotherapy versus rituximab immunotherapy for patients with relapsed or refractory low-grade, follicular, or transformed B-cell non-Hodgkin's lymphoma. *J Clin Oncol*. 2002 May 15; 20(10): 2453–2463.
65. Press OW, Unger JM, Braziel RM, Maloney DG, Miller TP, LeBlanc M, Gaynor ER, Rivkin SE, Fisher RI. A phase 2 trial of CHOP chemotherapy followed by tositumomab/iodine I 131 tositumomab for previously untreated follicular non-Hodgkin lymphoma: Southwest Oncology Group Protocol S9911. *Blood*. 2003; 102(5): 1606–1612.
66. Vose JM, Colcher D, Gobar L, Bierman PJ, Augustine S, Tempero M, Leichner P, Lynch JC, Goldenberg D, Armitage JO. Phase I/II trial of multiple dose 131Iodine-MAb LL2 (CD22) in patients with recurrent non-Hodgkin's lymphoma. *Leuk Lymphoma*. 2000 Jun; 38(1–2): 91–101.
67. Denardo GL, Lamborn KR, Goldstein DS, Kroger LA, Denardo SJ. Increased survival associated with radiolabeled lym-1 therapy for non-Hodgkins lymphoma and chronic lymphocytic leukemia. *Cancer* 1997; 80: 2706–2711.
68. Postema EJ, Raemaekers JM, Oyen WJ, Boerman OC, Mandigers CM, Goldenberg DM, van Dongen GA, Corstens FH. Final results of a phase I radioimmunotherapy trial using (186)Re-epratuzumab for the treatment of patients with non-Hodgkin's lymphoma. *Clin Cancer Res*. 2003; 9(10 Pt 2): 3995S–34002S.
69. Schilder RJ, Witzig TE, Gordon L, Emmanouilides CE, Bartlett NL, Flinn IW, Ugoretz R, Ding E, Czuczman MS, White CA. 90Y ibritumomab tiuxetan (Zevalin®) radioimmunotherapy does not preclude effective delivery of subsequent therapy for lymphoma. *Proc Am Soc Clin Oncol*. 2002.

70. Ansell SM, Ristow KM, Habermann TM, Wiseman GA, Witzig TE. Subsequent chemotherapy regimens are well tolerated after radioimmunotherapy with yttrium-90 ibritumomab tiuxetan for non-Hodgkin's lymphoma. *J Clin Oncol.* 2002; 20(18): 3885–3890.
71. Leung SO, Goldenberg DM, Dion AS, Pellegrini MC, Shevitz J, Shih LB, Hansen HJ. Construction and characterization of a humanized, internalizing, B-cell (CD22)-specific, leukemia/lymphoma antibody, LL2. *Mol Immunol.* 1995; 32(17–18): 1413–1427.
72. Leonard JP, Coleman M, Ketas JC, Chadburn A, Ely S, Furman RR, Wegener WA, Hansen HJ, Ziccardi H, Eschenberg M, Gayko U, Cesano A, Goldenberg DM. Phase I/II trial of epratuzumab (humanized anti-CD22 antibody) in indolent non-Hodgkin's lymphoma. *J Clin Oncol.* 2003; 21(16): 3051–3059.
73. Leonard JP, Coleman M, Matthews JC, et al. Epratuzumab (anti-CD22) and rituximab (anti-CD20) combination immunotherapy for non-Hodgkin's lymphoma: preliminary response data. *Proc Am Soc Clin Oncol.* 2002; 266a.
74. Coleman M, Goldenberg DM, Siegel AB, Ketas JC, Ashe M, Fiore JM, Leonard JP. Epratuzumab: Targeting B-cell malignancies through CD22. *Clin Cancer Res.* 2003; 9(10 Pt 2): 3991S–3994S.
75. Micallef IN, Kahl BS, Maurer MJ, Dogan A, Ansell SM, Colgan JP, Geyer S, Inwards DJ, White WL, Habermann TM. A pilot study of epratuzumab and rituximab in combination with cyclophosphamide, doxorubicin, vincristine, and prednisone chemotherapy in patients with previously untreated, diffuse large B-cell lymphoma. *Cancer.* 2006 Dec 15; 107(12): 2826–2832.
76. DiJoseph JF, Goad ME, Dougher MM, Boghaert ER, Kunz A, Hamann PR, Damle NK. Potent and specific antitumor efficacy of CMC-544, a CD22-targeted immunoconjugate of calicheamicin, against systemically disseminated B-cell lymphoma. *Clin Cancer Res.* 2004 Dec 15; 10(24): 8620–8629.
77. DiJoseph JF, Dougher MM, Kalyandrug LB, Armellino DC, Boghaert ER, Hamann PR, Moran JK, Damle NK. Antitumor efficacy of a combination of CMC-544 (inotuzumab ozogamicin), a CD22-targeted cytotoxic immunoconjugate of calicheamicin, and rituximab against non-Hodgkin's B-cell lymphoma. *Clin Cancer Res.* 2006 Jan 1; 12(1): 242–249.
78. Advani A, Giné E, Gisselbrecht C, Rohatiner A, Rosen S, Smith M, Boni J, Lejeune C, Patel H. Preliminary report of a phase 1 study of CMC-544, an antibody-targeted chemotherapy agent, in patients with B-cell non-Hodgkin's lymphoma (NHL). *Proc Am Soc Hematol.* 2005.
79. Bains SK, Mone A, Yun Tso J, Lucas D, Byrd JC, Weiner GJ, Green JM. Mitochondria control of cell death induced by anti-HLA-DR antibodies. *Leukemia.* 2003; 17(7): 1357–1365.
80. Link BK, Wang H, Byrd JC, Leonard JP, Davis TA, Flinn I, Hall WC, Turner JF, Levitt D, Weiner GJ. Phase I study of Hu1D10 monoclonal antibody in patients with B-cell lymphoma. *Proc Am Soc Clin Oncol.* 2001; 1135.
81. Link BK, Wang H, Byrd JC, Leonard JP, Davis TA, Flinn I, Hall WC, Turner JF, Bowles J, Shannon M, Levitt D, Weiner GJ. Prolonged clinical responses in patients with follicular lymphoma treated on a phase I trial of the anti-HLA-DR monoclonal antibody Remitogen™ (Hu1D10). *Proc Am Soc Hematol.* 2001; 98: 244b.
82. Link BK, Kahl B, Czuczman M, Powell BL, Bartlett N, Leonard JP, Ansell S, Porcu P, Byrd JC, Lazarus HM, Flinn I, Jones GJ, Levitt D, Hall WC, Weiner GJ. A phase II study of Remitogen™ (Hu1D10), a humanized monoclonal antibody in patients with relapsed or refractory follicular, small lymphocytic, or marginal zone/MALT B-cell lymphoma. *Proc Am Soc Hematol.* 2001.

83. Lundin J, Osterborg A, Brittinger G, Crowther D, Dombret H, Engert A, Epenetos A, Gisselbrecht C, Huhn D, Jaeger U, Thomas J, Marcus R, Nissen N, Poynton C, Rankin E, Stahel R, Uppenkamp M, Willemze R, Mellstedt H. CAMPATH-1H monoclonal antibody in therapy for previously treated low-grade non-Hodgkin's lymphomas: A phase II multicenter study. European Study Group of CAMPATH-1H Treatment in Low-Grade Non-Hodgkin's Lymphoma. *J Clin Oncol.* 1998; 16(10): 3257–3263.
84. Khorana A, Bunn P, McLaughlin P, Vose J, Stewart C, Czuczman MS. A phase II multicenter study of CAMPATH-1H antibody in previously treated patients with nonbulky non-Hodgkin's lymphoma. *Leuk Lymphoma.* 2001; 41(1–2): 77–87.
85. Gallamini A, Zaja F, Patti C, Billio A, Specchia MR, Tucci A, Levis A, Manna A, Secondo V, Rigacci L, Pinto A, Iannitto E, Zoli V, Torchio P, Pileri S, Tarella C. Alemtuzumab (Campath-1H) and CHOP chemotherapy as first-line treatment of peripheral T-cell lymphoma: Results of a GITIL (Gruppo Italiano Terapie Innovative nei Linfomi) prospective multicenter trial. *Blood.* 2007 Oct 1; 110(7): 2316–2323.
86. Kim JG, Sohn SK, Chae YS, Cho YY, Yang DH, Lee JJ, Kim HJ, Shin HJ, Chung JS, Cho GJ, Lee WS, Joo YD, Sohn CH, Oh SJ. Alemtuzumab plus CHOP as front-line chemotherapy for patients with peripheral T-cell lymphomas: A phase II study. *Cancer Chemother Pharmacol.* 2007 Jun; 60(1): 129–134.
87. Younes A, Hariharan K, Allen RS, Leigh BR. Initial trials of anti-CD80 monoclonal antibody (galiximab) therapy for patients with relapsed or refractory follicular lymphoma. *Clin Lymphoma*, 2003; 3: 257–259.
88. Czuczman MS, Thall A, Witzig TE, Vose JM, Younes A, Emmanouilides C, Miller TP, Moore JO, Leonard JP, Gordon LI, Sweetenham J, Alkuzweny B, Finucane DM, Leigh BR. Phase I/II study of galiximab, an anti-CD80 antibody, for relapsed or refractory follicular lymphoma. *J Clin Oncol.* 2005 Jul 1; 23(19): 4390–4398.
89. Leonard J, Friedberg J, Younes A, Fisher D, Gordon L, Moore J, Czuczman M, Miller T, Stiff P, Cheson B, Forero-Torres A, Chieffo N, McKinney B, Finucane D, Molina A. A phase I/II study of galiximab (an anti-CD80 monoclonal antibody) in combination with rituximab for relapsed or refractory, follicular lymphoma. *Ann Oncol.* 2007 Jul; 18(7): 1216–1223.
90. Zhang Z, Zhang M, Ravetch JV, Goldman C, Waldmann TA. Effective therapy for a murine model of adult T-cell leukemia with the humanized anti-CD2 monoclonal antibody, MEDI-507. *Blood.* 2003 Jul 1; 102(1): 284–288.
91. Janik JE, Morris JC, Stetler-Stevenson M, Gao WW, O'Hagan D, Moses LS, Taylor M, McEwen C, Hammershaimb LD, Waldmann TA. Phase I trial of siplizumab in CD2-positive lymphoproliferative disease. 2005 ASCO Annual Meeting Proceedings. Part I of II. *J Clin Oncol.* 2005; 23(16S, Jun 1 Suppl): 2533.
92. Casale DA, Bartlett NL, Hurd DD, Foss F, Pro B, Sokol L, Lieberman R, Luz H. A phase I open label dose escalation study to evaluate MEDI-507 in patients with CD2-positive T-cell lymphoma/leukemia. *Blood.* 2006; 108(11): abstr 2727.
93. Olsen E, Duvic M, Frankel A, Kim Y, Martin A, Vonderheid E, Jegasothy B, Wood G, Gordon M, Heald P, Oseroff A, Pinter-Brown L, Bowen G, Kuzel T, Fivenson D, Foss F, Glode M, Molina A, Knobler E, Stewart S, Cooper K, Stevens S, Craig F, Reuben J, Bacha P, Nichols J. Pivotal phase III trial of two dose levels of denileukin diftitox for the treatment of cutaneous T-cell lymphoma. *J Clin Oncol.* 2001 Jan 15; 19(2): 376–388.
94. Dang NH, Pro B, Hagemeister FB, Samaniego F, Jones D, Samuels BI, Rodriguez MA, Goy A, Romaguera JE, McLaughlin P, Tong AT, Turturro F, Walker PL, Fayad L. Phase II trial of denileukin diftitox for relapsed/refractory T-cell non-Hodgkin lymphoma. *Br J Haematol.* 2007 Feb; 136(3): 439–447.

95. Dang NH, Hagemeister FB, Pro B, McLaughlin P, Romaguera JE, Jones D, Samuels B, Samaniego F, Younes A, Wang M, Goy A, Rodriguez MA, Walker PL, Arredondo Y, Tong AT, Fayad L. Phase II study of denileukin diftitox for relapsed/refractory B-cell non-Hodgkin's lymphoma. *J Clin Oncol.* 2004; 22(20): 4095–4102.
96. Dang NH, Fayad L, McLaughlin P, Romaguara JE, Hagemeister F, Goy A, Neelapu S, Samaniego F, Walker PL, Wang M, Rodriguez MA, Tong AT, Pro B. Phase II trial of the combination of denileukin diftitox and rituximab for relapsed/refractory B-cell non-Hodgkin lymphoma. *Br J Haematol.* 2007 Aug; 138(4): 502–505.
97. Ansell SM, Horwitz SM, Engert A, Khan KD, Lin T, Strair R, Keler T, Graziano R, Blanset D, Yellin M, Fischkoff S, Assad A, Borchmann P. Phase I/II study of an anti-CD30 monoclonal antibody (MDX-060) in Hodgkin's lymphoma and anaplastic large-cell lymphoma. *J Clin Oncol.* 2007 Jul 1; 25(19): 2764–2769.
98. Forero A, Bernstein SH, Gopal A, Foss F, Leonard JP, Rosenblatt JD, Bartlett NL, Cheson BD, Epner E, Hernandez-Ilizaliturri FJ, Proctor SJ, Brice P, Tilly H, Haioun C, Jerkeman M, Bosly A, Gianni AM, Holte H, Mahendra P, Uharek L, Lorenz J, Barton J, Sievers E. SGN-30 (anti-CD30 mAb) has a single-agent response rate of 21 percent in patients with refractory or recurrent systemic anaplastic large cell lymphoma (ALCL). *Blood.* 2006; 108(11): abstr 2718.
99. Georgakis GV, Li Y, Humphreys R, Andreeff M, O'Brien S, Younes M, Carbone A, Albert V, Younes A. Activity of selective fully human agonistic antibodies to the TRAIL death receptors TRAIL-R1 and TRAIL-R2 in primary and cultured lymphoma cells: Induction of apoptosis and enhancement of doxorubicin- and bortezomib-induced cell death. *Br J Haematol.* 2005 Aug; 130(4): 501–510.
100. Younes A, Vose J, Zelenetz AD, Czuczman MS, et al. Results of a phase 2 trial of HGS-ETR1 (agonistic human monoclonal antibody to TRAIL receptor 1) in subjects with relapsed/refractory non-Hodgkin's lymphoma (NHL) (ETR1-HM01). 47th Annual Meeting of the American Society of Hematology, 2005.
101. Law CL, Gordon KA, Collier J, Klussman K, McEarchern JA, Cerveny CG, Mixan BJ, Lee WP, Lin Z, Valdez P, Wahl AF, Grewal IS. Preclinical antilymphoma activity of a humanized anti-CD40 monoclonal antibody, SGN-40. *Cancer Res.* 2005 Sep 15; 65(18): 8331-8338. Erratum in: *Cancer Res.* 2006 Feb 15; 66(4): 2495.
102. Forero-Torres A, Furman RR, Rosenblatt JD, et al. A humanized antibody against CD40 (SGN-40) is well tolerated and active in non-Hodgkin's lymphoma (NHL): Results of a phase I study. *ASCO Meeting Abstracts.* 2006; 24: 7534.
103. Vonderheide RH, Flaherty KT, Khalil M, Stumacher MS, Bajor DL, Hutnick NA, Sullivan P, Mahany JJ, Gallagher M, Kramer A, Green SJ, O'Dwyer PJ, Running KL, Huhn RD, Antonia SJ. Clinical activity and immune modulation in cancer patients treated with CP-870,893, a novel CD40 agonist monoclonal antibody. *J Clin Oncol.* 2007 Mar 1; 25(7): 876–883.
104. Munafo A, Priestley A, Nestorov I, Visich J, Rogge M. Safety, pharmacokinetics and pharmacodynamics of atacicept in healthy volunteers. *Eur J Clin Pharmacol.* 2007 Jul; 63(7): 647–656.
105. Ansell SM, Witzig TE, Novak A, Inwards DJ, Porrata L, Ythier A, Ferrande L, Nestorov I, DeVries T, Sievers E. Phase 1 clinical study of atacicept in patients with relapsed and refractory B-cell lymphoma. *Blood* 2006; 108: 770a (abstr 2722).
106. Lin WY, Gong Q, Seshasayee D, Lin Z, Ou Q, Ye S, Suto E, Shu J, Lee WP, Lee CW, Fuh G, Leabman M, Iyer S, Howell K, Gelzleichter T, Beyer J, Danilenko D, Yeh S, Deforge LE, Ebens A, Thompson JS, Ambrose C, Balazs M, Starovasnik MA, Martin F. Anti-BR3 antibodies—A new class of B cell immunotherapy combining cellular depletion and survival blockade. *Blood.* 2007 Aug 8; [Epub ahead of print].

107. Ganjoo KN, An CS, Robertson MJ, Gordon LI, Sen JA, Weisenbach J, Li S, Weller EA, Orazi A, Horning SJ. Rituximab, bevacizumab and CHOP (RA-CHOP) in untreated diffuse large B-cell lymphoma: Safety, biomarker and pharmacokinetic analysis. *Leuk Lymphoma*. 2006 Jun; 47(6): 998–1005.
108. Dupont J, Camastra D, Gordon MS, Mendelson D, Murren J, Hsu A, Lucarelli A, Cedarbaum JM, Spriggs DR. Phase 1 study of VEGF Trap in patients with solid tumors and lymphoma. *Proc Am Soc Clin Oncol*. 2003; 22: abstr 776.

12 Biologic Therapies for Chronic Lymphocytic Leukemia

Mohamed A. Kharfan-Dabaja and Edgardo S. Santos

CONTENTS

KEY WORDS: alemtuzumab, chronic lymphocytic leukemia (CLL), minimal residual disease (MRD), monoclonal antibodies (mAb), rituximab

ABSTRACT

Identification of cancer-specific antigens is changing the therapeutic algorithm of chronic lymphocytic leukemia (CLL). Monoclonal antibodies (mAb), as monotherapy or in combination with conventional chemotherapy, are becoming integral therapies for CLL. Alemtuzumab is a humanized anti-CD52 IgG_1 mAb that is approved by the U.S.

Food and Drug Administration (FDA) in patients with CLL in front-line as well as in fludarabine-refractory CLL also treated with prior alkylator chemotherapy. Rituximab is a mAb that targets CD20, but has limited activity as monotherapy and rarely induces complete remissions (CR) by itself. But, rituximab is feasible to combine with purine analogues, alkylators, and/or alemtuzumab-based regimens resulting in improved responses in CLL. Novel mAbs, currently undergoing clinical trials, are demonstrating activity in patients with relapsed or refractory CLL and other lymphoid malignancies. Modern methods to detect chromosomal abnormalities and biologic and molecular features of CLL are also helping to better prognosticate this disease, and in some cases tailor therapy accordingly. Traditional definitions of response, such as CR or partial responses (PR), are being superseded by novel definitions like minimal residual disease negativity (MRD-negativity). We here review the available body of published literature evaluating alemtuzumab, rituximab, and other newer mAbs across different stages of CLL.

MONOCLONAL ANTIBODIES IN THE TREATMENT OF CLL

Chronic lymphocytic leukemia (CLL) is the most common leukemia diagnosed in the Western world.[1] In 2008, there will be an estimated 15,110 cases of CLL in the United States.[1] In most cases, the disease is indolent and does not require immediate treatment; however, in a third of patients CLL progresses rapidly even with aggressive therapy.[2,3] The emergence of more effective therapies for CLL, including monoclonal antibodies (mAbs), has shifted our treatment goal from symptom palliation to eradication of disease. Development of fluorescence *in situ* hybridization (FISH) and DNA sequencing technology has helped identify chromosomal, biologic, and molecular markers that characterize high-risk CLL sufferers, even at early clinical stages, who may benefit from more aggressive interventions including combination chemoimmunotherapy or allogeneic hematopoietic cell transplantation.[4–9] Alemtuzumab is effective in CLL with associated p53 abnormalities (mutations or deletions), known to confer resistance to chemotherapy, particularly purine analogues and alkylators.[10,11] We hereby provide a comprehensive review of alemtuzumab, rituximab, and other novel biologics in the treatment of CLL.

Alemtuzumab

Alemtuzumab is approved by the U.S. Food and Drug Administration (FDA) for treatment of CLL at initial presentation as well as for fludarabine-refractory CLL previously treated with alkylator chemotherapy. Alemtuzumab is a recombinant humanized mAb that targets CD52, a cell surface protein with hitherto unknown function(s) which is expressed on CLL cells, normal and malignant B and T cells, and natural killer (NK) cells.[12,13] Alemtuzumab is capable of inducing cell death through various mechanisms including antibody-dependent cytotoxicity, complement-dependent cytolysis, and apoptosis, among others.[14–16]

Alemtuzumab in Previously Untreated CLL

The efficacy of alemtuzumab as first-line therapy for CLL has been demonstrated in randomized as well as single-arm clinical trials. Hillmen et al. conducted a randomized, multicenter, international trial comparing alemtuzumab IV versus oral chlorambucil as first-line therapy for patients with CLL.[17] A total of 297 patients were randomly assigned to intravenous (IV) alemtuzumab ($N = 149$) or oral (PO) chlorambucil ($N = 148$). Alemtuzumab was escalated (3 mg → 10 mg → 30 mg) until tolerated at 30 mg three times per week (TIW) for up to 12 weeks. Chlorambucil (PO) was prescribed at a dose of 40 mg/m^2 every 28 days for up to 12 cycles. Patients treated with alemtuzumab were required to receive prophylactic trimethoprin/sulfamethoxazole and famciclovir; but no antimicrobial prophylaxis was required for patients on chlorambucil. The median age of patients receiving alemtuzumab or chlorambucil was 59 (35 to 86) years and 60 (36 to 83) years, respectively. Bulky lymphadenopathy (defined as ≥5 cm) was present in 22.1 percent ($N = 33$) of patients enrolled in the alemtuzumab treatment arm versus 23 percent ($N = 34$) in the chlorambucil group. Presence of chromosomal abnormalities were analyzed by interphase FISH for patients treated with alemtuzumab—17p13.1 (p53) = 7.7 percent; any del 11q = 16.1 percent; trisomy 12 = 16.8 percent; normal = 17.5 percent; sole del 13q = 23.1 percent; others = 18.9 percent—or chlorambucil—17p13.1 (p53) = 7.2 percent; any del 11q = 22.3 percent; trisomy 12 = 7.2 percent; normal = 18.7 percent; sole del 13q = 24.5 percent; others = 20.1 percent. The primary end point of this trial was progression-free survival (PFS). The median length of exposure to alemtuzumab or chlorambucil was 11.7 (0 to 33) weeks and 28.3 (4 to 59) weeks, respectively. Alemtuzumab yielded superior median PFS (14.6 months vs. 11.7 months, $P = 0.0001$), overall response rates (ORR; 83.2 percent vs. 55.4 percent, $P < 0.0001$), complete remission (CR) rates (24.2 percent vs. 2.0 percent, $P < 0.0001$), and minimal residual disease negativity (MRD-negativity) by four-color flow cytometry for CD5/CD19/CD52/κ or λ in peripheral blood (PB) or bone marrow (BM) (7.4 percent vs. 0 percent, $P = 0.0008$). Interestingly, patients with 17p del treated with alemtuzumab had over threefold improved response rates (64 percent vs. 20 percent, $P = 0.0805$) and near fivefold improvement in median PFS (10.7 months vs. 2.2 months, $P = 0.4066$), but failed to reach statistical significance. Conversely, a significant improvement in ORR was seen in cases with 11q del (87 percent vs. 29 percent, $P < 0.0001$) or sole 13q del (91 percent vs. 62 percent, $P = 0.0087$) treated with alemtuzumab. The toxicity profile was different among the groups, particularly in regards to cytomegalovirus (CMV) reactivation (defined as asymptomatic cytomegalovirus (CMV) with PCR positivity) or CMV disease (symptomatic CMV with PCR positivity). The incidence of CMV reactivation (52.4 percent vs. 7.5 percent) and CMV disease (15.6 percent vs. 0 percent) was significantly higher following treatment with alemtuzumab. Nonetheless, 22 of 23 patients with CMV disease received antiviral therapy and all recovered without end organ damage. The authors conclude that alemtuzumab IV improves outcomes in CLL with predictable and manageable toxicity. One may argue that using chlorambucil as the comparator arm is not reflective of the current prescribing trend for patients with CLL because fludarabine-based regimens are nowadays more commonly used as first-line treatment for CLL.

Österborg et al. assessed the efficacy of alemtuzumab as front-line therapy in a small series of patients with progressive CLL.[18] Alemtuzumab was administered IV ($N = 5$) or subcutaneously (SQ) ($N = 4$), TIW, for 18 weeks. ORR was 89 percent (CR = 33 percent) and duration of responses ranged from 8 to more than 24 months. One (11 percent) patient developed CMV disease manifesting as pneumonitis. This study, albeit small in size, also demonstrates efficacy of alemtuzumab in untreated CLL.

A single-arm phase II trial by Lundin et al. evaluated front-line treatment with alemtuzumab SQ in 41 patients with symptomatic, previously untreated CLL.[19] The median age of enrolled patients was 66 (44 to 75) years; two-thirds had B-symptoms and 69 percent were Rai stage III or higher. Alemtuzumab SQ was administered at 30 mg TIW for 18 weeks. Antimicrobial prophylaxis with fluconazole, cotrimoxazole, and valacyclovir, was prescribed until 2 months after completing alemtuzumab. There were only 38 evaluable patients. ORR was 87 percent (CR = 19 percent). CR or nodular PR (nPR), in the BM, occurred in 66 percent, after 18 weeks. Localized skin reactions were seen in 90 percent, but resolved after 2 weeks in most cases. Infectious complications included CMV reactivation in four cases and *Pneumocystis carinii* pneumonia (PCP) in one case. This study shows that treatment with alemtuzumab SQ induces significant responses in patients with previously untreated CLL. Using the SQ route generally requires a longer treatment course compared to the IV route. These studies are summarized in Table 12.1.

TABLE 12.1
Selected Clinical Studies of Alemtuzumab in Patients with Untreated CLL

Author (Ref.)	Study Type	Study Design	N	Outcomes	Toxicities
Hillmen et al.[17]	Randomized, multicenter	Alemtuzumab (IV) vs. Chlorambucil (PO)	$N = 149$ $N = 148$	ORR Alemtuzumab = 83.2% Chlorambucil = 55.4% $P < 0.0001$ CR Alemtuzumab = 24.2% Chlorambucil = 2.0% $P < 0.0001$ PFS (median) in months Alemtuzumab = 14.6 Chlorambucil = 11.7 $P = 0.0001$	CMV reactivation (%) Alemtuzumab = 52.4% Chlorambucil = 7.5% CMV disease (%) Alemtuzumab = 15.6% Chlorambucil = 0%
Österborg et al.[18]	Small series (non-randomized)	Alemtuzumab IV or SQ TIW × 18 weeks	IV = 5 SQ = 4	ORR = 89% CR = 33%	CMV disease = 11%

TABLE 12.1 (CONTINUED)
Selected Clinical Studies of Alemtuzumab in Patients with Untreated CLL

Author (Ref.)	Study Type	Study Design	*N*	Outcomes	Toxicities
Lundin et al.[19]	Phase II (single arm)	Alemtuzumab SQ TIW × 18 weeks	*N* = 41	ORR = 87% CR = 19%	Localized skin reactions = 90% CMV reactivation = 11%
GCLLSG[20]	Randomized	F or FC Followed by alemtuzumab 30 mg IV TIW vs. observation	Alemtuzumab = 11 Observation = 10	MRD-negativity Alemtuzumab = 45.5% Observation = 0% *P* = 0.048 PFS (months) Alemtuzumab = NR Observation = 24.7	CMV reactivation (%) Alemtuzumab = 36%
Montillo et al.[25]	Single arm	Various fludarabine-based regimens → alemtuzumab 10 mg SQ TIW × 6 wks	*N* = 34	MRD-negativity = 56%	CMV reactivation = 53%
O'Brien[26]	Single arm	Various regimens → alemtuzumab 10 mg IV TIW or alemtuzumab 30 mg IV TIW	*N* = 41	Molecular responses = 38%	CMV reactivation = 22%

Abbreviations: *N*, number of patients; ORR, overall response rate; CR, complete remissions; PFS, progression-free survival; TIW, thrice a week; CMV, cytomegalovirus; MRD, minimal residual disease; F, fludarabine; FC, fludarabine and cyclophosphamide.

Alemtuzumab as Consolidation Therapy to Induce Molecular Remission in CLL

Eradication of MRD represents an important step toward improving survival in CLL.[20] A variety of techniques are currently available to assess MRD-negativity in patients with CLL.[21–23] Commonly used assays include multiparametric flow cytometry which targets expression of CLL cell surface markers or PCR-based tests using consensus primer immunoglobulin gene rearrangement. A more sensitive technique that targets patient-specific primers, namely, allele-specific oligonucleotide PCR

(ASO-PCR), is very costly, laborious, and not yet practical for universal use. It is important to note that MRD assays differ in their sensitivities; as a result, treating oncologists should avoid using them interchangeably because measurements may vary among them.

The German CLL Study group (GCLLSG) conducted a randomized, multicenter phase III clinical trial that evaluated the concept of inducing MRD-negativity using alemtuzumab as consolidative therapy after achieving a response to fludarabine-based regimens.[20] A total of 21 patients with a median age of 60 (37 to 66) years completed initial therapy with fludarabine or fludarabine plus cyclophosphamide (FC). Of the patients, 11 were randomized to receive alemtuzumab 30 mg IV TIW for 12 weeks, and 10 patients were assigned to observation. Patients' demographics, among both groups, were similarly distributed for age, gender, disease stage, presence of chromosomal abnormalities, and response to fludarabine-based therapies, among others. Interestingly, alemtuzumab was capable of inducing MRD-negativity, in the PB (45.5 percent vs. 0 percent, $P = 0.048$). Patients treated with alemtuzumab achieved longer PFS (not reached, NR, vs. 24.7 months, $P = 0.036$). However, patients who received alemtuzumab had higher incidence of infections (64 percent vs. 20 percent), including four cases of CMV reactivation. Using alemtuzumab as a consolidative approach to induce MRD-negativity is promising and needs to be further evaluated in larger randomized studies.

Other trials using a similar approach have yielded similar outcomes.[24,25] Montillo et al. treated 9 patients with alemtuzumab SQ, at a lower dose of 10 mg TIW for 6 weeks, after previous response to fludarabine-based therapy.[2] Monoclonal rearrangement of immunoglobulin H (IgH) was still present in all patients after completing fludarabine-based regimens. Patients received prophylactic antimicrobials with cotrimozaxole and acyclovir. MRD-negativity was achieved in 3 (33 percent) patients. These results are interesting because they show that molecular responses are still achievable using lower doses of alemtuzumab for a shorter period of time. A follow-up study in 34 patients resulted in improved CR rates, from 35 percent after fludarabine to 79.4 percent after alemtuzumab-consolidation, and MRD-negativity in 56 percent.[25] Reactivation of CMV was seen in 18 (53 percent), all successfully treated with ganciclovir. Interestingly, using alemtuzumab for consolidation did not prevent hematopoietic stem cell mobilization and subsequent autografting.

O'Brien et al. evaluated different doses of alemtuzumab IV as a treatment for residual disease in 41 patients with CLL with a median age of 60 (44 to 79) years.[26] Of the patients, 24 received alemtuzumab 10 mg IV TIW for 4 weeks and 17 patients 30 mg IV TIW for 4 weeks. ORR was 46 percent. Molecular responses were seen in 11 (38 percent) of 29 evaluable cases using PCR for specific IgVH rearrangement. These studies are summarized in Table 12.1.

Flowers et al. evaluated a regimen of alemtuzumab IV at a dose of 30 mg for 5 days and fludarabine 25 mg/m^2 for 5 days on a 28-day cycle for four cycles in 56 patients, median age of 62 (37 to 86) years, with relapsed or refractory CLL.[27] Patients who achieved less than a CR with four cycles were eligible to receive two additional cycles for a maximum of six cycles. The ORR was 64 percent in 28 evaluable patients. MRD-negativity using four-color flow cytometry was achieved in 8 (47 percent) of 17 assessable patients. These findings demonstrate that combining alemtuzumab and fludarabine induces molecular remissions, even in refractory heavily pretreated patients.

There are several unanswered questions when using alemtuzumab as consolidative therapy to eradicate CLL: (1) What is the optimal dose and length of therapy with alemtuzumab? (2) What is the optimal route of administration (SQ vs. IV)? (3) What is the best technique to measure MRD-negativity?

Alemtuzumab as Salvage Therapy for Relapsed or Refractory CLL

Several studies reviewed below have shown that alemtuzumab is effective in patients with relapsed or refractory CLL.

Single Agent Alemtuzumab

Österborg et al. evaluated alemtuzumab IV in 29 heavily pretreated patients with relapsed or refractory CLL (48 percent had received three or more therapies).[28] The dose of alemtuzumab was escalated to 30 mg TIW for 12 weeks as previously reported.[17] Antimicrobial prophylaxis with acyclovir or trimethoprim/sulfamethoxazole was not mandatory. ORR was 42 percent (CR: 4 percent) with a median duration of response of 12 months. CR rates varied by disease sites; for instance, CR rate in the BM was 36 percent and in the lymph nodes was 7 percent. Neutropenia and thrombocytopenia were reported in 10 and 7 percent, respectively. Infectious complications included herpes simplex virus (HSV) in 11 patients, oral candidiasis in 5 patients, PCP in 2 cases, and bacteremia in 4 patients. Observed responses to alemtuzumab in the salvage setting are encouraging, but treating oncologists should educate patients about increased risks of developing opportunistic infections, particularly in heavily pretreated cases. Antimicrobial prophylaxis with trimethoprim/sulfamethoxazole and acyclovir (or comparable) is imperative to consider.

Keating et al. treated 93 patients with relapsed or refractory CLL who had failed fludarabine, among others therapies, in a multicenter, international study.[29] Of the patients, 55 (59 percent) had Rai stage IV disease. Dose of alemtuzumab was escalated to 30 mg TIW for 12 weeks as previously reported.[17,28] In this case, antimicrobial prophylaxis was mandatory, consisting of trimethoprim/sulfamethoxazole and famciclovir, or equivalent, until 2 months after completing alemtuzumab. ORR, on intent-to-treat, was 33 percent (CR: 2 percent), median time-to-response (TTR) was 1.5 (0.4 to 3.7) months, and median time-to-progression (TTP) was 4.7 months. Overall median survival was 16 months when all patients were analyzed; and 32 months in patients who demonstrated a response to alemtuzumab. More than half of the patients (55 percent) developed at least one infection during the treatment phase. Grade III/IV sepsis was reported in 10 cases. Treatment associated death was 6.5 percent. Interestingly, fewer grade III/IV infections were noted among patients who responded to alemtuzumab (10 percent vs. 36 percent, $P < 0.01$). This study demonstrates the efficacy of alemtuzumab in patients with relapsed or refractory CLL, consistent with a previous study.[28] Serious infections may still occur despite antimicrobial prophylaxis.

An open-label multicenter phase II trial by Rai et al. evaluated the efficacy of alemtuzumab in 24 patients with previously treated CLL.[30] Of the 24 patients, 13 (54 percent) had failed to respond to previous fludarabine, and 6 (25 percent) relapsed within 6 months of fludarabine. Alemtuzumab was administered to a target dose of 30 mg IV TIW for 16 weeks. Antimicrobial prophylaxis was not mandatory in this study. ORR was 33 percent (CR: 0 percent) and the median duration of response was 15.4 months. Median TTR was 3.9 months. Median survival was 35.8 months. Opportunistic infections occurred in 10 (41.7 percent) patients; 5 of 10 patients developed opportunistic infections despite antimicrobial prophylaxis. This study supports alemtuzumab as effective salvage therapy for relapsed CLL. These and other studies[27] are summarized in Table 12.2.

Alemtuzumab is effective in the setting of relapsed or refractory CLL resulting in rapid responses in the PB and BM.[26,28] Responses are generally less remarkable in cases of bulky lymphadenopathy.[26,29,31] It is very important to institute prophylactic antimicrobial therapy.

Feasibility of Combining Alemtuzumab with Chemotherapy and/or Other mAbs

Alemtuzumab is feasible to combine with other mAb such as rituximab. A single-arm, single institution study by Faderl et al. treated 48 patients with relapsed or refractory lymphoproliferative disorders with a combination of alemtuzumab plus rituximab.[32] More than two-thirds of the patients had CLL. Interestingly, 32 (67 percent) patients had received rituximab and 1 (2 percent) had received prior alemtuzumab. Rituximab 375 mg/m^2 was administered weekly for 4 weeks. Alemtuzumab

TABLE 12.2
Selected Clinical Studies of Alemtuzumab as Salvage Therapy in CLL

Author (Ref.)	Study Type	Treatment	*N*	Outcomes
Flowers et al.[27]	Phase II, single arm, multicenter	Alemtuzumab 30 mg IV × 5 days plus Fludarabine 25 mg/m^2 × 5 days	*N* = 56	ORR = 64% MRD-negativity = 47%
Österborg et al.[28]	Phase II, single arm, multicenter	Alemtuzumab 30 mg IV x 12 weeks	*N* = 29	ORR = 42% CR = 4%
Keating et al.[29]	Phase II, single arm, multicenter, international	Alemtuzumab 30 mg IV × 12 weeks	*N* = 93	ORR = 33% CR = 2% Median TTP = 4.7 months
Rai et al.[30]	Phase II, single arm, multicenter	Alemtuzumab 30 mg IV × 16 weeks	*N* = 24	ORR = 33% CR = 0%
Faderl et al.[32]	Phase II, single arm, single institution	Alemtuzumab IV plus rituximab	*N* = 48	ORR = 52% CR = 8%
Nabhan et al.[33]	Pilot trial, single arm	Alemtuzumab IV plus rituximab	*N* = 12	ORR = 9%
Wierda et al.[34,35]	Single arm, single institution	Alemtuzumab plus FCR	*N* = 78	ORR = 65% CR = 24%

Abbreviations: ORR, overall response rate; MRD, minimal residual disease; CR, complete remission; TT, time-to-progression; FCR, fludarabine, cyclophosphamide, and rituximab.

was escalated (3 mg → 10 mg → 30 mg) on three consecutive days during the first week, followed by 30 mg on days 3 and 5 of weeks 2, 3, and 4. Additional cycles of therapy were allowed based on responses or permissible toxicities. Antimicrobial prophylaxes were instituted, and continued for 2 months after completing treatment. ORR was 52 percent (CR = 8 percent, nPR = 4 percent). The median TTP was 6 (1 to 20) months. Median survival was higher for patients with clinical responses (11+ months vs. 6 months). CMV reactivation was reported in 27 percent. This study shows that combining alemtuzumab and rituximab is feasible and responses are achieved even in patients who have failed each mAb as single therapy.

Nabhan et al. also demonstrated feasibility to combine alemtuzumab and rituximab in 12 patients, median age 69.5 (53 to 73) years, with relapsed or refractory CLL.[33] In this study, all patients received prior alkylator chemotherapy and failed fludarabine-containing regimens. The majority of patients (75 percent) had advanced disease (Rai stage IV). Patients were excluded, however, if they received alemtuzumab or rituximab within 6 months of enrollment. Rituximab 375 mg/m^2 was prescribed weekly and alemtuzumab 30 mg IV TIW for 4 weeks. This study failed to demonstrate significant responses, showing an ORR of only 9 percent.

Alemtuzumab has also been demonstrated to be feasible to combine with a potent chemoimmunotherapy combination of fludarabine, cyclophosphamide, and rituximab (FCR) in 78 patients with heavily pretreated CLL.[34,35] The regimen consisted of fludarabine IV 25 mg/m^2/day (days 3 to 5), cyclophosphamide IV 250 mg/m^2/day (days 3 to 5), rituximab 375 to 500 mg/m^2 (day 2), and alemtuzumab IV 30 mg TIW; and was repeated every 4 weeks for a total of six cycles. This combination yielded an impressive ORR of 67 percent (CR = 22 percent). Responses were more pronounced in CLL cases with low β2-microglobulin and without high-risk chromosomal abnormalities. Major infections still occur despite antimicrobial prophylaxis. These studies are summarized in Table 12.2.

Alemtuzumab and p53 Mutations

Novel genetic and biologic prognostic factors vary in their ability to predict specific clinical outcomes. Patients with CLL with associated 17p del, through p53 mutations or deletions, are likely to be resistant to fludarabine and akylator chemotherapy.[36–38] Alemtuzumab is effective therapy in patients with CLL with associated p53 mutations or deletions who have failed prior chemotherapy regimens. Studies have shown ORR ranging from 40 to 53.8 percent.[39–42] These findings suggest a role for screening for p53 abnormalities, at initial diagnosis of CLL, to guide therapy accordingly to avoid preventable toxicities.

Rituximab

Rituximab is a genetically engineered chimeric murine/human IgG$_1$ kappa anti-CD20 mAb that is approved by the FDA for treatment of relapsed or refractory low-grade or follicular B-cell non-Hodgkin lymphoma (NHL), as first-line treatment for diffuse large B-cell (DLBC) NHL in combination with cyclophosphamide, doxorubicin, vincristine, and prednisone (CHOP) or other anthracycline-based chemotherapy regimens, and as first-line treatment of low-grade or follicular B-cell NHL either combined with cyclophosphamide, vincristine, and prednisone (CVP) or following CVP. Rituximab has become a cornerstone in the management of B-cell malignancies. ORRs using rituximab as monotherapy are usually less remarkable in patients with CLL,[43–47] even at high doses,[48] than in B-cell follicular NHL (10 to 15 percent vs. 40 to 60 percent). This may be explained in part by the lower density of CD20 expression in CLL cells.[43–49]

Rituximab as Part of Chemoimmunotherapy Regimens

Rituximab has also been shown to be feasible to combine with conventional chemotherapy, resulting in improved responses in patients with previously untreated, relapsed, or refractory CLL.

Untreated CLL

Investigators from the Cancer and Leukemia Group B (CALGB) evaluated a regimen of fludarabine and rituximab (FR), comparing a concurrent (FR→R) versus a sequential (F→R) approach in 104 patients with untreated symptomatic CLL.[50] Fludarabine IV was administered at a dose of 25 mg/m²/day for 5 days every 4 weeks for six cycles concurrently with rituximab 375 mg/m² on days 1 and 4 during the first cycle, and on day 1 of subsequent cycles, followed by rituximab consolidation ($N = 51$); or a sequential fludarabine IV at same dose and schedule for six cycles followed by rituximab consolidation (F→R) ($N = 53$). Rituximab was well tolerated in both arms. ORRs were better with the concurrent regimen—90 percent (CR: 47 percent) vs. 77 percent (CR: 28 percent)—but 2-year PFS was 70 percent for each regimen. Neutropenia (grade III/IV) was higher in the concurrent arm (37 percent vs. 18 percent).

Egle et al. evaluated a novel approach consisting of a chemoimmunotherapy combination of fludarabine, cyclophosphamide, and rituximab (FCR) for three cycles followed by three cycles of FR, and subsequent maintenance therapy with rituximab every 3 months for 2 years in 43 patients, median age of 63 (36 to 81) years, with previously untreated CLL.[51] The primary end point was CR rate. Secondary end points included ORR, toxicity, and MRD assessment using four-color flow cytometry. A planned interim analysis showed a CR rate, in 23 evaluable patients, of 83 percent after three cycles of FCR; increased to 94 percent, in 6 evaluable patients, after completion of three cycles of FR. MRD-negativity was reported in 10 (59 percent) of 17 evaluated patients after completion of FR. Moreover, 12 evaluable cases achieved CR after 6 months of maintenance therapy. Grade III/IV hematologic toxicities were observed in 43 percent. Grade III infections were reported in 13 percent. One case of fatal herpes was reported 1 year after initiation of therapy. This study demonstrates that limiting cyclophosphamide to three cycles only does not compromise efficacy. But, toxicities, hematologic and nonhematologic, remain significant despite the deintensification.

A phase II study by Keating et al. evaluated FCR in 224 patients with progressive or advanced chemonaïve CLL.[52] Rituximab 375 mg/m² was prescribed on day 1 of first cycle and increased to 500 mg/m² on additional cycles for a total of six cycles.

ORR was 95 percent (CR = 70 percent; nPR = 10 percent). Grade III/IV neutropenia occurred in 52 percent, but only 2.6 percent resulted in major infections. These results demonstrate that FCR is an active regimen in patients with untreated symptomatic CLL.

In a novel approach to reduce toxicity of FCR, Tarhini et al. evaluated an FCR-lite regimen in 50 patients with untreated CLL with a median age of 58 (36 to 85) years.[53] Treatment consisted of fludarabine IV 20 mg/m^2/day for 3 days, cyclophosphamide IV 150 mg/m^2/day for 3 days, and rituximab 500 mg/m^2 (on days 1 and 14) every 4 weeks. Rituximab 500 mg/m^2 was administered as maintenance therapy every 3 months until progression. Of 50 enrolled patients, 40 were evaluable for response. ORR in 40 subjects was 100 percent (CR = 85; PR = 15 percent). Authors conclude that FCR-lite is effective in untreated CLL. Preliminary results suggest that hematologic toxicities are less pronounced when compared with historical toxicities seen with standard FCR, but the sample size is relatively small to draw significant conclusions.

A phase II by Kay et al. evaluated rituximab in combination with pentostatin, and cyclophosphamide in 64 patients with symptomatic untreated CLL.[54] Rituximab 100 mg/m^2 was prescribed on day 1, and 375 mg/m^2 on days 3 and 5 during the first cycle. Rituximab 375 mg/m^2 was used for subsequent cycles. ORR was 91 percent (CR = 41 percent; nPR = 22 percent). Poor-risk factors such as overexpression of CD38, ZAP-70, unmutated IgVH, and 11q del did not have an effect on responses with PCR regimen. These and other studies[55,56] are summarized in Table 12.3.

Rituximab plus Chemotherapy in Relapsed or Refractory CLL

Wierda et al. treated 177 patients with CLL, who had received a median of two prior therapies (fludarabine-refractory = 21 percent), with FCR.[57] The ORR was 73 percent (CR = 25 percent; nPR = 16 percent; PR = 32 percent). CR rates were less pronounced in those patients previously refractory to fludarabine (6 percent vs. 33 percent). Overall survival (OS) at 4 years was 50 percent. This shows that FCR is effective in relapsed or refractory CLL.

Lamanna et al. evaluated the combination of pentostatin 4 mg/m^2, cyclophosphamide IV 600 mg/m^2, and rituximab 375 mg/m^2 on day 1 of cycles 2 to 6, for six cycles, in 32 patients with relapsed or refractory CLL.[58] ORR was 75 percent (CR = 25 percent; nPR = 3 percent; PR = 47 percent). No distinction in ORR (75 percent) was reported for previously fludarabine-refractory patients ($N = 8$). Median survival was 44 months; 3-year OS was 100 percent for patients in CR or nPR, versus 28 percent for patients in PR or those without response. Incidence of grade III/IV infections was 28 percent. Combining rituximab, pentostatin, and cyclophosphamide in relapsed/refractory CLL is feasible and yields significant responses. This PCR regimen is also feasible to combine with mitoxantrone (PCRM) in patients previously treated with PCR-regimen or FCR.[59] ORR, in 16 evaluable patients, was 94 percent (CR = 25 percent; PR = 69 percent).

In 2008, the FDA approved bendamustine for first-line treatment of CLL. Bendamunstine is also feasible to combine with rituximab (BR) for relapsed CLL.[60] Fischer et al.[60] treated 81 cases of relapsed or refractory CLL with bendamustine 70 mg/m^2 IV on days 1 and 2, and rituximab 375 mg/m^2 during the first cycle, and

TABLE 12.3
Selected Clinical Studies of Rituximab plus Chemotherapy in Untreated CLL

Author (Ref.)	Study Type	Treatment	*N*	Outcomes
CALGB[50]	Phase II, randomized, multicenter	FR → R vs. F → R	FR → R = 51 F → R = 53	ORR (%) FR → R = 90% F → R = 77% CR(%) FR → R = 47% F → R = 28%
Egle et al.[51]	Phase II, non-randomized	FCRx 3 → FR × 3 → rituximab q3 months for 24 months	*N* = 43 *N* = 23 (evaluable)	CR = 94% MRD-negativity = 59%
Keating et al.[52]	Phase II, single arm, single institution	FCR	*N* = 224	ORR = 95% CR = 70% nPR = 10%
Tarhini et al.[53]	Phase II, single arm, single institution	FCR-Lite	*N* = 50 *N* = 40 (evaluable)	ORR = 100% CR = 85% PR = 15%
Kay et al.[54]	Phase II, single arm, multicenter	PCR	*N* = 64	ORR = 91% CR = 41% nPR = 22%
Bosch et al.[55]	Phase II, single arm	R-FCM → maintenance rituximab	*N* = 69 Evaluable = 39	ORR = 92%. MRD-negativity = 36%
Faderl et al.[56]	Phase II, single arm, single institution	FCM-R	*N* = 31 Treated = 30 Evaluable = 29	ORR = 97% CR = 41% MRD-negativity = 34%

Abbreviations: CALGB, Cancer and Leukemia Group B; *N*, number of patients; F, fludarabine; R, rituximab; FR, fludarabine plus rituximab; ORR, overall response rate; CR, complete remission; FCR, fludarabine, cyclophosphamide, and rituximab; MRD, minimal residual disease; nPR, nodular partial response; FCR-Lite, F 20 mg/m^2 d1 → 3 q 4 wk; C150 mg/m^2 d1 → 3 q 4 wk; R 500 mg/m^2 d1 and d 14 q 4 wks; maintenance R 500 mg/m^2 x1 q 3 months until progression; PCR, pentostatin, cyclophosphamide, and rituximab; R-FCM, rituximab, fludarabine, cyclophosphamide, and mitoxantrone, followed by maintenance rituximab q3 months × 2 years; FCM-R, fludarabine, cyclophosphamide, mitoxantrone plus rituximab.

500 mg/m^2 during subsequent 28-day cycles for six courses. ORR in 23 evaluable patients was 65 percent (CR = 13 percent; PR = 52 percent). Reponses were observed in patients with 11q del (4 of 5 patients); but none with 17p del. This study shows that BR is an active regimen in relapsed or refractory CLL. These studies are summarized in Table 12.4.

TABLE 12.4
Selected Studies of Rituximab Plus Chemotherapy in Relapsed or Refractory CLL

Author (Ref.)	Study Type	Treatment	*N*	Outcomes
Wierda et al.[57]	Phase II, single arm, single institution	FCR	*N* = 177	ORR = 73% CR = 25% 4-year OS = 50%
Lamanna et al.[58]	Phase II, single arm, multicenter	PCR	*N* = 32	ORR = 75% CR = 25% nPR = 3%
Lamanna et al.[59]	Phase II, single arm, single institution	PCRM	*N* = 16 (evaluable)	ORR = 94% CR = 25% PR = 69%
Fischer et al. (GCLLSG)[60]	Phase II, single arm, multicenter	BR	*N* = 81 *N* = 23 (evaluable)	ORR = 65% CR = 13%

Abbreviations: *N*, number of patients; FCR, fludarabine, cyclophosphamide, and rituximab; ORR, overall response rate; CR, complete remission; nPR, nodular partial response; OS, overall survival; PCR, pentostatin, cyclophosphamide, and rituximab; PCRM, pentostatin, cyclophosphamide, rituximab, and mitoxantrone; GCLLSG, German Chronic Lymphocytic Leukemia Study Group; BR, bendamustine plus rituximab.

Other Monoclonal Antibodies

New monoclonal antibodies that target CD20 or other surface antigens are being evaluated for the treatment of CLL and other lymphoid neoplasms.

Ofatumomab (HuMax-CD20)

Ofatumumab is a fully human mAb that targets a distinct small loop epitope on the CD20 molecule. The safety and efficacy of ofatumumab was evaluated in a three-cohort, multicenter dose-escalating study of 33 patients with relapsed or refractory CLL.[61] Patients in the first cohort (*N* = 3) received one 100 mg and three 500 mg doses; patients in the second cohort (*N* = 3) received one 300 mg and three 1,000 mg doses; and those in the third cohort (*N* = 27) received one 500 mg and three 2,000 mg doses. Two-thirds of patients were Binet stage B with a median of three previous treatments. The maximum tolerated dose (MTD) was not reached. Most of the related adverse events occurred during the first infusion, decreasing with subsequent infusion. The incidence of infections was 53 percent. ORR for patients treated on the third cohort was 50 percent (nPR = 4 percent; PR = 46 percent). These encouraging findings support future studies of ofatumumab in CLL.

Epratuzumab

Epratuzumab (hLL2), is a 90 to 95 percent humanized anti-CD22 mAb with antitumor effect mostly through antibody-dependent cytotoxicity.[62,63] Clinical studies support a synergistic role for combining epratuzumab with rituximab (ER) in various lymphoid malignancies.[64,65] Strauss et al. evaluated ER in 65 patients with $CD20^+$ lymphoid malignancies (follicular (FL) = 34; DLBC = 15; CLL/SLL = 4; others = 12).[64] ORR, on 64 assessable patients, was 47 percent. Patients with FL exhibited the best responses (ORR = 64 percent, CR = 24 percent). None of the CLL/SLL ($N = 4$) responded to ER.[64] On the other hand, a recent study by Leonard et al. using the same combination (ER) showed an ORR of 57 percent ($N = 4/N = 7$) in patients with CLL/SLL.[65] Further studies combining epratuzumab, rituximab, and conventional chemotherapies are needed to better establish a role for epratuzumab in patients with CLL.

IMMU-106

IMMU-106 is a humanized anti-CD20 mAb with mechanisms of cytotoxicity similar to rituximab, including direct apoptosis, antibody-dependent cellular cytotoxicity, and complement-mediated cytotoxicity.[66] Preclinical studies in the SU-DHL-6 tumor cell line suggest that combining IMMU-106 and epratuzumab appears to be more effective than either mAb alone.[66]

Lumiliximab

Lumiliximab, a macaque-human primatized mAb, targets CD23 which has limited expression on activated B cells including CLL.[67] Byrd et al. conducted a phase I study evaluating the role of lumiliximab in 46 cases, median age 62 (47 to 80) years, with refractory CLL.[67] There were no CR or PR observed, but 17 (52 percent) of 33 patients had a decrease in lymph node size and 42 (91 percent) of 46 patients had a decrease in lymphocytosis. These improvements were transient, with most patients progressing within 2 months.[67]

Apolizumab (Hu1D10)

Mone et al. demonstrated that apolizumab is capable of inducing apoptosis concurrent with activation of the AKT survival pathway in primary CLL cells.[68] A phase II open label clinical trial evaluating the efficacy of apolizumab in patients with CLL is currently ongoing.

Anti-CD40 Antibodies

HCD122

The CD40 antigen is expressed on early B cells. HCD122 is a fully humanized IgG_1 antibody capable of binding the CD40 receptor and inhibiting CD40-ligand-mediated survival signals. A standard design phase I study by Byrd et al. treated 14 patients, median age of 65 (41 to 74) years, with relapsed/refractory (median of 3 (2 to 11) prior therapies) CLL to evaluate MTD, pharmacokinetics, and toxicity.[69] Patients were treated at three different doses: (1) 0.3 mg/kg ($N = 3$), (2) 1 mg/kg ($N = 4$), and (3) 3 mg/kg ($N = 7$). Dose limiting toxicity (DLT) occurred at 3 mg/kg after two infusions.

SGN-40

SGN-40 is another anti-CD40 antibody that is cytotoxic against CLL.[70] Advani et al. reported preliminary results of a phase I trial using SGN-40 in various $CD20^+$

lymphoid malignancies treated in various cohorts at doses ranging from 2 to 6 mg/kg/week for 4 weeks.[71] Four (67 percent) of six patients treated with planned infusions at 2 mg/kg/week did not report grade IV toxicities. Preliminary results suggest antitumor activity at the 2 mg/kg dose.

CONCLUSIONS

Recent introduction of mAbs, particularly alemtuzumab, to the therapeutic armamentarium against CLL has shifted treatment goals from palliation of symptoms to eradication of CLL. Other mAbs like rituximab appear to have limited activity against CLL when used as a single agent, even above standard doses. However, rituximab is feasible to combine with purine analogues, alkylators, alemtuzumab, and other newer mAbs resulting in improved clinical and molecular responses.

Better understanding of genetic, biologic, and molecular markers and their incorporation into risk-stratification strategies in CLL are helping recognize patients who may benefit from more specific therapies, and avoid preventable toxicities from less useful treatments. One such case is exemplified by using alemtuzumab in patients with CLL with associated p53 abnormalities known to be resistant to purine analogues and alkylators.

Development of mAbs remains a research area of high priority in this era of targeted therapies. The future will continue to bring effective biologics from the bench to the bedside which will continue to improve outcomes and overall prognosis of CLL and other neoplasms.

REFERENCES

1. Jemal A, Siegel R, Ward E, Hao Y, Xu J, Murray T, Thun MJ. 2008. Cancer statistics, 2008. *CA Cancer J Clin* 58(2): 71–96.
2. Rai KR, Sawitsky A, Cronkite EP, Chanana AD, Levy RN, Pasternack BS. 1975. Clinical staging of chronic lymphocytic leukemia. *Blood* 46(2): 219–34.
3. Binet JL, Auquier A, Dighiero G, et al. 1981. A new prognostic classification of chronic lymphocytic leukemia derived from a multivariate survival analysis. *Cancer* 48(1): 198–206.
4. Döhner H, Stilgenbauer S, Benner A, et al. 2000. Genomic aberrations and survival in chronic lymphocytic leukemia. *N Engl J Med* 343(26):1910–16.
5. Damle RN, Wasil T, Fais F, et al. 1999. Ig V gene mutation status and CD38 expression as novel prognostic indicators in Chronic Lymphocytic Leukemia. *Blood* 94 (6): 1840–47
6. Crespo M, Bosch F, Villamor N, et al. 2003. ZAP-70 Expression as a Surrogate for Immunoglobulin-Variable-Region Mutations in Chronic Lymphocytic Leukemia. *N Engl J Med* 348:1764–75.
7. Oscier D, Gardiner AC, Mould SJ, et al. 2002. Multivariate analysis of prognostic factors in CLL: clinical stage, IGVH gene mutational status, and loss or mutation of p53 gene are independent prognostic factors. *Blood* 100(4):1177–84.
8. Dreger P, Corradini P, Kimby E, et al. 2007. Indications for allogeneic stem cell transplantation in chronic lymphocytic leukemia: the EBMT transplant consensus. *Leukemia* 21: 12–17.

9. Kharfan-Dabaja MA, Anasetti C, Santos ES. 2007. Hematopoietic cell transplantation for chronic lymphocytic leukemia: an evolving concept. *Biol Blood Marrow Transplant* 13: 373–85.
10. Dohner H, Fischer K, Bentz M, et al. 1995. p53 gene deletion predicts for poor survival and non-response to therapy with purine analogs in chronic B-cell leukemias. *Blood* 85(6): 1580–89.
11. El Rouby S, Thomas A, Costin D, et al. 1993. p53 mutation in B-cell chronic lymphocytic leukemia is associated with drug resistance and is independent of MDR1/MDR3 gene expression. *Blood* 82(11): 3452–59.
12. Hale G, Dyer MJ, Clark MR, et al. 1988. Remission induction in non-Hodgkin lymphoma with reshaped human monoclonal antibody CAMPATH-1H. *Lancet* 2(8625): 1394–99.
13. Treumman A, Lifely MR, Schneider P, Ferguson MA. 1995. Primary structure of CD52. *J Biol Chem* 270: 6088–99.
14. Bindon CI, Hale G, Waldmann H. 1988. Importance of antigen specificity for complement-mediated lysis by monoclonal antibodies. *Eur J Immunol* 18(10): 1507–14.
15. Dyer MJ, Hale G, Hayhoe FG, Waldmann H. 1989. Effects of CAMPATH-1 antibodies in vivo in patients with lymphoid malignancies: influence of antibody isotype. *Blood* 73(6):1431–39.
16. Stanglmaier M, Reis S, Hallek M. 2004. Rituximab and alemtuzumab induce a non-classic caspase-independent apoptotic pathway in B-lymphoid cell lines and in chronic lymphocytic leukemia cells. *Ann Hematol* 83: 634–45.
17. Hillmen P, Skotnicki AB, Robak T, et al. 2007. Alemuzumab compared with chlorambucil as first-line therapy for chronic lymphocytic leukemia. *J Clin Oncol* 25(35): 5616–23.
18. Österborg A, Fassas AS, Anagnostopoulos A, Dyer MJ, Catovsky D, Mellstedt H. 1996. Humanized CD52 monoclonal antibody Campath-1H as first line treatment in chronic lymphocytic leukaemia. *Br J Haematol* 93(1): 151–53.
19. Lundin J, Kimby E, Björkholm M, et al. 2002. Phase II trial of subcutaneous anti-CD52 monoclonal antibody alemtuzumab (Campath-1H) as first-line treatment for patients with B-cell chronic lymphocytic leukemia (B-CLL). *Blood* 100(3): 768–73.
20. Wendtner C-M, Ritgen M, Schweighofer CD, et al. 2004. Consolidation with alemtuzumab in patients with chronic lymphocytic leukemia (CLL) in first remission- experience on safety and efficacy within a randomized multicenter phase III trial of the German CLL Study Group (GCLLSG). *Leukemia* 18(6): 1093–1101.
21. Rawstron AC, Kennedy B, Evans PA, et al. 2001. Quantitation of minimal disease levels in chronic lymphocytic leukemia using a sensitive flow cytometric assay improves the prediction of outcome and can be used to optimize therapy. *Blood* 98(1): 29–35.
22. Pfitzner T, Reiser M, Barth S, et al. 2002. Quantitative molecular monitoring of residual tumor cells in chronic lymphocytic leukemia. *Ann Hematol* 81(5): 258–66.
23. Rawstron AC, Villamor N, Ritgen M, et al. 2007. International standardized approach for flow cytometric residual disease monitoring in chronic lymphocytic leukaemia. *Leukemia* 21(5): 1–9.
24. Montillo M, Cafro AM, Tedeschi A, et al. 2002. Safety and efficacy of subcutaneous Campath-1H for treating residual disease in patients with chronic lymphocytic leukemia responding to fludarabine. *Haematologica* 87(7): 695–700.
25. Montillo M, Tedeschi A, Miqueleiz S, et al. 2006. Alemtuzumab as consolidation after a response to fludarabine is effective in purging residual disease in patients with chronic lymphocytic leukemia. *J Clin Oncol* 24(15): 2337–42.
26. O'Brien SM, Kantarjian HM, Thomas DA, et al. 2003. Alemtuzumab as treatment for residual disease after chemotherapy in patients with chronic lymphocytic leukemia. *Cancer* 98(12): 2657–63.

27. Flowers C, Rosenthal H, Brown J, et al. 2007. Analysis of Minimal Residual Disease (MRD) from the Phase 2 Multicenter Study of Subcutaneous (SC) Alemtuzumab Combined with Fludarabine for Treatment of Relapsed/Refractory B-Cell Chronic Lymphocytic Leukemia (CLL). *Blood* 110 (abstr # 3111): 914A.
28. Österborg A, Dyer MJ, Bunjes D, et al. 1997. Phase II multicenter study of human CD52 antibody in previously treated chronic lymphocytic leukemia. *J Clin Oncol* 15(4): 1567–74.
29. Keating MJ, Flinn I, Jain V, et al. 2002. Therapeutic role of alemtuzumab (Campath-1H) in patients who failed fludarabine: results of a large international study. *Blood* 99(10): 3554–61.
30. Rai KR, Freter CE, Mercier RJ, et al. 2002. Alemtuzumab in previously treated chronic lymphocytic leukemia patients who also had received fludarabine. *J Clin Oncol* 20(18): 3891–97.
31. Fiegl M, Falkner A, Hopfinger G, et al. 2006. Routine clinical use of alemtuzumab in patients with heavily pretreated B-cell chronic lymphocytic leukemia: a nation-wide retrospective study in Austria. *Cancer* 107(10): 2408–16.
32. Faderl S, Thomas DA, O'Brien S, et al. 2003. Experience with alemtuzumab plus rituximab in patients with relapsed and refractory lymphoid malignancies. *Blood* 101 (9): 3413–15.
33. Nabhan C, Patton D, Gordon LI, et al. 2004. A pilot trial of rituximab and alemtuzumab combination therapy in patients with relapsed and/ or refractory chronic lymphocytic leukemia (CLL). *Leuk Lymphoma* 45 (11): 2269–73.
34. Wierda WG, O'Brien S, Ferrajoli A, et al. 2005. Salvage therapy with combined cyclophosphamide, fludarabine, alemtuzumab, and rituximab (CFAR) for heavily pre-treated patients with CLL. *Blood* 106 (11): (abs. # 719)
35. Wierda W, O'Brien S, Faderl S, et al. 2006. Combined cyclophosphamide, fludarabine, alemtuzumab, and rituximab (CFAR) is active for heavily pretreated patients with CLL. *Blood* 108(11): (Abs. # 32)
36. El Rouby S, Thomas A, Costin D, et al. 1993. p53 gene mutation in B-cell chronic lymphocytic leukemia is associated with drug resistance and is independent of MDR1/ MDR3 gene expression. *Blood* 82(11): 3452–59.
37. Döhner H. 1995. Fischer K, Bentz M, et al. p53 gene deletion predicts for poor survival and non-response to therapy with purine analogs in chronic B-cell leukemias. *Blood* 85(6): 1580–89.
38. Sturm I, Bosanquet AG, Hermann S, Güner D, Dörken B, Daniel PT. 2003. Mutation of p53 and consecutive selective drug resistance in B-CLL occurs as a consequence of prior DNA-damaging chemotherapy. *Cell Death Differ* 10(4): 477–84.
39. Stilgenbauer S, Döhner H. 2002. Campath-1H-iinduced complete remission of chronic lymphocytic leukemia despite p53 gene mutation and resistance to chemotherapy. *N Engl J Med* 347(6): 452–53.
40. Lozanski G, Heerema NA, Flinn IW, et al. 2004. Alemtuzumab is an effective therapy for chronic lymphocytic leukemia with p53 mutations and deletions. *Blood* 103 (9): 3278–81.
41. Osuji NC, Del Giudice I, Matutes E, Wotherspoon AC, Dearden C, Catovsky D. 2005. The efficacy of alemtuzumab for refractory chronic lymphocytic leukemia in relation to cytogenetic abnormalities of p53. *Haematologica* 90(10): 1435–36.
42. Stilgenbauer S, Winkler D, Krober A, et al. 2004. Subcutaneous campath-1H (alemtuzumab) in fludarabine-refractory CLL: interim analysis of the CLL2h study of the German CLL Study Group (GCLLSG). *Blood* 104: (Abs. # 478).
43. Maloney DG, Grillo-López AJ, White CA, et al. 1997. IDECC2B8 (Rituximab) anti-CD20- monoclonal antibody therapy in patients with relapsed low-grade non-Hodgkin's lymphoma. *Blood* 90(6): 2188–95.

44. O'Brien SM, Kantarjian H, Thomas DA, et al. 2001. Rituximab dose-escalation trial in chronic lymphocytic leukemia. *J Clin Oncol* 19(8): 2165–70.
45. Byrd JC, Murphy T, Howard RS, et al. 2001. Rituximab using a thrice weekly dosing schedule in B-cell chronic lymphocytic leukemia and small lymphocytic lymphoma demonstrates clinical activity and acceptable toxicity. *J Clin Oncol* 19(8): 2153–2164.
46. Hainsworth JD, Litchy S, Barton JH, et al. 2003. Single-agent rituximab as first-line and maintenance treatment for patients with chronic lymphocytic leukemia or small lymphocytic lymphoma: a phase II trial of the Minnie Pearl Cancer Research Network. *J Clin Oncol* 21(9):1746–51.
47. Huhn D, von Schilling C, Wilhelm M, et al. 2001.Rituximab therapy of patients with B-cell chronic lymphocytic leukemia. *Blood* 98(5): 1326–31.
48. O'Brien SM, Kantarjian H, Thomas DA, et al. 2001. Rituximab dose-escalation trial in chronic lymphocytic leukemia. *J Clin Oncol* 19(8): 2165–70.
49. McLaughlin P, Grillo-López AJ, Link BK, et al. 1998. Rituximab chimeric anti-CD20 monoclonal antibody therapy for relapsed indolent lymphoma: Half of patients respond to a four-dose treatment program. *J Clin Oncol* 16(8): 2825–33.
50. Byrd JC, Peterson BL, Morrison VA, et al. 2003. Randomized phase 2 study of fludarabine with concurrent versus sequential treatment with rituximab in symptomatic,untreated patients with B-cell chronic lymphocytic leukemia: results from Cancer and Leukemia Group B 9712 (CALGB 9712). *Blood* 101(1): 6–14.
51. Egle A, Weiss L, Russ G, et al. 2007. Planned first efficacy and safety analysis of de-intensified induction with fludarabine, cyclophosphamide plus rituximab (FCR) followed by fludarabine plus rituximab (FR) and remission maintenance therapy with rituximab in previously untreated B-chronic lymphocytic leukemia (B-CLL): The Chairos Study. *Blood* 110(11): (abs. # 2045).
52. Keating MJ, O'Brien S, Albitar M, et al. 2005. Early results of a chemoimmunotherapy regimen of fludarabine, cyclophosphamide, and rituximab as initial therapy for chronic lymphocytic leukemia. *J Clin Oncol* 23(18): 4079–88.
53. Tarhini AA, Land S, Pietragallo L, et al. 2007. Final results of lower dose fludarabine (F) and cyclophosphamide (C), and high dose rituximab (R), (FCR-Lite) for patients with untreated chronic lymphocytic leukemia (CLL). *Blood* 110(11): (abst# 2037).
54. Kay NE, Geyer SM, Call TG, et al. 2007. Combination chemoimmunotherapy with pentostatin, cyclophosphamide, and rituximab shows significant clinical activity with low accompanying toxicity in previously untreated B chronic lymphocytic leukemia. *Blood* 109(2): 405–11.
55. Bosch F, Muntañola A, Villamor N, et al. 2007. Preliminary results of the combination rituximab, fludarabine, cyclophosphamide and mitoxantrone (R-FCM) followed by rituximab maintenance in previously untreated chronic lymphocytic leukemia (CLL). *Blood* 110(11): (abs. # 626).
56. Faderl S, Wierda W, Ferrajoli A, et al. 2007. Update of experience with fludarabine, cyclophosphamide, mitoxantrone plus rituximab (FCM-R) in frontline therapy for chronic lymphocytic leukemia (CLL). *Blood* 110(11): (abs# 627).
57. Wierda W, O'Brien S, Wen S, et al. 2005. Chemoimmunotherapy with fludarabine, cyclophosphamide, and rituximab for relapsed and refractory chronic lymphocytic leukemia. *J Clin Oncol* 23(18): 4070–78.
58. Lamanna N, Kalaycio M, Maslak P, et al. 2006. Pentostatin, cyclophosphamide, and rituximab is an active,well-tolerated regimen for patients with previously treated chronic lymphocytic leukemia. *J Clin Oncol* 24(10): 1575–81.
59. Lamanna N, Heaney ML, Brentjens RJ, Jurcic JG, Weiss MA. 2007. Pentostatin, cyclophosphamide, rituximab, and mitoxantrone (PCRM): a new highly active regimen for patients with chronic lymphocytic leukemia (CLL) previously treated with PCR or FCR. *Blood* 110(11): (abs.# 3115).

60. Fischer K, Stilgenbauer S, Schweighofer CD, et al. 2007. Bendamustine in combination with rituximab (BR) for patients with relapsed chronic lymphocytic leukemia (CLL): a multicentre phase II trial of the German CLL study group (GCLLSG). *Blood* 110(11): (abs. # 3106).
61. Coiffier B, Lepretre S, Pedersen LM, et al. 2008. Safety and efficacy of ofatumumab, a fully human monoclonal anti-CD20 antibody, in patients with relapsed or refractory B-cell chronic lymphocytic leukemia: a phase 1-2 study. *Blood* 11(3): 1094–1100.
62. Leung SO, Goldenberg DM, Dion AS, et al. 1995. Construction and characterization of a humanized, internalizing, B-cell (CD22)-specific, leukemia/lymphoma antibody, LL2. *Mol Immunol* 32: 1413–27.
63. Stein R, Qu Z, Chen S, et al. 2004. Characterization of a new humanized anti-CD20 monoclonal antibody, IMMU-106, and its use in combination with the humanized anti-CD22 antibody, epratuzumab, for the therapy of non-Hodgkin's lymphoma. *Clin Cancer Res* 10: 2868–78.
64. Strauss SJ, Morschhauser F, Rech J, et al. 2006. Multicenter phase II trial of immunotherapy with the humanized anti-CD22 antibody, epratuzumab, in combination with rituximab, in refractory or recurrent non-Hodgkin's lymphoma. *J Clin Oncol* 24(24): 3880–86.
65. Leonard JP, Schuster SJ, Emmanouilides C, et al . 2007. Durable Complete Responses Following Therapy with Epratuzumab Plus Rituximab: Final Efficacy Results of a Multicenter Study in Recurrent Indolent Non-Hodgkin's Lymphoma (NHL). *Blood* 110: (abs.# 3419).
66. Stein R, Qu Z, Chen S, et al. 2004. Characterization of a New Humanized Anti-CD20 Monoclonal Antibody, IMMU-106, and Its Use in Combination with the Humanized Anti-CD22 Antibody, Epratuzumab, for the Therapy of Non-Hodgkin's Lymphoma. *Clin Cancer Res* 10(8): 2868–78.
67. Byrd JC, O'Brien S, Flinn I, et al. 2007. Phase 1 Study of Lumiliximab with Detailed Pharmacokinetic and Pharmacodynamic Measurements in Patients with Relapsed or Refractory Chronic Lymphocytic Leukemia. *Clin Cancer Res* 13: 4448–55.
68. Mone AP, Huang P, Pelicano H, et al. 2004. Hu1D10 induces apoptosis concurrent with activation of the AKT survival pathway in human chronic lymphocytic leukemia cells. *Blood* 103(5): 1846–54.
69. Byrd JC, Flinn IW, Khan KD, et al. 2006. Pharmacokinetics and pharmacodynamics from a first-in-human phase 1 dose escalation study with antagonist anti-CD40 antibody, HCD122 (formerly CHIR-12.12), in patients with relapsed and refractory chronic lymphocytic leukemia. *Blood* 108(11): (abs. # 2837).
70. Gowda AC, Zhao XB, Cheney C, et al. 2005. Humanized anti CD-40 antibody SGN-40 effectively induces cytotoxicity against chronic lymphocytic leukemia (CLL) cells through antibody mediated cytotoxicity and demonstrates modest biologic evidence of CD40 activation. *Blood* 106(11):832a (abs. # 2966).
71. Advani RH, Furmann RR, Rosenblatt JD, et al . 2005. A phase I study of humanized anti-CD40 immunotherapy with SGN-40 in non-Hodgkin's lymphoma. *Blood* 106(11):433a (abs.# 1504).

13 Biologic Therapies for Multiple Myeloma and Plasma Cell Disorders

Shuji Ozaki

CONTENTS

KEY WORDS: multiple myeloma, monoclonal antibody therapy, antibody engineering

ABSTRACT

Multiple myeloma (MM) is a hematological malignancy characterized by the accumulation of neoplastic plasma cells mainly in the bone marrow. This disease causes clinical symptoms resulting from end-organ damage such as anemia, destructive bone lesions, renal impairment, and immunodeficiency. The introduction of chemotherapy and hematopoietic stem cell transplantation has improved overall survival, but unfortunately, MM remains incurable in most patients. Basic and clinical studies have provided new insights into the mechanism of MM pathogenesis within MM cells and the relationship between the tumor and the bone marrow microenvironment. Novel therapeutic agents such as thalidomide, lenalidomide, and bortezomib have been shown to significantly improve patient outcomes. These agents have multiple biologic activities against both MM cells and the bone marrow microenvironment, and combination therapy has emerged as a promising treatment paradigm. Additionally, numerous efforts have been made to identify fundamental molecular targets and establish more specific therapy by using biologic agents such as cytokines, monoclonal antibodies (mAbs), and vaccines. This review summarizes the key issues of mAb therapy and the recent advances in the development of therapeutic mAbs for MM and plasma cell disorders.

CONSTRUCTION OF THERAPEUTIC MONOCLONAL ANTIBODIES

Since Köhler and Milstein have reported that homogenous antibody could be prepared by the fusion of mouse myeloma cells to spleen cells derived from immunized mice,[1] numerous hybridomas were established to obtain monoclonal antibodies (mAbs) for specific antigens. These mAbs are able to bind to tumor cells selectively *in vivo*, and were expected to be a "magic bullet" for cancer therapy. However, mouse mAbs induced human antimouse antibody (HAMA) response, which limited the efficacy of the administered mAb in clinical settings. Therefore, additional approaches were necessary for the establishment of mAb therapy.

Subsequently, advances in molecular biology and antibody technology have been enabling the design of antibody-based reagents for specific purposes in cancer diagnosis and therapy.[2,3]

First, to minimize the immunogenicity and enhance the efficacy in human use, mouse antibodies were engineered to mouse/human chimeric antibodies by grafting of the variable regions to the human constant regions (Figure 13.1). As a next step, humanized antibodies were constructed by grafting of the complementarity-determining regions (CDR) to the human antibody framework. Moreover, fully human antibodies were developed by the use of transgenic animals or phage-display technology.[4] These mAbs exert Fc-mediated antitumor activities such as antibody-dependent cell-mediated cytotoxicity (ADCC) and complement-dependent cytotoxicity (CDC).

Second, mAbs were designed as immunoconjugates to deliver cytotoxic agents such as chemotherapeutic drugs, toxins, enzymes, and radionuclides.[5,6] These immunoconjugates specifically bind to tumor cells with highly active agents and induce tumor cell death independently on the host immune functions.

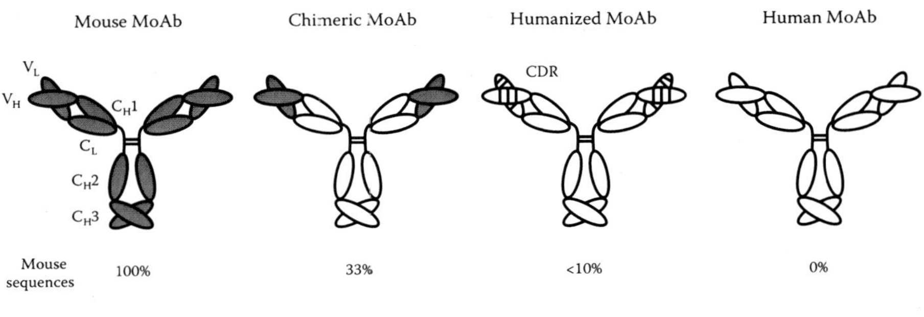

FIGURE 13.1 ***A color version of this figure follows page 176.*** Structures of therapeutic mAbs. Mouse antibodies are obtained by hybridoma technology following immunization of mice. Mouse/human chimeric antibodies are derived by the variable domains (VH and VL) of the parent mouse antibody and constant domains (CH1, CH2, CH3, and CL) of human antibody (retained 33 percent mouse sequences). Humanized antibodies are constructed by grafting the CDR from the parent mouse antibody to a human antibody (<10 percent mouse sequences). Human antibodies can be produced using transgenic animals or phage display technologies (0 percent mouse sequence).

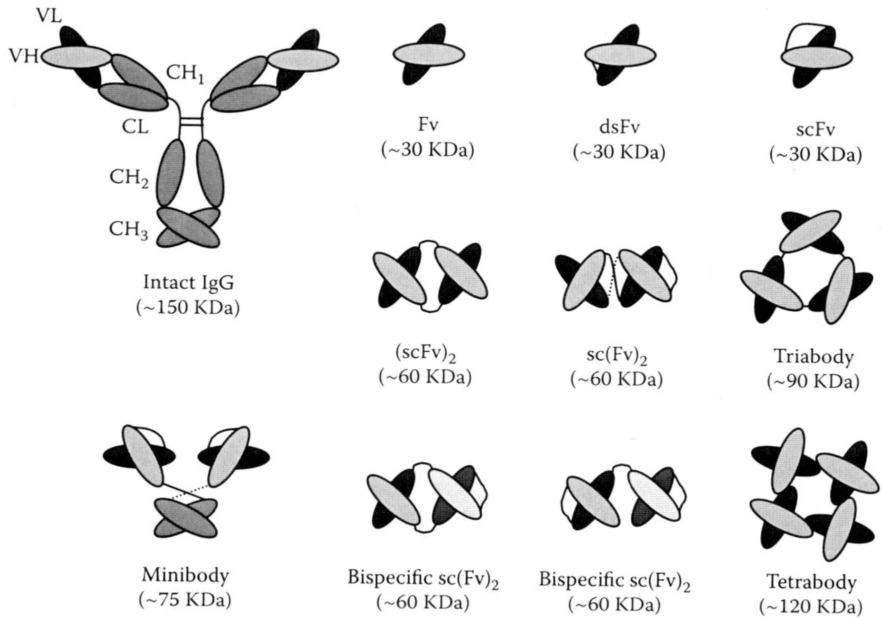

FIGURE 13.2 ***A color version of this figure follows page 176.*** Schematic structures of engineered antibody fragments. The variable regions of heavy (VH) and light chains (VL) contribute to the antigen binding (Fv fragment). Disulfide-stabilized Fv fragments are molecules in which the VH and VL are stabilized by an interchain disulfide bond. The VH and VL domains can be connected by a peptide linker to form single-chain Fv (scFv). The scFv fragments can form multimers such as $(scFv)_2$ diabody, triabody, and tetrabody depending on the linker length and the V-domain orientation. The bivalent $sc(Fv)_2$ can be generated by connecting two scFvs covalently. Bispecific diabodies can also be engineered by using two different Fv domains.

Third, engineered antibody fragments were investigated as alternative agents because of their unique properties resulting from the structure (Figure 13.2).[7,8] For example, single chain Fv and diabody are small molecule antibodies that have marked agonistic or cross-linking activities against their target molecules. Pharmacokinetically, low-molecular-weight antibodies (below 60 kDa) are subject to excretion by renal clearance, resulting in a shorter serum half-life than intact immunoglobulin G (IgG).[9] Such characteristics are important advantages for the imaging and targeted drug delivery because of the lower background levels and less undesired toxicity in normal tissues.

MONOCLONAL ANTIBODY THERAPY FOR HEMATOLOGICAL MALIGNANCIES

In hematological malignancies, mAb therapy has become an important strategy in clinical practice. CD20 antigen is specifically expressed in mature B cells, and an anti-CD20 chimeric antibody, rituximab, has emerged as a potent agent for the treatment of B-cell lymphoma.[10–12] More recently, anti-CD20 mAbs have been used for the selective delivery of radioactive isotopes. Clinical studies of radioisotope-conjugated anti-CD20 mAbs (^{90}Y ibritumomab tiuxetan and ^{131}I tositumomab) have shown the more significant response rate.[13,14] Similarly, CD33 is a specific marker for myeloid cells, and toxin-conjugated anti-CD33 mAb (gemtuzumab ozogamicin) has been successfully introduced for the treatment of acute myeloid leukemia.[15] However, no therapeutic mAb has been approved for the treatment of MM.

TARGET ANTIGENS FOR MONOCLONAL ANTIBODY THERAPY OF MULTIPLE MYELOMA

The selection of suitable target antigens is a key issue for the establishment of mAb-based strategies. Cellular and molecular studies have revealed several cell surface molecules specifically expressed on MM cells. Moreover, recent studies have identified specific growth factors and their downstream cell signaling such as Ras/MAPK, JAK/STAT3, PI3K/Akt, nuclear factor (NF)-κB, and Wnt/β-catenin pathways.[16] The interactions between MM cells and the bone marrow microenvironment have been shown to play an important role for MM pathogenesis through adherent stimulation and cytokine production.[17] In particular, MM cells activate osteoclasts in the bone marrow, and MM cell–osteoclast interactions form a vicious cycle, which leads to extensive bone destruction and MM cell expansion.[18] Therefore, biologic therapies are designed to target specific molecules in not only MM cells but also the bone marrow microenvironment (Figure 13.3). Several mAbs against these antigens are under preclinical investigation or in clinical trials (Table 13.1).

GROWTH FACTORS AND SIGNAL TRANSDUCTION PATHWAYS

Because interleukin 6 (IL-6) is a major growth factor for MM cells,[19] efficacy of blocking of IL-6 signaling has been investigated using anti-IL-6 mAb or anti-IL-6 receptor (IL-6R) mAb. Initial clinical studies of mouse anti-IL-6 mAb showed the decrease of M-protein and inhibition of C-reactive protein synthesis, but no significant response was observed as a single agent.[20,21] The potential of anti-IL-6 mAb was evaluated as maintenance therapy after autologous tandem stem cell transplantation, but the addition of the mAb did not improve progression-free survival or overall survival.[22] A chimeric mAb to IL-6 (CNTO 328) also failed to show significant anti-MM activity in a phase I trial,[23] and is currently under evaluation in combination with dexamethasone. Recent studies have shown that inhibition of IL-6 signaling with CNTO 328 enhances the activity of bortezomib in MM cells, and this combination warrants further investigation.[24] A humanized anti-IL-6R mAb (tocilizumab) also showed an inhibitory effect of IL-6 in clinical trials, resulting in disease stability in patients with refractory MM.[25] This agent is now used for the treatment of rheumatoid arthritis and Castleman's disease characterized by polyclonal activation of plasma cells and overexpression of IL-6.[26]

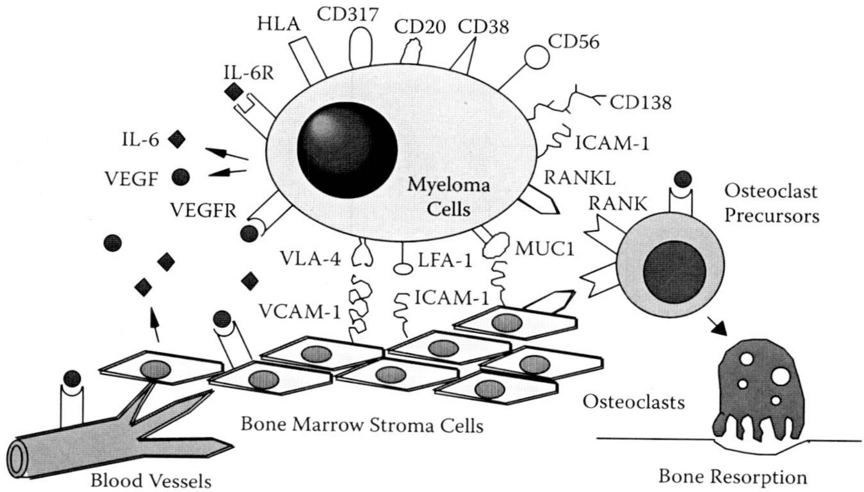

FIGURE 13.3 Target molecules for therapy of MM. MM cells express many lineage-specific antigens on the cell surface. MM cells adhere to extracellular matrix or bone marrow stroma cells, which stimulates the production of cytokines and regulates survival of MM cells by autocrine and paracrine fashions. Osteoclasts are markedly activated through the action of RANK ligand in the bone marrow. These cell surface molecules and cytokines are ideal targets for mAb therapy.

Vascular endothelial growth factor (VEGF) is implicated in the pathogenesis of MM by stimulating IL-6 production and angiogenesis in the bone marrow microenvironment.[27] Clinical studies of anti-VEGF mAb (bevacizumab) are ongoing in combination with chemotherapy.

Insulin-like growth factor (IGF)-1 also contributes to growth and survival of MM cells through activation of Ras/MAPK and PI3K/Akt pathways.[28] A phase I study of a fully human anti-IGF-1 mAb (CP-751, 871) has shown acceptable tolerability and objective responses in 9 of 27 patients when combined with dexamethasone.[29]

Cytogenetic and molecular studies have demonstrated the overexpression of fibroblast growth factor receptor (FGFR)-3 in MM patients with t(4;14).[30] Signals from FGFR3 activate Ras/MAPK and PI3K/Akt pathways, and this chromosomal abnormality relates to poor prognosis in clinical course. A human mAb against FGFR3 (PRO-001) is expected to be an agent for blocking of FGFR3 signaling in this group of patients.[31]

Tumor necrosis factor-related apoptosis-inducing ligand (TRAIL) is a potent activator of apoptotic pathway in a variety of tumor cells including MM cells, and therefore agonistic mAbs to its receptors are considered as possible therapeutic agents in cancer treatment. Several studies have shown that MM cells mostly express TRAIL receptor-1 and that anti-TRAIL receptor-1 mAbs induce apoptosis in MM cells.[32] Clinical trials of an anti-TRAIL receptor-1 mAb (mapatumumab) are ongoing in MM and other neoplastic diseases.

TABLE 13.1
Monoclonal Antibodies for the Treatment of Multiple Myeloma

Name	Target	Phase[a]
Rituximab	CD20	Phase II
Tositumomab	CD20	Phase II
CNTO 328	IL-6	Phase II
Bevacizumab	VEGF	Phase II
Cetuximab	EGFR	Phase II
Mapatumumab, TRM-1	TRAIL-R1	Phase II
Denosumab, AMG 162	RANKL	Phase II
Milatuzumab, hLL1	CD74	Phase I/II
Elotuzumab, HuLuc63	CD319	Phase I/II
HuMax-CD38	CD38	Phase I/II
Natalizumab	VLA-4	Phase I/II
Ibritumomab tiuxetan	CD20	Phase I
SGN-40	CD40	Phase I
HCD122, CHIR-12.12	CD40	Phase I
BB-10901, huN901-DM1	CD56	Phase I
CP-751, 871	IGF-1R	Phase I
1-7F9	KIR	Phase I
1D09C3	HLA class II	Preclinical
Anti-DKK1	DKK1	Preclinical
MABL	CD47	Preclinical
B-B4-DM1	CD138	Preclinical
AHM	CD317	Preclinical
PRO-001	FGFR3	Preclinical
2D7 diabody	HLA class I	Preclinical
D1 and E6	β2-Microglobulin	Preclinical

[a] Phase information of the trial is according to clinicaltrials.gov (http://www.clinicaltrials.gov/).

Osteolytic bone lesion is a hallmark of MM due to activation of osteoclasts and suppression of osteoblasts in the bone marrow. MM cells and bone marrow stroma cells secrete soluble factors and express functional ligands that affect the growth and differentiation of osteoclasts and osteoblasts. Among these, macrophage inflammatory protein (MIP)-1 produced by MM cells induces expression of receptor activator of NF-κB (RANK) ligand by stroma cells, which stimulates the essential signals for osteoclast maturation of osteoclast precursors.[33] A human anti-RANKL mAb (denosumab) has shown the improvement of bone metabolic markers in patients with breast cancer by blocking the RANK signaling.[34] Its clinical efficacy is now being investigated in patients with MM.

Dickkopf-1 (DKK1) is a soluble inhibitor of Wnt signaling secreted by MM cells. This factor disrupts the differentiation of osteoblasts and contributes to the development of osteolytic bone disease.[35] A DKK1-neutralizing mAb has been shown to reduce bone resorption and increase bone formation in mouse models.[36]

Cell Surface Antigens and Adhesion Molecules

Several differentiation antigens specifically expressed on MM cells have been considered as target molecules for mAb therapy, including CD20, CD38, CD40, CD47, CD56, CD74, CD138 (syndecan-1), CD317 (HM1.24/BST2), and HLA molecules. More recently, CD2 subset-1 (CS-1, CD319) has been studied as an attractive therapeutic target for MM.

CD20 is a specific marker for B cells and is expressed in approximately 20 percent of patients with MM, especially in the case of t(11;14).[37] It has been speculated that the cancer stem cells of MM express CD20, and several clinical trials of anti-CD20 mAb (rituximab) have been conducted.[38] Treon et al. have treated 19 patients with rituximab, and reported that this therapy was effective in only 6 patients whose MM cells were positive for CD20.[39] To date, radioactive anti-CD20 mAbs, ^{131}I tositumomab and ^{90}Y ibritumomab tiuxetan, are under evaluation in clinical trials for eradicating residual MM cells after chemotherapy and with autologous stem cell transplantation.

CD38 is an ectoenzyme involved in transmembrane signaling and cell adhesion, and is also used as a disease marker for leukemia and myeloma. CD38 is overexpressed in MM cells, and several clinical trials were conducted by using chimeric mAbs, humanized mAbs, or immunotoxins.[40] However, this antigen is also expressed in vascular endothelial cells, activated T cells, and hematopoietic stem cells. Therefore, the possibility of adverse events should be carefully considered.

CD40 is the receptor for CD40 ligand that belongs to the TNF gene family. This antigen is widely expressed on B cells and MM cells. CD40 signaling induces NF-κB activation as a downstream target of Ras/MAPK and PI3K/AKT pathways, and stimulates growth of MM cells, secretion of cytokines, and expression of adhesion molecules.[41] Thus, inhibition of CD40 signaling is considered as a therapeutic option for MM. Two mAbs against CD40 have been established and evaluated in clinical trials. A humanized mAb (SGN-40) binds to CD40 on MM cells, and induces ADCC against MM cells without activation of downstream signaling through CD40.[42] A clinical trial of SGN-40 has shown the disappearance of peripheral B cells and reduction of M-protein in 4 out of 16 patients.[43] In addition, combination therapy with lenalidomide has shown that lenalidomide augments the activity of SGN-40 by killing MM cells and enhancing the potential of host effector cells.[44] In contrast, a human mAb (HCD122, formerly CHIR-12.12) has a potential to block CD40 signaling in addition to ADCC activity.[45] This mAb also inhibits adhesion of MM cells to bone marrow stroma cells, and prevents secretion of cytokines such as IL-6 and VEGF.

CD47 is a member of the immunoglobulin superfamily containing a five-pass transmembrane attachment, and is also known as Rh-associated protein. Its expression is widely distributed in hematopoietic cells including T cells, B cells, monocytes, platelets, and erythrocytes as well as epithelial cells, endothelial cells, fibroblasts, and neural tissues. Recent studies have shown that a single-chain Fv diabody against

CD47 cross-links these molecules on the cell surface and induces caspase-independent apoptosis in lymphoid neoplastic cells and MM cells without host effector mechanisms.[46] Thus, the direct activity of small molecule antibodies might benefit even in MM patients with immune deficiency.

CD56 is also known as a neural cell adhesion molecule (NCAM) that consists of five immunoglobulin-like domains and two fibronectin type III domains. This molecule is expressed on the surface of neurons, skeletal muscle, natural killer cells, and MM cells. An immunoconjugate composed of a humanized anti-CD56 mAb (huN901) and a potent antimicrotubule agent (maytansinoid DM1) has been developed. This immunoconjugate has shown *in vitro* and *in vivo* cytotoxic activity against $CD56^+$ MM cells in a mouse model.[47]

CD74 was originally identified as the invariant chain that is associated with the alpha and beta chains of human leukocyte antigen (HLA)-DR (major histocompatibility complex, MHC, class II). CD74 is frequently expressed on B cells and MM cells, and is an important target for the development of immunoconjugates because this molecule internalizes rapidly into the target cells following antibody binding. IMMU-110, an immunoconjugate of a humanized anti-CD74 mAb (hLL1, milatuzumab) and doxorubicin, has shown cytotoxic efficacy against MM cells both *in vitro* and *in vivo*.[48] Clinical trials are planned for MM and other B-cell malignancies.

CD138 is a heparan sulfate proteoglycan that is a member of the syndecan proteoglycan family. Altered syndecan-1 expression has been detected in several different tumor types including MM. A mouse anti-CD138 mAb (B-B4) conjugated with maytansinoid DM1 has shown anti-MM activity against $CD138^+$ mature MM cells.[49] However, $CD138^-$ clonogenic population of MM stem cells might not be affected by this agent.

CD317 (HM1.24) was originally identified as a cell surface protein that is preferentially overexpressed on MM cells,[50] and later was found to be identical to bone marrow stromal cell antigen 2 (BST2).[51,52] This antigen is also expressed at low levels in selected tissues, but is specifically upregulated in metastatic tumor cells and chemoresistant cancer cells.[53,54] Subsequent studies have revealed that CD317 is a type II transmembrane glycoprotein with a unique topology that is present in lipid rafts via a GPI anchor at the C-terminus although its biologic function is unknown.[55] A phase I trial of humanized anti-HM1.24 mAb (AHM) showed an acceptable toxicity profile; however, its clinical efficacy was limited in these heavily treated patients with relapsed or refractory MM because of suppression of immune effector functions.[56,57]

CD2 subset-1 (CS-1, CRACC, SLAMF7, and CD319) is a cell surface glycoprotein that is a member of the immunoglobulin gene superfamily. This antigen is highly expressed on MM cells, and a humanized anti-CD319 mAb (HuLuc63, elotuzumab) has been shown to induce ADCC activity against MM cells.[58,59] Clinical trial of HuLuc63 is currently ongoing.

HLA class I molecule is a heterodimer consisting of a heavy chain and a light chain (β2-microglobulin). Several studies have shown that HLA class I is overexpressed on lymphoid neoplastic cells and that anti-HLA class I heavy chain mAb mediates tumor cell death.[60,61] Recently, anti-β2-microglobulin mAb has been shown to induce the recruitment of HLA class I molecules into lipid rafts and exclude cytokine receptors such as IL-6R and IGF-1R from lipid rafts, resulting in inhibition of their signaling pathways.[62] Moreover, a single-chain Fv diabody against HLA-A heavy chain has been shown to bind to HLA-A molecules closely, and specifically induces Rho activation and actin aggregation that leads to caspase-independent cell death in MM cells.[63]

HLA class II molecule is expressed on MM cells in one-fourth of patients with MM. A fully human mAb against HLA-DR (1D09C3) has been shown to induce apoptosis of MM cells correlated with HLA-DR expression.[64] Interferon gamma (IFNγ) upregulates HLA-DR expression on MM cells and enhances the anti-MM activity of this mAb.

CONCLUSION AND FUTURE DIRECTIONS

Numerous approaches have been employed to develop mAb therapy specific for MM. However, unfortunately their efficacy has not been satisfactory in clinical trials. One of the reasons is that immune effector functions such as ADCC activity are mostly suppressed in patients with MM.[65] Therefore, antibody modifications are required to improve therapeutic potential in patients with immunodeficiency. For example, engineered mAbs such as low fucosylated antibodies have an increased affinity of the Fc domain for Fcγ receptors on immune effector cells.[66] Bispecific antibodies that target simultaneously two antigens can be designed to recruit effector cells more efficiently.[67] Recombinant antibodies such as immunoconjugates and small molecule antibodies are also promising candidates because these agents directly induce MM cell death without the need for host immune functions.[6,7] The evolving technologies of antibody engineering will continue to provide alternative approaches to establish more specific and effective therapeutic opportunities for MM.

mAb therapy is likely to be effective in minimal residual disease but not a large tumor burden. Recent studies have suggested the existence of cancer stem cells within MM cell population in the bone marrow.[68] These MM stem cells contribute to the relapse of the disease even after treatment with chemotherapy and novel agents. Therefore, mAb therapy focusing on MM stem cells might be a suitable approach to improve long-term prognosis.

Because MM is a highly heterogeneous disease with respect to genetic alterations, there is no single target molecule responsible for tumorigenesis of MM cells. Moreover, blocking of a single cytokine pathway by mAbs seems to be insufficient to suppress the growth and survival of MM cells. Novel agents such as thalidomide, lenalidomide, and bortezomib have multiple biologic activities against both MM cells and the bone marrow microenvironment, and are now part of standard treatment.[69] Therefore, further studies are needed to evaluate the efficacy of therapeutic mAbs in combination with chemotherapy and novel agents to establish a promising strategy for MM and plasma cell disorders.

REFERENCES

1. Kohler G, Milstein C. Continuous cultures of fused cells secreting antibody of predefined specificity. *Nature* 1975;256(5517):495–97.
2. Carter P. Improving the efficacy of antibody-based cancer therapies. *Nat Rev Cancer* 2001;1(2):118–29.

3. Adams GP, Weiner LM. Monoclonal antibody therapy of cancer. *Nat Biotechnol* 2005;23(9):1147–57.
4. Ishida I, Tomizuka K, Yoshida H, et al. Production of human monoclonal and polyclonal antibodies in TransChromo animals. *Cloning Stem Cells* 2002;4(1):91–102.
5. Pastan I, Hassan R, Fitzgerald DJ, Kreitman RJ. Immunotoxin therapy of cancer. *Nat Rev Cancer* 2006;6(7):559–65.
6. Schrama D, Reisfeld RA, Becker JC. Antibody targeted drugs as cancer therapeutics. *Nat Rev Drug Discov* 2006;5(2):147–59.
7. Holliger P, Hudson PJ. Engineered antibody fragments and the rise of single domains. *Nat Biotechnol* 2005;23(9):1126–36.
8. Jain M, Kamal N, Batra SK. Engineering antibodies for clinical applications. *Trends Biotechnol* 2007;25(7):307–16.
9. Batra SK, Jain M, Wittel UA, Chauhan SC, Colcher D. Pharmacokinetics and biodistribution of genetically engineered antibodies. *Curr Opin Biotechnol* 2002;13(6):603–8.
10. Maloney DG, Grillo-Lopez AJ, White CA, et al. IDEC-C2B8 (Rituximab) anti-CD20 monoclonal antibody therapy in patients with relapsed low-grade non-Hodgkin's lymphoma. *Blood* 1997;90(6):2188–95.
11. Coiffier B, Lepage E, Briere J, et al. CHOP chemotherapy plus rituximab compared with CHOP alone in elderly patients with diffuse large-B-cell lymphoma. *N Engl J Med* 2002;346(4):235–42.
12. Hiddemann W, Kneba M, Dreyling M, et al. Frontline therapy with rituximab added to the combination of cyclophosphamide, doxorubicin, vincristine, and prednisone (CHOP) significantly improves the outcome for patients with advanced-stage follicular lymphoma compared with therapy with CHOP alone: Results of a prospective randomized study of the German Low-Grade Lymphoma Study Group. *Blood* 2005;106(12):3725–32.
13. Witzig TE, Gordon LI, Cabanillas F, et al. Randomized controlled trial of yttrium-90-labeled ibritumomab tiuxetan radioimmunotherapy versus rituximab immunotherapy for patients with relapsed or refractory low-grade, follicular, or transformed B-cell non-Hodgkin's lymphoma. *J Clin Oncol* 2002;20(10):2453–63.
14. Kaminski MS, Tuck M, Estes J, et al. 131I-Tositumomab therapy as initial treatment for follicular lymphoma. *N Engl J Med* 2005;352(5):441–49.
15. Sievers EL, Larson RA, Stadtmauer EA, et al. Efficacy and safety of gemtuzumab ozogamicin in patients with CD33-positive acute myeloid leukemia in first relapse. *J Clin Oncol* 2001;19(13):3244–54.
16. Hideshima T, Mitsiades C, Tonon G, Richardson PG, Anderson KC. Understanding multiple myeloma pathogenesis in the bone marrow to identify new therapeutic targets. *Nat Rev Cancer* 2007;7(8):585–98.
17. Mitsiades CS, Mitsiades NS, Munshi NC, Richardson PG, Anderson KC. The role of the bone microenvironment in the pathophysiology and therapeutic management of multiple myeloma: Interplay of growth factors, their receptors and stromal interactions. *Eur J Cancer* 2006;42(11):1564–73.
18. Abe M, Hiura K, Wilde J, et al. Osteoclasts enhance myeloma cell growth and survival via cell-cell contact: a vicious cycle between bone destruction and myeloma expansion. *Blood* 2004;104(8):2484–91.
19. Kawano M, Hirano T, Matsuda T, et al. Autocrine generation and requirement of BSF-2/IL-6 for human multiple myelomas. *Nature* 1988;332(6159):83–85.
20. Klein B, Wijdenes J, Zhang XG, et al. Murine anti-interleukin-6 monoclonal antibody therapy for a patient with plasma cell leukemia. *Blood* 1991;78(5):1198–204.
21. Bataille R, Barlogie B, Lu ZY, et al. Biologic effects of anti-interleukin-6 murine monoclonal antibody in advanced multiple myeloma. *Blood* 1995;86(2):685–91.

22. Moreau P, Hullin C, Garban F, et al. Tandem autologous stem cell transplantation in high-risk de novo multiple myeloma: Final results of the prospective and randomized IFM 99-04 protocol. *Blood* 2006;107(1):397–403.
23. van Zaanen HC, Lokhorst HM, Aarden LA, et al. Chimaeric anti-interleukin 6 monoclonal antibodies in the treatment of advanced multiple myeloma: A phase I dose-escalating study. *Br J Haematol* 1998;102(3):783–90.
24. Voorhees PM, Chen Q, Kuhn DJ, et al. Inhibition of interleukin-6 signaling with CNTO 328 enhances the activity of bortezomib in preclinical models of multiple myeloma. *Clin Cancer Res* 2007;13(21):6469–78.
25. Nishimoto N, Shima Y, Sasai M, Danno N, Yoshizaki K. [Clinical application of interleukin-6 receptor antibody]. *Nihon Rinsho Meneki Gakkai Kaishi* 1997;20(2):87–94.
26. Nishimoto N, Kanakura Y, Aozasa K, et al. Humanized anti-interleukin-6 receptor antibody treatment of multicentric Castleman disease. *Blood* 2005;106(8):2627–32.
27. Podar K, Tai YT, Davies FE, et al. Vascular endothelial growth factor triggers signaling cascades mediating multiple myeloma cell growth and migration. *Blood* 2001;98(2):428–35.
28. Tai YT, Podar K, Catley L, et al. Insulin-like growth factor-1 induces adhesion and migration in human multiple myeloma cells via activation of beta1-integrin and phosphatidylinositol 3′-kinase/AKT signaling. *Cancer Res* 2003;63(18):5850–58.
29. Lacy MQ, Alsina M, Fonseca R, et al. Phase I, pharmacokinetic and pharmacodynamic study of the anti-insulinlike growth factor type 1 receptor monoclonal antibody CP-751,871 in patients with multiple myeloma. *J Clin Oncol* 2008;26(19):3196–203.
30. Chesi M, Nardini E, Brents LA, et al. Frequent translocation t(4;14)(p16.3;q32.3) in multiple myeloma is associated with increased expression and activating mutations of fibroblast growth factor receptor 3. *Nat Genet* 1997;16(3):260–64.
31. Trudel S, Stewart AK, Rom E, et al. The inhibitory anti-FGFR3 antibody, PRO-001, is cytotoxic to t(4;14) multiple myeloma cells. *Blood* 2006;107(10):4039–46.
32. Menoret E, Gomez-Bougie P, Geffroy-Luseau A, et al. Mcl-1L cleavage is involved in TRAIL-R1- and TRAIL-R2-mediated apoptosis induced by HGS-ETR1 and HGS-ETR2 human mAbs in myeloma cells. *Blood* 2006;108(4):1346–52.
33. Abe M, Hiura K, Wilde J, et al. Role for macrophage inflammatory protein (MIP)-1alpha and MIP-1beta in the development of osteolytic lesions in multiple myeloma. *Blood* 2002;100(6):2195–202.
34. Body JJ, Facon T, Coleman RE, et al. A study of the biological receptor activator of nuclear factor-kappaB ligand inhibitor, denosumab, in patients with multiple myeloma or bone metastases from breast cancer. *Clin Cancer Res* 2006;12(4):1221–28.
35. Tian E, Zhan F, Walker R, et al. The role of the Wnt-signaling antagonist DKK1 in the development of osteolytic lesions in multiple myeloma. *N Engl J Med* 2003;349(26):2483–94.
36. Yaccoby S, Ling W, Zhan F, Walker R, Barlogie B, Shaughnessy JD, Jr. Antibody-based inhibition of DKK1 suppresses tumor-induced bone resorption and multiple myeloma growth in vivo. *Blood* 2007;109(5):2106–11.
37. Robillard N, Avet-Loiseau H, Garand R, et al. CD20 is associated with a small mature plasma cell morphology and t(11;14) in multiple myeloma. *Blood* 2003;102(3):1070–71.
38. Kapoor P, Greipp PT, Morice WG, Rajkumar SV, Witzig TE, Greipp PR. Anti-CD20 monoclonal antibody therapy in multiple myeloma. *Br J Haematol* 2008; 141(2):135–48.
39. Treon SP, Pilarski LM, Belch AR, et al. CD20-directed serotherapy in patients with multiple myeloma: Biologic considerations and therapeutic applications. *J Immunother* 2002;25(1):72–81.
40. Stevenson GT. CD38 as a therapeutic target. *Mol Med* 2006;12(11–12):345–46.

41. Tai YT, Podar K, Mitsiades N, et al. CD40 induces human multiple myeloma cell migration via phosphatidylinositol 3-kinase/AKT/NF-kappa B signaling. *Blood* 2003;101(7):2762–69.
42. Tai YT, Catley LP, Mitsiades CS, et al. Mechanisms by which SGN-40, a humanized anti-CD40 antibody, induces cytotoxicity in human multiple myeloma cells: Clinical implications. *Cancer Res* 2004;64(8):2846–52.
43. Hussein M, Berenson J, Niesvizky R, et al. A phase I humanized anti-CD40 monoclonal antibody (SGN-40) in patients with multiple myeloma [abstr]. *Blood* 2005;106(11):723a.
44. Tai YT, Li XF, Catley L, et al. Immunomodulatory drug lenalidomide (CC-5013, IMiD3) augments anti-CD40 SGN-40-induced cytotoxicity in human multiple myeloma: Clinical implications. *Cancer Res* 2005;65(24):11712–20.
45. Tai YT, Li X, Tong X, et al. Human anti-CD40 antagonist antibody triggers significant antitumor activity against human multiple myeloma. *Cancer Res* 2005;65(13):5898–906.
46. Kikuchi Y, Uno S, Kinoshita Y, et al. Apoptosis inducing bivalent single-chain antibody fragments against CD47 showed antitumor potency for multiple myeloma. *Leuk Res* 2005;29(4):445–50.
47. Tassone P, Gozzini A, Goldmacher V, et al. In vitro and in vivo activity of the maytansinoid immunoconjugate huN901-N2′-deacetyl-N2′-(3-mercapto-1 -oxopropyl)-maytansine against CD56+ multiple myeloma cells. *Cancer Res* 2004;64(13):4629–36.
48. Sapra P, Stein R, Pickett J, et al. Anti-CD74 antibody-doxorubicin conjugate, IMMU-110, in a human multiple myeloma xenograft and in monkeys. *Clin Cancer Res* 2005;11(14):5257–64.
49. Tassone P, Goldmacher VS, Neri P, et al. Cytotoxic activity of the maytansinoid immunoconjugate B-B4-DM1 against CD138+ multiple myeloma cells. *Blood* 2004;104(12):3688–96.
50. Goto T, Kennel SJ, Abe M, et al. A novel membrane antigen selectively expressed on terminally differentiated human B cells. *Blood* 1994;84(6):1922–30.
51. Ishikawa J, Kaisho T, Tomizawa H, et al. Molecular cloning and chromosomal mapping of a bone marrow stromal cell surface gene, BST2, that may be involved in pre-B-cell growth. *Genomics* 1995;26(3):527–34.
52. Ohtomo T, Sugamata Y, Ozaki Y, et al. Molecular cloning and characterization of a surface antigen preferentially overexpressed on multiple myeloma cells. *Biochem Biophys Res Commun* 1999;258(3):583–91.
53. Walter-Yohrling J, Cao X, Callahan M, et al. Identification of genes expressed in malignant cells that promote invasion. *Cancer Res* 2003;63(24):8939–47.
54. Becker M, Sommer A, Kratzschmar JR, Seidel H, Pohlenz HD, Fichtner I. Distinct gene expression patterns in a tamoxifen-sensitive human mammary carcinoma xenograft and its tamoxifen-resistant subline MaCa 3366/TAM. *Mol Cancer Ther* 2005;4(1):151–68.
55. Kupzig S, Korolchuk V, Rollason R, Sugden A, Wilde A, Banting G. Bst-2/HM1.24 is a raft-associated apical membrane protein with an unusual topology. *Traffic* 2003;4(10):694–709.
56. Ozaki S, Kosaka M, Wakahara Y, et al. Humanized anti-HM1.24 antibody mediates myeloma cell cytotoxicity that is enhanced by cytokine stimulation of effector cells. *Blood* 1999;93(11):3922–30.
57. Powles R, Sirohi B, Morgan G, et al. A phase I study of the safety, tolerance, pharmacokinetics, antigenicity and efficacy of a single intravenous dose of AHM followed by multiple doses of intravenous AHM in patients with multiple myeloma [abstr]. *Blood* 2001;98(11):165a.
58. Tai YT, Dillon M, Song W, et al. Anti-CS1 humanized monoclonal antibody HuLuc63 inhibits myeloma cell adhesion and induces antibody-dependent cellular cytotoxicity in the bone marrow milieu. *Blood* 2008;112(4):1329–37.

59. Hsi ED, Steinle R, Balasa B, et al. CS1, a potential new therapeutic antibody target for the treatment of multiple myeloma. *Clin Cancer Res* 2008;14(9):2775–84.
60. Genestier L, Meffre G, Garrone P, et al. Antibodies to HLA class I alpha1 domain trigger apoptosis of CD40-activated human B lymphocytes. *Blood* 1997;90(2):726–35.
61. Woodle ES, Smith DM, Bluestone JA, Kirkman WM, 3rd, Green DR, Skowronski EW. Anti-human class I MHC antibodies induce apoptosis by a pathway that is distinct from the Fas antigen-mediated pathway. *J Immunol* 1997;158(5):2156–64.
62. Yang J, Zhang X, Wang J, et al. Anti beta2-microglobulin monoclonal antibodies induce apoptosis in myeloma cells by recruiting MHC class I to and excluding growth and survival cytokine receptors from lipid rafts. *Blood* 2007;110(8):3028–35.
63. Sekimoto E, Ozaki S, Ohshima T, et al. A single-chain Fv diabody against human leukocyte antigen-A molecules specifically induces myeloma cell death in the bone marrow environment. *Cancer Res* 2007;67(3):1184–92.
64. Carlo-Stella C, Guidetti A, Di Nicola M, et al. IFN-gamma enhances the antimyeloma activity of the fully human anti-human leukocyte antigen-DR monoclonal antibody 1D09C3. *Cancer Res* 2007;67(7):3269–75.
65. Harrison SJ, Cook G. Immunotherapy in multiple myeloma—possibility or probability? *Br J Haematol* 2005;130(3):344–62.
66. Shinkawa T, Nakamura K, Yamane N, et al. The absence of fucose but not the presence of galactose or bisecting *N*-acetylglucosamine of human IgG1 complex-type oligosaccharides shows the critical role of enhancing antibody-dependent cellular cytotoxicity. *J Biol Chem* 2003;278(5):3466–73.
67. Kontermann RE. Recombinant bispecific antibodies for cancer therapy. *Acta Pharmacol Sin* 2005;26(1):1–9.
68. Matsui W, Wang Q, Barber JP, et al. Clonogenic multiple myeloma progenitors, stem cell properties, and drug resistance. *Cancer Res* 2008;68(1):190–97.
69. Hideshima T, Anderson KC. Molecular mechanisms of novel therapeutic approaches for multiple myeloma. *Nat Rev Cancer* 2002;2(12):927–37.

Section IV

Development of Biotherapies for Solid Tumors

14 Antibody-Based Therapies for Carcinomas

David Schrama and Jürgen C. Becker

CONTENTS

KEY WORDS: antibodies, carcinoma, radioimmunoconjugates, chemoimmunoconjugates, immunotoxins

ABSTRACT

The major problem for treatment of cancer is that treatment should be as aggressive as possible to kill all tumor cells while at the same time this aggressiveness often causes severe side effects. Indeed, due to the side effects many therapeutics cannot be applied at the desired doses, or some even cannot be applied systemically. In addition, therapeutics like cytokines require a locally enhanced level to exert their effect appropriately. Hence, recent therapeutic strategies try to target the tumor more specifically. In this regard, monoclonal antibodies can be used to achieve this aim, either as naked antibodies or as drug-carriers, that is, fusion of drugs to tumor-specific antibodies. In fact, all different kinds of agents have already been fused to tumor-specific antibodies. Here, we discuss recent preclinical and clinical data of both naked antibodies and drug–antibody-fusion proteins.

INTRODUCTION

Until improved techniques and histological control made surgery more efficient, cancer was more or less regarded as incurable. By now, chemotherapy and radiation have become standard cancer therapies in addition to surgery. Nevertheless, cancer remains to be one of the most life-threatening diseases. A reason for the failure of most therapies can be ascribed to the fact that it is difficult to differentiate tumor cells from healthy cells, and thus most therapies cause severe side effects. Indeed, despite intensive research on cancer and cancer therapy over the past decades, prognosis of metastatic cancer has not been satisfactorily improved; however, the research has unraveled many pathways and characteristics of different tumor entities. Based on this information, specific tumor therapies are conducted either to target the proteins involved in the neoplastic process directly or to target drugs toxic to the tumor.

Importantly, monoclonal antibodies (mAb) can be used to realize both strategies because of their attributes of specificity and high affinity. Notably, Paul Ehrlich envisioned antibodies as "magic bullets" at the end of the nineteenth century. Nevertheless, it took until 1975 when Köhler and Milstein described the generation of murine mAbs that a technique was available to pursue this vision.[1] For the development of mAb as therapeutic agents, however, several other problems had to be solved by technical advances in the process of antibody generation. For example, that patients' immune response readily inactivated antibodies of nonhuman origin limited the use of such antibodies. Moreover, not all mAb isotypes have the capacity to mediate antibody-dependent cell-mediated cytotoxicity through human peripheral blood lymphocytes.[2]

The first mAb for treatment of cancer was approved in Germany in 1995. This murine, anti-EPCAM mAb (edrecolomab) was used to treat colorectal cancer after surgery. However, this mAb was withdrawn from the market in 2000 due to inferior performance compared to standard therapy with fluorouracil. The first therapeutic antibody for cancer therapy approved by the U.S. Food and Drug Administration (FDA) was rituximab in 1997 (Table 14.1) and mAb-based therapies became a major strategy in medicine.[3]

mAb can be used therapeutically as naked mAb, or to target therapeutic measures to tumor cells (Figure 14.1). Indeed, conjugates of radionuclides, cytotoxic drugs, toxins, or cytokines and tumor-specific mAbs have been evaluated. This strategy should increase specificity of the therapeutic intervention thereby minimizing side effects while maximizing desired effects. Here, we discuss recent preclinical and clinical data on naked mAb and immunoconjugates with a focus on the latter.

NAKED MAB

Therapeutic mAbs have been clinically studied for almost 30 years. During the last 25 years approximately 200 mAbs for the treatment of cancer entered clinical trials of which around 110 were discontinued.[4] Nevertheless, from the 11 anticancer mAbs approved in at least one country, 7 are naked mAbs. In this regard, almost 60 percent of mAb-based anticancer therapies are naked mAbs. A common primary mode of action is the destruction of targeted cells through activation of components of the human immune system involving mechanisms such as antibody-dependent cell cytotoxicity (ADCC) and complement-dependent cytotoxicity (CDC). For example, ADCC and CDC are the main modes of action of rituximab and alemtuzumab. Alternative primary modes of action include blockade of growth factor-receptor interaction, receptor down-modulation, and signal transduction.[5]

TABLE 14.1
Therapeutic Antibodies Approved for Cancer Treatment

	Type	Isotype	Target	Indication	Year of Approval[a]
mAbs					
Rituximab (Rituxan)	Chimeric	IgG_1	CD20	Low-grade B-cell NHL	1997
Trastuzumab (Herceptin)	Humanized	IgG_1	Her2/*neu*	Metastatic breast cancer	1998
Alemtuzumab (Campath)	Humanized	IgG_1	CD52	CLL	2001
Bevacizumab (Avastin)	Chimeric	IgG_1	VEGF	Metastatic colorectal cancer	2004
Cetuximab (Erbitux)	Chimeric	IgG_1	EGF receptor	Metastatic colorectal cancer	2003 (Switzerland); 2004 (USA)
Nimotuzumab (TheraCIM)	Humanized	IgG_1	EGF receptor	Squamous cell carcinoma	2005 (China)
Panitumumab (Vectibix)	Human	IgG_2	EGF receptor	Metastasized colorectal cancer following standard chemotherapy	2006
Immunoconjugates		**Isotype/ Conjugate**			
Gemtuzumab ozogamicin (Mylotarg)	Humanized	IgG_4/ calicheamicin	CD33	AML	2000
Ibritumomab tiuxetan (Zevalin)	Murine	$IgG_1/^{90}Y$	CD20	Relapsed or refractory NHL	2002
^{131}I ch-TNT (Cotara)	Chimeric	$IgG_1/^{131}I$	DNA/ histone complex	Advanced lung cancer	2003 (China)
Tositumomab and ^{131}I tositumomab (Bexxar)	Murine	$IgG_{2a}/^{131}I$	CD20	NHL refractory to rituximab and relapsed following chemotherapy	2003

[a] Year of approval is given for United States by the FDA if not otherwise stated.
Abbreviations: AML, acute myelogenous leukemia; CLL, chronic lymphocytic leukemia; NHL, non-Hodgkin's lymphoma.

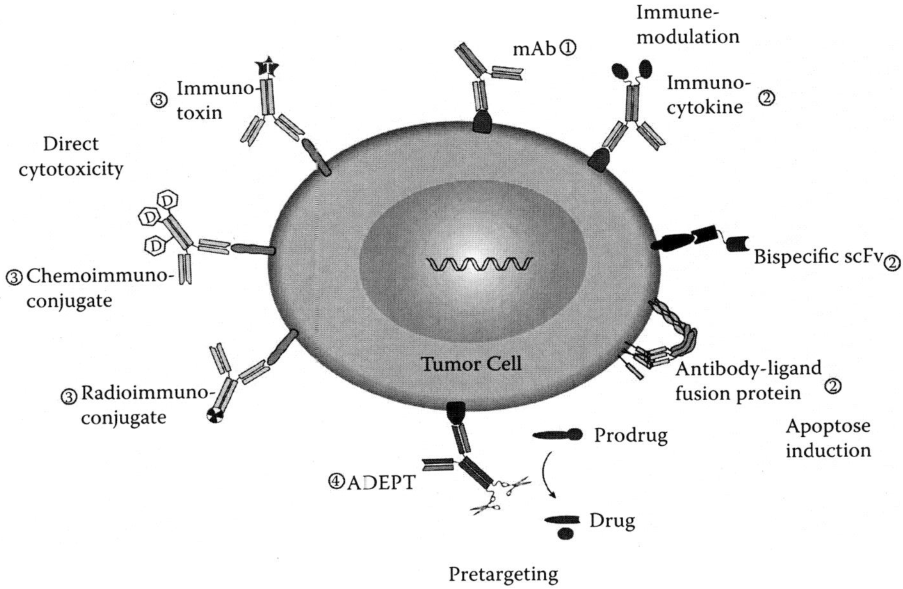

FIGURE 14.1 ***A color version of this figure follows page 176.*** Concepts of therapeutic antibodies. Binding of naked mAbs to tumor cells may result in the destruction of the tumor cells by ADCC or CDC. Similarly, the immune response can be modulated by targeting cytokines or immunomodulatory molecules either by diabodies or antibody–ligand fusion proteins to the tumor. In addition, antibody–ligand fusion proteins can induce apoptosis to targeted cell as well as bystander cell by, for example, presenting FasL. The targeted cell can be killed directly by conjugation of cytotoxic drugs (D), toxins (T), or radionucleotides to the mAbs. The direct targeting approach might require the homogeneous expression of antigen in the tumor cell population although dependent on the radionucleotides used, radioimmunoconjugates can have a bystander effect, e.g., when beta emitters are used. The ADEPT approach specifically aims at this effect by targeting an enzyme to the tumor cell and subsequently delivers a prodrug which can be converted to a chemotherapeutic drug by the targeted enzyme.

In addition, mAbs can be used to enhance T-cell responses. Indeed, several therapeutic approaches to target T-cell costimulatory and inhibitory pathways for turning on or preventing the turning off of immune responses in patients with cancer have been tested, that is, antagonizing receptors that suppress immune responses or activating others that amplify immune responses. Binding of CD137 (4-1BB), for example, leads to T-cell growth and activation. Hence, agonistic anti-CD137 mAb can enhance proliferation and survival of activated cytotoxic T-cell lymphocytes (CTLs).[6] In contrast, CD152—CTL antigen-4 (CTLA-4) and CD279 (programmed death-1, PD-1—are powerful in defusing T-cell responses. Thus, anti-CTLA-4 and anti-PD-1 mAbs directly assist T-cell activation or prevent T-cell apoptosis, respectively (Figure 14.2). It should be noted that complications such as autoimmunity and systemic inflammation can be problematic side effects associated with these agents.

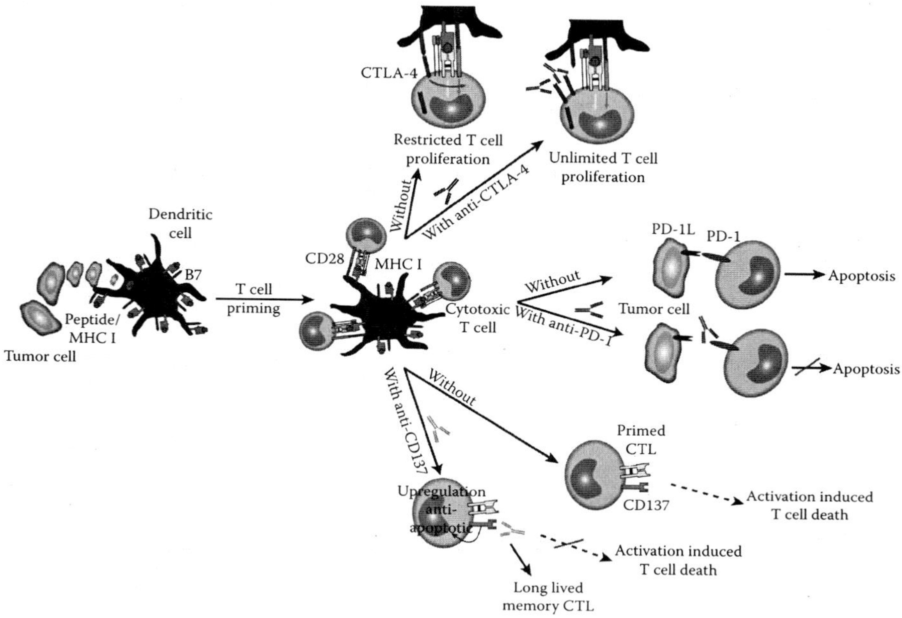

FIGURE 14.2 ***A color version of this figure follows page 176.*** Immunostimulatory mAb. Dendritic cells pick up tumor cells and present peptide/MHC class I complexes to T cells. Cytotoxic T cells (CTLs) which possess the appropriate T-cell receptor recognizing this peptide/MHC class I complex will be primed. In the normal setting the following T-cell proliferation will be impaired by CTLA-4/CD28 interactions. In the presence of anti-CTLA-4 mAb, however, T-cell proliferation will be unrestricted (upper part). In addition, PD-1/PD-1L interaction between T cell and cancer cell will lead to apoptosis of the T cells. This can be prevented by the administration of anti-PD-1 mAb (middle part). Primed CTLs also express CD137. Ligand binding, which can be substituted by anti-CD137 mAb, prevents activation-induced T-cell death by leading to upregulation of antiapoptotic proteins (lower part). In addition, CD137 is required for the stable differentiation of CTLs and the establishment of long-lived memory CTLs.

Due to their multiple capabilities for disrupting cellular function and survival, unconjugated mAbs are attractive therapeutics. Naked mAbs, however, are often not very potent and therefore they are commonly used in combinations with drugs to increase the potency. Similarly, mAbs can be conjugated to other cytotoxic compounds to increase their potency.

RADIOIMMUNOCONJUGATES

One strategy to increase the efficacy of antibody therapies is to link tumor-specific antibodies to highly toxic radioisotopes. This approach thus enables one to target the killing power of these isotopes directly on tumor cells. In this regard, some radiolabeled mAbs are being evaluated in phase II or III studies. Moreover, two anti-CD20 mAb regimens for the treatment of non-Hodgkin's lymphoma (NHL) have been approved by the FDA (Table 14.1).

A central point for the design of radiolabeled mAbs is the choice of the radioisotope. Certainly, no single radionuclide will be able to address every therapeutic need, especially as cancer does not exist in an exclusive form. The characteristics that should be considered for the selection include emission type, energy and range of emission, and the half-life of the radionuclide.[7] Currently, from the three types of emission (α- or β-particles, or Auger electrons) β-emitting radionuclides are almost exclusively used in the clinic. The main advantage of β-emitters is the relatively long path lengths in biologic tissue. In addition, β-emitters have a relatively low linear energy transfer; that is, energy deposition takes place some distance from the actual decay event. These characters enable β-transmitters to irradiate cancer cells that have not bound radiolabeled mAb (crossfire) bypassing tumor antigen heterogeneity.[8] This advantage, however, is accompanied with the disadvantage that β-emitters are not very efficient in inducing lethal lesions in small tumors and single cell metastatic disease.

Because of the availability, extensive clinical experience in the treatment of thyroid disorders, the convenient half-life, and radiolabeling chemistry (iodine-131, ^{131}I, is incorporated into antibody by direct iodination of tyrosine residues), ^{131}I (half-life: 8 days; maximum range β particle in tissue: 3.0 mm) is quite often used for labeling mAb. Other clinically relevant β-emitters include yttrium-90 (^{90}Y; half-life: 64 hours; maximum range β particle in tissue: 12 mm), copper-67 (^{67}Cu; 61.9 hours; 1.8 mm), rhenium-186 (^{186}Re; 90.7 hours; 5.1 mm), and lutetium-177 (^{177}Lu; 6.7 days; 2.5 mm). Due to the imagable γ emission, ^{131}I-labeled antibody is used to derive patient-specific doses based on the uptake of the tracer antibody. However, this high energy results in undesirable nonspecific irradiation of the whole body. The convenient incorporation of ^{131}I into antibody, however, is associated with the problem of rapid deiodination of the protein after cellular internalization. This issue can be bypassed when the radioisotope such as ^{90}Y is conjugated to antibodies via a chelator in a process requiring more elaborate chemistry.[9] The longer emission range of ^{90}Y, however, may result in elevated toxicity by irradiation of normal neighboring cells.[10] The characteristics of the radionuclides are also important with respect to the eventual radiation energy dose that is delivered to the tumor. For example, minimal disease consisting of tumor nodules with a diameter of only a few millimeters seems not to be suitable for targeting with ^{90}Y-labeled mAb, since the β-particles' energy is so high that 70 percent of the radiation energy is deposited outside of these small tumors. In fact, several preclinical studies have shown that medium-energy β-emitters, such as ^{131}I and ^{177}Lu are more effective for the treatment of small tumor nodules.[11] In contrast, high-energy β-emitters such as ^{90}Y are more suitable for radioimmunotherapy of larger tumors.[12]

For small tumor cell clusters, α-emitters are prime candidates and are, for example, being explored for the treatment of myeloid leukemia.[13] They bear the advantage that their path length will traverse several cell diameters and that their high linear energy transfer increases the probability for cell killing even when targeting low-density antigens. The third class of radioisotopes, Auger-emitters, exert their activity mostly to the targeted cell and, therefore, could be ideally suited for treating micrometastatic disease, even in the bone marrow as nontargeted cells would be unaffected. However, their low energy requires that a substantial number of Auger particles must be delivered to kill a cell. Although antibodies radiolabeled with α- or Auger-emitters are typically considered for applications such as leukemia, they may have a role in the treatment of other hematologic malignancies and maybe even solid tumors. Radioimmunotherapy with α- or Auger-emitters could also be combined with chemotherapy or radioimmunotherapy with β-emitters, to destroy minimal residues of cancer cells that might not otherwise be as effectively treated with β-emitters or drugs.

Until now, radioimmunotherapy has emerged as an accepted treatment for patients with NHL, for which two radiolabeled mAbs have been approved. The therapeutic application of radiolabeled mAbs in solid cancers, however, has met considerable difficulties, mainly because tumoricidal radiation doses often cannot be achieved due to low uptake of the radiolabeled mAb in tumor tissue. A solution to this problem might be pretargeting: in this procedure a tumor-specific antibody conjugate is administered, and after this conjugate is bound and cleared from the blood, the radionuclide attached to a compound that can bind to the antibody conjugate is given. Thus, this strategy allows administration of the radionuclide when optimal targeting has been accomplished.[14] Moreover, myelosuppression, which is the major side effect of radioimmunotherapy, can be reduced with this approach. It should be noted that responses in lymphoma are most certainly related to the radiosensitivity of these tumors, as the measured uptake of radiolabeled antibodies in lymphoma is not higher than in solid tumors.[15] The few clinical studies performed on radioimmunotherapy for the treatment of patients with solid cancers suggest that this therapy may be much more effective in patients with small volume disease than in advanced patients with bulky disease. In this regard, patient selection may be an important factor determining the eventual success of the therapy for patients with solid cancers. Moreover, radioimmunotherapy might be an appropriate adjuvant therapy for patients with solid tumors, for example, after resection of primary tumors and for patients with a high risk of distant metastases.

CHEMOIMMUNOCONJUGATES

In addition to radiation, traditional chemotherapy is still the standard therapy for most cancers. The systemic administration of cytotoxic agents results in the death of actively dividing cells within the body, specifically cancer cells, but also has the undesired side effect of killing oral and intestinal mucosa, hair follicles, and bone marrow, increasing the potency of the drug. Consequently, anticancer chemotherapeutics are often given at suboptimal doses to reduce this risk. To achieve high efficacy and low toxicity, cytotoxic agents can be conjugated to tumor-specific antibodies facilitating the transport of a drug cargo in the body. Such chemoimmunoconjugates indeed demonstrate selectivity toward the cells which express the respective antigens.[16] After binding, the conjugate is internalized via receptor-mediated endocytosis and the parent drug is released from the lysosome into the cell to restore its cytotoxic activity (Figure 14.3).

In the first preclinical and clinical trials, standard chemotherapeutic agents belonging to the antifolates, vinca alkaloids, or anthracyclines have been conjugated

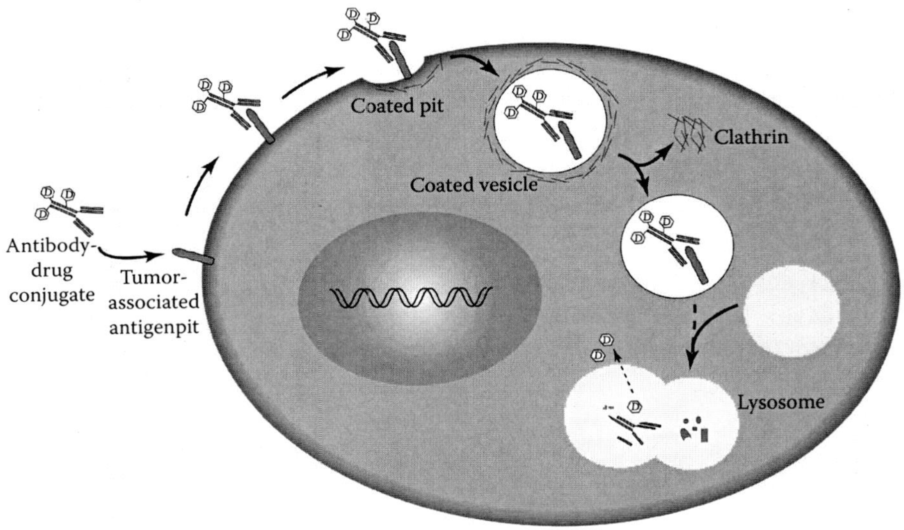

FIGURE 14.3 ***A color version of this figure follows page 176.*** Internalization of antibody–drug conjugates. To regain their cytotoxic activity, the cytotoxic agent has to be cleaved from the chemoimmunoconjugate. Uptake of antibodies predominantly occurs via the clathrin-mediated endocytosis pathway. After binding the respective antigen associated with coated pits, the antibody–drug conjugates will readily be endocytosed initiating their transit via several stages of transport and endosomal vesicles and finally end up in a lysosome. Subsequently, linkers and antibody will be cleaved releasing the cytotoxic agent which—after exit from the lysosomal compartment—exerts its cytotoxic effect.

to mAbs.[17] Such immunoconjugates demonstrated efficiency in preclinical models,[18] but not in the clinical situation. This was because therapeutic levels of the cytotoxic agents within the cells were not achieved as these drugs possessed only moderate cytotoxic potency. In fact, a high cytotoxic potency is important as both the amount of cytotoxic drug which can be conjugated to the antibody and the expression of antigens on the tumor cell are limited.[19] These initial studies, however, were important to gather crucial information concerning the influence of the targeting device on the pharmacokinetics of the conjugates as well as the different strategies to couple cytotoxic agents to the antibody. In this regard, peptide linkers which are stable in serum but can be readily degraded in intracellular compartments were found to be superior to hydrazone linkers. In this respect, disulfide linkers are suitable for conjugation of drug and mAb and are cleavable by disulfide exchange with an intracellular thiol such as gluthatione; this is of particular advantage as the gluthatione concentration in tumor cells is generally higher than in normal cells.[20] The development of molecular engineering techniques and of recombinant DNA technology allowed the production of chimeric, humanized, and human mAbs solving the major problem of early antibody–drug conjugates, that is, the rather high clearance rate of the immunoconjugates due to the development of human antimouse antibody (HAMA) responses in the treated patients.[21] Similarly, modifications of the crystalizable fragment (Fc) modify recruitment of Fc-dependent effector mechanisms and thereby increase the half-life in humans.

The cytotoxic potency of antibody–drug conjugates can be improved by (1) increasing the number of molecules delivered per targeting moiety by means of carriers like liposomes and polymers, and/or (2) using highly cytotoxic compounds. Indeed, targeting extremely cytotoxic drugs, such as calicheamicin, the maytansine derivative DM1, or monomethyl auristatin E, to the tumor results in a marked antitumor activity *in vivo*. Nevertheless, to date only one immunoconjugate containing a cytotoxic drug has been approved by the FDA for the treatment of cancer. This immunoconjugate (Mylotarg) consists of a humanized anti-CD33 mAb (gemtuzumab) linked to the cytotoxic antibiotic *N*-acetyl-γ–calicheamicin. Mylotarg is licensed for the treatment of acute myeloid leukemia (AML) in patients over 60 years of age and considered unsuitable for other chemotherapy.[22] In a multicenter phase II trial the combined response rate was around 30 percent with a median relapse-free survival of 7 months. A recent phase II trial demonstrated that Mylotarg can also be safely applied as first-line therapy with objective responses in 27 percent of the patients.[23] Notably, preliminary studies suggest that the efficacy of Mylotarg treatment may be increased by concomitant intense chemotherapy. Consequently, several phase III trials are currently testing the impact of Mylotarg on standard chemotherapy treatment in leukemia patients (Reference 24 and http://www.clinicaltrials.gov).

Other cytotoxic drugs currently investigated for use in drug–antibody conjugates include doxorubicin, DM1, CC-1065, second-generation taxane, monomethyauristatin, and geldanamycin E (Table 14.2).[25] For example, SGN-15, an antibody–doxorubicin conjugate in combination with docetexal was well tolerated and could serve as an active second- and third-line treatment for patients with non-small cell lung carcinoma (NSCLC).[26] It should be noted that the efficacy of therapeutic mAbs may be significantly improved by chemical conjugation with cytotoxic drugs; for example, the Herceptin–geldanamycin conjugate demonstrated in a xenograft tumor model a much greater antitumor effect than the anti-HER2 mAb alone.[27]

It should be noted, however, that only tumor cells presenting the respective antigen are affected by chemoimmunoconjugates. This limitation can be bypassed by the antibody-directed enzyme prodrug therapy (ADEPT) expanding the antitumor effect toward cells not expressing the respective ligand. ADEPT is the selective activation of anticancer prodrugs by enzymes conjugated to a delivery antibody converting a weakly toxic prodrug into a very toxic agent.[28]

Two characteristics are important for antibody–enzyme conjugates: (1) they have to remain on the cell surface after binding the respective antigen and (2) the enzyme–antibody conjugate has to be cleared rapidly from the circulation to prevent toxicity. The latter can be realized by mannose glycosylation of the targeting moiety.[29] The enzymes used for ADEPT can be divided into three classes:[30] (1) enzymes of nonmammalian origin with no mammalian homologues, (2) enzymes of nonmammalian origin with a mammalian homologue, and (3) enzymes of mammalian origin. Certainly, each class has both advantages and disadvantages. For example, the use of prodrugs cleaved by class I enzymes abates toxicity against normal cells because they are not cleavable by endogenous enzymes. However, class I enzymes cause a strong immune response due to their foreign nature. In contrast, class III

TABLE 14.2
Antibodies and Immunoconjugates in Clinical Development

Name	Targeting Device	Conjugate	Tumor	Phase
Naked Antibodies[a]				
GC1008	Human anti-TGFβ		Renal cell carcinoma; melanoma	I
HGS-ETR2 (lexatumumab)	Human anti-TRAIL-R2 mAb		Refractory pediatric solid tumors	I
MORAb-009	Chimeric antimesothelin mAb		Pancreatic; ovarian, NSCLC; mesothelioma	I[b]
RAV12	Chimeric anti-RAAG12 mAb		Adenocarcinoma	I
MDX-1106	Human anti-PD-1 mAb		NSCLC, colorectal cancer, melanoma, and others	I
BMS-663513	Humanized agonistic anti-CD137 mAb		Advanced solid malignancies	I/II
SGN-33 (HuM195; lintuzumab)	Humanized anti-CD33 mAb		AML	I/II
MDX-CTLA-4 (ipilimumab)	Human anti-CTLA-4 mAb		Melanoma; pancreatic cancer, and others	II
MDX-060	Human anti-CD30 mAb		Hodgkin's disease	II
SGN-30	Chimeric anti-CD30 mAb		Lymphoma	II
SGN-40	Humanized anti-CD40 mAb		Lymphoma	II
CP-675206 (ticilimumab)	Human anti-CTLA-4 mAb		Melanoma, NSCLC, and others	II
CS-1008	Humanized anti-DR5 mAb		Pancreatic cancer	II
cG250 (Rencarex)	Chimeric anti-CAIX/MN mAb		Kidney cancer	III[c]
HuMax-EGFr (Zalutumumab)	Human anti-EGFR mAb		Head and neck cancer	III

TABLE 14.2 (CONTINUED)
Antibodies and Immunoconjugates in Clinical Development

Name	Targeting Device	Conjugate	Tumor	Phase
Radioconjugates				
^{213}Bi-HuM19	Humanized anti-CD33	^{213}Bi	Leukemia	I/II
^{213}Bi-HuM19	Humanized anti-CD33	^{213}Bi	Leukemia	I/II
^{131}I mAb 3F8	Murine anti-GD2 mAb	^{131}I	Central nervous system tumors, melanoma, retinoblastoma	II
^{90}Y-muHMFG1	Murine anti-Muc1 mAb	^{90}Y	Ovarian, gastric carcinoma	II/III
Epratuzumab (Lymphocide)	Humanized anti-CD22 mAb	^{90}Y	NHL	III
Immunoconjugates				
AVE9633 (huMy9-6-DM4)	Humanized anti-CD33 mAb	DM4	AML	I[d]
SGN-35	Chimeric anti-CD30 mAb	Monomethyl auristatin E	Lymphoma	I
CMC-544 (Inotuzumab Ozogamicin)	Humanized anti-CD22 mAb	Calicheamicin	B-cell NHL	I
huC242-DM4	Humanized anti-CanAg mAb	DM4	CRC, pancreatic cancer	I
BB-10901 (huN901-DM1)	Humanized anti-CD56 mAb	DM1	Recurrent or refractory lung cancer or other $CD56^{+}$ solid tumors	II
MLN2704	Humanized anti-PSMA mAb	DM1	Prostate cancer	II
Trastuzumab-MCC-DM1	Humanized anti-Her2 mAb	DM1	Metastatic breast cancer	II
Gemtuzumab Ozogamicin	Humanized anti-CD33 mAb	Calicheamicin	Older patients with relapsed or untreated AML	II/III
ADEPT				
A5CP + ZD2767P	Murine anti-CEA F(ab)2 fragment fused to CPG2	Prodrug ZD2767P	Advanced CRC	I
MFECP1 + ZD2767P	Murine anti-CEA scFv fragment fused to CPG2	Prodrug ZD2767P	CEA expressing tumors	I

TABLE 14.2 (CONTINUED)
Antibodies and Immunoconjugates in Clinical Development

Name	Targeting Device	Conjugate	Tumor	Phase
Immunotoxins				
Hum-195/rGel	Humanized anti-CD33 antibody	Recombinant gelonin	Advanced myeloid malignancies	I
LMB-9	Murine anti-Le(Y) dsFv fragment	Truncated *Pseudomonas* exotoxin A	Advanced pancreatic, esophageal, stomach cancer, or CRC	I Completed
RFB4-dgA	Murine anti-CD22 dsFv fragment	Deglycosylated ricin A chain	Leukemia, lymphoma	I
HD37-dgA + RFB4-dgA [Combotox]	Murine anti-CD19 and anti-CD22	Deglycosylated ricin A chain	NHL	I
CAT-8015	Murine anti-CD22 dsFv fragment	Truncated *Pseudomonas* exotoxin A	Leukemia, lymphoma	I
SS1(dsFv)-PE38 [SS1P]	Murine anti-mesothelin dsFv fragment	Truncated *Pseudomonas* exotoxin	Mesothelin-expressing tumors like mesothelioma, ovarian and pancreatic adenocarcinoma	I Completed[78]
BL22 [CAT-3888]	Murine anti-CD22 dsFv fragment	Truncated *Pseudomonas* exotoxin A	Leukemia and lymphoma	I/II
LMB-2	Murine anti-CD25 scFv fragment	Truncated *Pseudomonas* exotoxin A	Leukemia and lymphoma	I/II
RFT5-dgA	Murine anti-CD25 mAb	Deglycosylated ricin A chain	Melanoma	II
Immunocytokines				
huKS-IL2 (EMD 273066)	Humanized anti-EpCAM mAb	IL-2	Ovarian, prostate, CRC or NSCL cancers	I
hu14.18-IL2 fusion protein (EMD 273063)	Humanized anti-GD2 mAb	IL-2	Melanoma	II
Bispecific double scFv				
BiTE MT103	Rabbit anti-CD19 scFv fragment	scFv fragment of a murine anti-CD3 mAb	Relapsed NHL	I

TABLE 14.2 (CONTINUED)
Antibodies and Immunoconjugates in Clinical Development

Name	Targeting Device	Conjugate	Tumor	Phase
rM28	Murine anti-M-AP scFv fragment	scFV fragment of a murine anti-CD28 mAb	Metastatic melanoma	I/II
Catumaxomab (Removab)	Murine anti-EPCAM IgG_{2a}	Rat anti-CD3 IgG_{2b}	Ovarian cancer	II
Ertumaxomab (Rexomun)	Murine anti-HER2 IgG_{2a}	Rat anti-CD3 IgG_{2b}	Breast cancer	II

Note: Information on ongoing trials was gathered from http://www.clinicaltrials.gov (except as marked). Status quo September 2007.

[a] Many naked antibodies are in clinical trials in combination with cytotoxic drugs.

[b] Received FDA orphan drug status in November 2006.

[c] Also clinical trials phase I/II conjugated to different radionuclides.

[d] Information on ongoing trials gathered from http://www.immunogen.com/wt/page/other. Status quo September 2007.

Abbreviations: AML, acute myelogenous leukemia; CPG2, carboxypeptidase 2; CRC, colorectal cancer; ds, disulfide-stabilized; M-AP, melanoma-associated proteoglycan; NHL, non-Hodgkin's lymphoma; NSCLC, non-small cell lung cancer; PSMA, prostate-specific membrane antigen; ZD2767P, bis-iodo phenol mustard.

enzymes are generally not immunogenic, but endogenous enzymes may cleave prodrugs designed for class III enzymes at inappropriate sites. Preclinical studies have proved and are still proving the feasibility of ADEPT;[31] clinical data, however, are limited and largely restricted to phase I trials.[32] Although the clinical efficiency for ADEPT remains to be established, the preclinical data demonstrating the capacity of this approach to reach high concentrations of cytotoxic drugs in the tumor and to kill antigen-negative tumor cells through a bystander effect without severe systemic toxicity warrant further testing of this approach.

IMMUNOTOXIN

Immunotoxins are proteins that contain a toxin, that is, poisonous substance produced by living cells or organisms, along with an antibody or peptide that binds specifically to target cells. Indeed, toxins or toxin subunits derived from microorganisms, plants, insects, and vertebrates, such as diphtheria toxin, ricin, golin, saporin, pokeweed, and pseudomonas exotoxin A, have been conjugated to mAb and tested for antitumor therapy efficacy.[33] Toxins are enzymes which exert their cytotoxic activity inside the cell, mostly by inhibiting protein synthesis enzymatically, and in most cases one single molecule is sufficient to kill the cell. To be used therapeutically, toxins have to be modified to remove their tissue binding sites.[34] Furthermore, deglycosylation of toxins avoids their rapid clearance by liver cells expressing mannose receptors. Tumor targeting devices include mAbs (immunotoxins) or growth factors, cytokines, and peptide hormones (fusion protein toxins). Immunotoxins bind to specific antigens on the surface of cancer cells and enter cells by endocytosis. Within intracellular acidic compartments, the toxin will be separated from the antibody and will translocate in the cytosol where it exerts its function. The affinity of immunotoxins for antigen determines the quantity and quality of association with the cell surface antigen. Notably, an increased affinity is generally associated with a higher killing capacity of the immunotoxin. However, high local concentrations of immunotoxin may allow nonadsorptive or pinocytic uptake into nontarget cells causing undesirable side effects. In addition to antigen-binding affinity, internalization rate, intracellular processing, and the intrinsic potency of the toxin domain determine the cytotoxic potency of immunotoxins.[35] Importantly, toxicities are not as high as would be concluded from the toxicity of the toxin per se, as the uptake of the immunotoxins into the intracellular compartment is often rather inefficient.

Several clinical trials addressed or currently address the efficiency of toxin conjugates for therapy of hematological and solid tumors (Table 14.2). In this regard, some demonstrated an impressive clinical efficiency with response rates greater than 30 percent. As examples, (1) denileukin diftitox (ONTAK) for the treatment of patients with therapy-refractory cutaneous T-cell lymphoma, (2) LMB-2 and (3) BL-22 both indicated for treatment of hairy cell leukemia, as well as (4) transferrin-CRM107 for therapy of patients with high-grade glioma were tested.[34] To date, however, only denileukin diftitox consisting of diphtheria toxin fragments fused to interleukin-2 (IL-2) as targeting device is approved by the FDA. Observed side effects of targeted-toxin-based therapies include vascular leak syndrome and hepatocyte injury. Notably, to date most success has been achieved treating hematologic tumors.

A major obstacle of both immuno- and fusion-protein toxins is the immunogenicity of the toxin, which limits the number of cycles that can be given.[36] This immunological response to the toxin component not only reduces the serum half-life but also significantly inhibits the cytotoxic activity. Thus, several approaches to decrease the immunogenicity of toxins are being pursued, namely, modifications of the toxin or coadministration of immunosuppressive agents. To date, however, these strategies either have not been effective in patients or have not yet been tested in clinical trials.[37] PEGylation of the toxin, that is, its modification by conjugation with polyethylene glycol, a method most useful for other biologic agents, which not only blocks immunogenicity but also prolongs half-life, might solve the problem. However, PEGylating a toxin is challenging because disturbing sites on a toxin reduces toxin activity. Another solution approach is the usage of humanized toxins,[38] for example, human RNase which downregulates gene expression in the targeted cell. Fusion proteins of RNase with humanized or fully human antibodies should possess only limited immunogenicity. In this regard, two independent research groups generated fusion proteins consisting of either human pancreatic RNase and a fully human anti-ErbB-2 single chain antibody[39] or angiogenin and a humanized anti-CD22 single chain antibody.[40] Both constructs demonstrated specific binding and exhibited cytotoxic/cytostatic activity toward these cells. The anti-ErbB-2-human pancreatic RNase fusion protein also demonstrated an antitumor effect *in vivo*. This effect was dependent on its RNase activity inducing cell death by apoptosis after reaching the cytosol directly from the endosomal compartment.[39] From the wide variety of immunotoxins tested in cell culture, in animal models, and in patients, the most useful appear to be relatively small fusion toxins that contain either growth factors or Fv fragments as ligands. Nevertheless, the expected low toxicity and any possible therapeutic effect have still to be analyzed in clinical trials.

ANTIBODY–CYTOKINE FUSION PROTEINS

In addition to surgery, chemotherapy and radiation immunotherapy evolved rapidly as a treatment of neoplastic disease since the identification of tumor specific or associated antigens.[41] In this regard, antitumor immune responses can be improved by immunregulatory cytokines. In fact, the systemic administrations of IL-2, IL-12, or granulocyte–macrophage colony-stimulating factor (GM-CSF) enhance the immunogenicity of some tumors thereby inducing or improving the immune responses in such a way that they are effective in killing the tumor. Moreover, cytokines can exert direct effects on tumor or tumor stroma cells. For example, tumor necrosis factor α (TNFα) directly damages the tumor-associated vasculature.[42] Because cytokines function physiologically as auto- or paracrine factors with high concentrations only around the producing cell, systemic administration of cytokines neglects this working mechanism. Therefore, it is not surprising that clinically effective doses are frequently accompanied by severe side effects.[43] One solution to this problem is the direct injection of the cytokines into solid tumors to achieve sufficient concentrations of the cytokines at the tumor site while reducing generalized toxicity.[44] Similarly, locoregional treatment is applicable for locally disseminated tumors, that is, by isolated limb perfusion.[45] Generally, however, advanced tumors are neither localized nor accessible, which are the prerequisites for both approaches. An alternative strategy that allows us to address even micrometastases is the genetic fusion of the cytokine to tumor specific antibodies, which allows us to achieve sufficient cytokine concentration at the tumor site (Figure 14.4). This strategy was published in 1991 by two independent research groups.[46,47] Importantly, the fusion of cytokines to antibodies neither impaired the binding capacity of the mAb nor—with the exception of a GM-CSF–antibody fusion protein—the biologic activity of the cytokine. Furthermore, the serum half-life of the cytokine is clearly increased since the clearance rate of these constructs in most cases lies between that of the mAbs and the cytokine.

The best-studied antibody–cytokine fusion proteins are those based on IL-2. This cytokine is known to promote T-cell proliferation and to activate the cytotoxic capacity of T and natural killer (NK) cells. The FDA has approved IL-2 for therapy of advanced melanoma or renal carcinoma. An antibody[ch14.18]–IL-2 fusion protein specific for the disialogangliosid GD2 commonly overexpressed on human melanoma and neuroblastoma cells demonstrated antitumor efficacy in preclinical models *in vitro* and *in vivo*. To this end, the ch14.18–IL2 fusion protein led to the eradication of established experimental and spontaneous metastases both in xenogeneic and syngeneic tumor models.[48] The therapeutic effect was dependent on the fusion of antibody and cytokine since neither the combined treatment of parental antibody and cytokine nor the application of an antibody-IL-2 fusion protein

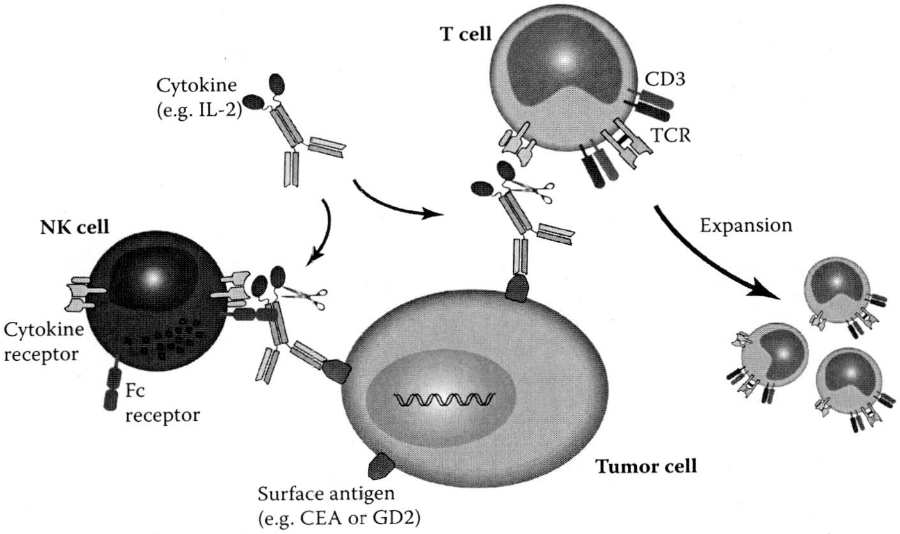

FIGURE 14.4 ***A color version of this figure follows page 176.*** Working mechanism of immunocytokines exemplified for tumor-targeted IL-2. Cytokines fused to mAb specific for a tumor-associated antigen allows the enrichment of these in the tumor microenvironment. In the case of IL-2 it enhances ADCC mediated by F_c-receptor positive effector cells like NK cells. In addition, tumor-targeted IL-2 stimulates the expansion of T cells. Because the linker contains a plasmin-cutting site, the high concentrations of plasmin at the tumor site enable the cleavage of IL-2 from the fusion protein.

directed to an irrelevant antigen displayed any therapeutic effects. Indeed, tumor-specific fusion protein accumulated within the tumor-bearing organs in the preclinical models.[48] The antibody–IL2 fusion protein mediates its therapeutic effect either by innate or adaptive immunity, demonstrating the broad spectrum of activity of IL-2: tumor eradication in the melanoma and colon carcinoma models was dependent on T cells, whereas in the neuroblastoma model the efficacy relied on NK cells. In fact, in the melanoma model, the antibody–IL-2 fusion protein seems to boost a preexisting T-cell response and requires $CD4^+$ T cell help for the therapeutic efficacy. Consequently, in lymphotoxin-α (LT-α) knockout mice where the induction of immune responses are impaired due to the absence of secondary lymphoid tissue, antibody–IL-2 fusion protein treatment is only effective if primed tumor-specific T cells are adoptively transferred prior to IL-2 fusion protein therapy.[49]

In addition, antibody–IL-2 fusion proteins are also effective in adjuvant settings. For example, antitumor immune responses induced by DNA vaccination are clearly enhanced by tumor-targeted IL-2. Similarly, therapeutic vaccination with tumor antigen loaded dendritic cells induced a therapeutic effect only in a murine melanoma model when vaccination was followed by IL-2 administration. Notably, tumor-targeted IL-2 instead of systemic IL-2 treatment clearly improved the therapeutic effect.[50] Interestingly, antibody–IL-2 fusion proteins not only boost vaccination-induced immune responses, but also increase the immunogenicity of antigens:

targeting IL-2 to a soluble, poorly immunogenic antigen triggered an immune response which led to significant tumor growth retardation.[51]

Based on the encouraging preclinical data, antibody–IL-2 fusion proteins were tested in the clinic. To this end, antibody–IL-2 fusion proteins were generally well tolerated; drug-related toxicities equal or larger than grade 3 occurred only in very few patients.[52,53] These immunocytokines led to an increase in total lymphocyte and NK cell counts as well as enhanced NK cell and antibody-dependent cellular cytotoxic activity. Although these trials were designed to evaluate only safety and the maximum tolerated dose, a clinical benefit was observed for the patients with melanoma.[52] Consequently, a phase II trial investigating the therapeutic activity is currently recruiting patients.

Other antibody–cytokine fusion proteins were generated with GM-CSF, IL-12, interferon gamma (IFNγ), TNFα, or LT-α, but have not yet been tested clinically. This may be at least in part be because their generation encountered some difficulties. For example, antibody–GM-CSF fusion proteins are cleared from the plasma much faster than the parental antibody.[54] Moreover, the biologic activity of TNFα, LT-α, and IL-12 within the fusion proteins is markedly lower than the recombinant cytokines.[55] This can be largely ascribed to the nature of the cytokines: TNFα and LT-α are trimeric structures and IL-12 a heterodimer. Nevertheless, all these constructs demonstrated potent antitumor efficiency in preclinical models. Interestingly, LT-α fusion protein seems to evoke a peritumoral tertiary lymphoid tissue which allows the priming of tumor-specific T cell directly at the tumor in a syngeneic melanoma model. Thus, the antitumor effect of LT-α fusion protein is based on the induction of a tumor-specific immune response. Consequently, the antitumor effect is also evident in LT-α knockout mice.[49]

In conclusion, targeting cytokines to the site of interest by an antibody prevents the severe side effects of systemic administrations of high doses of cytokines. Importantly, this approach demonstrated therapeutic efficiency in several preclinical models. To date, however, the clinical characterization is limited to antibody–IL-2 fusion proteins and thus it will be interesting to see if the preclinical therapeutic potential of targeted cytokines can be confirmed in clinical trials.

EVOLVING APPROACHES

In this section several new approaches to fight cancer are discussed. These strategies are based on disrupting aberrant signaling in the tumor cell or enhancing processes enabling the eradication of tumors. Notably, however, most of these attempts have been tested only in *in vitro* culture and in preclinical models so far. Thus, their clinical efficacy has yet to be determined.

In most cancers, tumor cells have an altered threshold, but are still able to undergo programmed cell death; that is, the mechanisms of apoptosis are dysregulated. Consequently, induction of programmed cell death is believed to be a powerful strategy to treat cancer. In this regard, extrinsic factors like FasL (CD95L), TNF-related apoptosis-inducing ligand (TRAIL), or TNF can trigger death receptors initiating apoptosis. Indeed, the feasibility of Fas ligation for cancer therapy has been demonstrated.[56] However, systemic treatment with Fas agonists such as anti-Fas antibodies or multimeric recombinant FasL causes severe systemic toxicities.[57] A solution to this problem was presented by Samel and colleagues who targeted soluble FasL (sFasL) to the tumor using a tumor-specific single chain antibody; by binding to the respective cell surface molecule, these constructs aggregate mimicking the affect of membrane-bound FasL.[58] Importantly, treatment with this antibody–sFasL fusion protein did not cause any systemic toxicity, but prevented tumor growth in a xenogeneic tumor model. Similarly, an anti-CD7–sFasL fusion protein revealed potent CD7-restricted induction of apoptosis in acute T-cell leukemia cell lines.[59] Analogously, a single chain antibody–sTRAIL fusion protein with specificities for CD7,[60] CD19,[61] or the EGFR[62] demonstrated preclinical efficacy. This approach offers the advantage that cancer cells are more sensitive to TRAIL-induced apoptosis than are normal cells. Interestingly, targeting sTRAIL to antigen-expressing tumor cells also induced apoptosis in antigen-negative bystander tumor cells.[62]

Another method to harm the tumor cells directly is to hamper the translation of messenger RNA and thereby to block the expression of proteins. This can be achieved by the use of ribozymes, antisense RNA, RNA interference, or more broadly by RNases. RNA-based strategies hold the potential for exquisite selectivity; rapid plasma elimination, poor intracellular delivery, and nuclease sensitivity, however, hamper their use as therapeutic agents. Their circulation lifetime, however, can be substantially prolonged by use of carriers and chemical modifications. In addition, fusion to targeting devices such as mAbs and specific ligands enables the selective delivery of these agents. For instance, tumor targeting of chemically modified antisense oligonucleotides specific for c-myc[63] or antisense oligonucleotides specific for c-myb was increased by encapsulation within liposomes.[64] Because small interfering (si)RNA became a frequently used tool for specific gene silencing *in vitro*, siRNA is now also seized as a drug candidate. Treatment based on siRNA faces the same problems as antisense oligonucleotides, that is, poor stability, short elimination half-life, and inefficient mechanisms for delivery of siRNA to respective cells.[65] One way to deliver siRNA to tumor cells is by ligand-targeted nanoparticles:[66] such constructs demonstrated tumor selectivity as well as specific silencing of the targeted gene both *in vitro* and *in vivo* in preclinical models. Nevertheless, improvement of tumor-directed delivery and cellular uptake remains crucial to translate the concept of targeted gene expression into clinical success.

Naturally occurring immune processes should be enhanced by targeting either costimulatory or immunmodulatory molecules to the tumor microenvironment. In the early 1990s a large series of reports demonstrated that triggering of costimulatory pathways augments the initiation and/or efficacy of immune responses against the tumor. Thus, it was hypothesized that aiming costimulatory molecules to the cell surface of tumors renders them immunogenic. Indeed, recombinant fusion proteins containing the costimulatory molecule B7 were able to mediate antitumor effects if directed to the tumor.[67] Grosse-Hovest and colleagues improved this approach by generating a bispecific Ab which allowed the targeting of an anti-CD28 antibody to melanoma cells.[68] This bispecific antibody mediated effective tumor eradication by activation of nonspecific cytotoxic cells. Other bispecific diabodies are currently

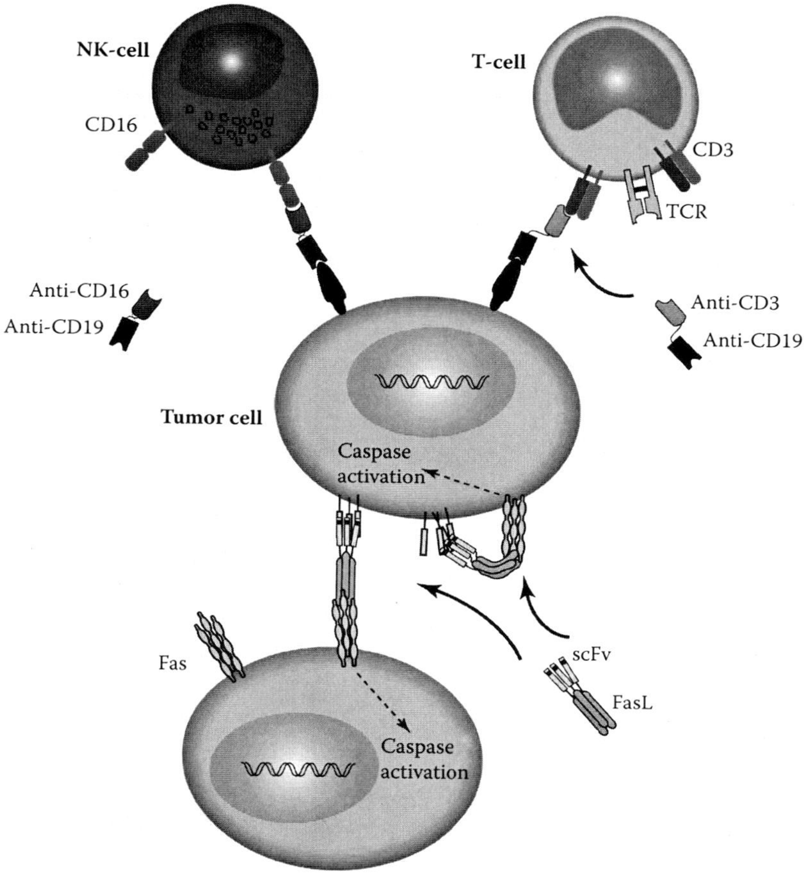

FIGURE 14.5 ***A color version of this figure follows page 176.*** Mode of action of immune modulating or apoptosis-inducing antibody fusion proteins. Bispecific antibody fragments allow the targeting of immunomodulatory molecules to the tumor cell. As exemplified, anti-CD16/CD19 diabody or anti-CD19/CD3 bispecific scFv can enhance the NK cell or T-cell response, respectively, to B-cell lymphomas. Triggering of death receptors by targeting extrinsic factors such as FasL or TRAIL to the cell surface of tumor cells can induce apoptosis in an autocrine as well as a paracrine way.

being tested *in vitro* and *in vivo* in preclinical models for their therapeutic potential. Diabodies specific for CD19 and CD3 or CD19 and CD16 mediated lysis of tumor cells by T cells or mononuclear cells, respectively (Figure 14.5).[69] Interestingly, a synergistic antitumor effect of anti-CD19/CD3 and anti-CD19/CD16 diabodies was observed in a murine NHL model.[70] Furthermore, anti-p-glycoprotein/CD3 or PSMA/CD3 diabodies in combination with activated PBLs inhibited the growth of xenograft tumors.[71,72] The use of multivalent antibodies is likely to further improve this strategy; a trispecific antibody targeting simultaneously CD3 and CD28 and the tumor abolished the requirement of preactivated T cells for tumor lysis.[73] From the evolving approaches described in this section, diabodies present the only ones already in clinical evaluation (Table 14.2).

CONCLUSIONS

mAbs are coming to have an impact on the management of malignant disease. Indeed, they have proved to be well tolerated and effective for the treatment of different cancers and several were consequently approved by the FDA. In addition to their role as cancer therapeutics per se, mAbs can be used as a targeting device increasing the selectivity of different kinds of therapeutic measures. For some of these drugs, such as toxins, fusion to such a targeting moiety is necessary in order to use them therapeutically. However, for their broader use as a targeting device a few problems still have to be solved, namely, (1) immunogenicity, (2) selectivity, and (3) penetration into solid tumors. The immunogenicity of earlier constructs can already be decreased by the generation of fully human antibodies. To increase the selectivity of antibodies and thus reduce unspecific toxicity of fusion proteins, new tumor antigens and the respective mAbs have to be identified. New targets will probably be identified through technologies like gene or protein profiling analysis, enabling us to generate respective mAbs. Most of the targeted therapies approved so far are for lympho-hematopoetic as these are generally easier to access. In contrast, the targeted delivery of macromolecules to solid tumors is difficult, as the therapeutic agent must overcome several obstacles including the vascular endothelium, stromal and epithelial barriers, and high interstitial pressure.[74] In this regard, smaller recombinant mAb constructs like single chain antibodies might penetrate better into the tumor,[75] but have the disadvantage that those are more rapidly cleared from the plasma.[76] An alternative approach for solid tumors is to target the tumor microenvironment in general and especially the tumor blood vessels.[77] The tumor vasculature is very accessible, relatively genetically stable (thereby avoiding antigen loss variants), and by damaging the tumor vasculature, many tumor cells, irrespective their antigen expressing profile, are affected.

Several antibody-based therapies are available for the treatment of cancer. Certainly, the role of mAb therapies should continue to expand, as clinical trials are performed in patients with early stage disease alone and in combination with other therapies. Indeed, antibody-based therapies might be most effective in the treatment of minimal residual disease. Thus, it will be important to select appropriate clinical settings where these agents have the greatest impact on the course of the disease.

REFERENCES

1. Köhler G, Milstein C. Continuous cultures of fused cells secreting antibody of predefined specificity. 1975. *Biotechnology* 1992;24:524–46.
2. Kipps TJ, Parham P, Punt J, Herzenberg LA. Importance of immunoglobulin isotype in human antibody-dependent, cell-mediated cytotoxicity directed by murine monoclonal antibodies. *J Exp Med* 1985 Jan;161(1):1–17.
3. Grillo-Lopez AJ, Hedrick E, Rashford M, Benyunes M. Rituximab: Ongoing and future clinical development. *Semin Oncol* 2002 Feb;29(1 Suppl 2):105–12.

4. Reichert JM, Rosensweig CJ, Faden LB, Dewitz MC. Monoclonal antibody successes in the clinic. *Nat Biotechnol* 2005 Sep;23(9):1073–78.
5. Schaedel O, Reiter Y. Antibodies and their fragments as anti-cancer agents. *Curr Pharm Des* 2006;12(3):363–78.
6. Melero I, Hervas-Stubbs S, Glennie M, Pardoll DM, Chen L. Immunostimulatory monoclonal antibodies for cancer therapy. *Nat Rev Cancer* 2007 Feb;7(2):95–106.
7. Milenic DE, Brady ED, Brechbiel MW. Antibody-targeted radiation cancer therapy. *Nat Rev Drug Discov* 2004 Jun;3(6):488–99.
8. Karagiannis TC. Comparison of different classes of radionuclides for potential use in radioimmunotherapy. *Hell J Nucl Med* 2007 May;10(2):82–88.
9. Gordon LI, Witzig TE, Wiseman GA, Flinn IW, Spies SS, Silverman DH, Emmanouilides C, Cripe L, Saleh M, Czuczman MS, Olejnik T, White CA, Grillo-Lopez AJ. Yttrium 90 ibritumomab tiuxetan radioimmunotherapy for relapsed or refractory low-grade non-Hodgkin's lymphoma. *Semin Oncol* 2002 Feb;29(1 Suppl 2):87–92.
10. Koppe MJ, Postema EJ, Aarts F, Oyen WJ, Bleichrodt RP, Boerman OC. Antibody-guided radiation therapy of cancer. *Cancer Metastasis Rev* 2005 Dec;24(4):539–67.
11. Koppe MJ, Bleichrodt RP, Soede AC, Verhofstad AA, Goldenberg DM, Oyen WJ, Boerman OC. Biodistribution and therapeutic efficacy of (125/131)I-, (186)Re-, (88/90)Y-, or (177)Lu-labeled monoclonal antibody MN-14 to carcinoembryonic antigen in mice with small peritoneal metastases of colorectal origin. *J Nucl Med* 2004 Jul;45(7):1224–32.
12. Cardillo TM, Ying Z, Gold DV. Therapeutic advantage of (90)yttrium- versus (131) iodine-labeled PAM4 antibody in experimental pancreatic cancer. *Clin Cancer Res* 2001 Oct;7(10):3186–92.
13. Jurcic JG, Larson SM, Sgouros G, McDevitt MR, Finn RD, Divgi CR, Ballangrud AM, Hamacher KA, Ma D, Humm JL, Brechbiel MW, Molinet R, Scheinberg DA. Targeted alpha particle immunotherapy for myeloid leukemia. *Blood* 2002 Aug;100(4):1233–39.
14. Goldenberg DM, Sharkey RM. Novel radiolabeled antibody conjugates. *Oncogene* 2007 May;26(25):3734–44.
15. Sharkey RM, Goldenberg DM. Perspectives on cancer therapy with radiolabeled monoclonal antibodies. *J Nucl Med* 2005 Jan;46 Suppl 1:115S–27S.
16. Johnson JR, Ford CH, Newman CE, Woodhouse CS, Rowland GF, Simmonds RG. A vindesine-anti-CEA conjugate cytotoxic for human cancer cells in vitro. *Br J Cancer* 1981 Sep;44(3):472–75.
17. Schrama D, Reisfeld RA, Becker JC. Antibody targeted drugs as cancer therapeutics. *Nat Rev Drug Discov* 2006 Feb;5(2):147–59.
18. Deguchi T, Chu TM, Leong SS, Horoszewicz JS, Lee CL. Effect of methotrexate-monoclonal anti-prostatic acid phosphatase antibody conjugate on human prostate tumor. *Cancer Res* 1986 Aug;46(8):3751–55.
19. Allen TM. Ligand-targeted therapeutics in anticancer therapy. *Nat Rev Cancer* 2002 Oct;2(10):750–63.
20. Jaracz S, Chen J, Kuznetsova LV, Ojima I. Recent advances in tumor-targeting anticancer drug conjugates. *Bioorg Med Chem* 2005 Jun.
21. Carter P. Improving the efficacy of antibody-based cancer therapies. *Nat Rev Cancer* 2001 Nov;1(2):118–29.
22. Bross PF, Beitz J, Chen G, Chen XH, Duffy E, Kieffer L, Roy S, Sridhara R, Rahman A, Williams G, Pazdur R. Approval summary: Gemtuzumab ozogamicin in relapsed acute myeloid leukemia. *Clin Cancer Res* 2001 Jun;7(6):1490–96.
23. Nabhan C, Rundhaugen LM, Riley MB, Rademaker A, Boehlke L, Jatoi M, Tallman MS. Phase II pilot trial of gemtuzumab ozogamicin (GO) as first line therapy in acute myeloid leukemia patients age 65 or older. *Leuk Res* 2005 Jan;29(1):53–57.

24. Tallman MS, Gilliland DG, Rowe JM. Drug therapy of acute myeloid leukemia. *Blood* 2005 Aug 15;106(4):1154–63.
25. Sanderson RJ, Hering MA, James SF, Sun MM, Doronina SO, Siadak AW, Senter PD, Wahl AF. In vivo drug-linker stability of an anti-CD30 dipeptide-linked auristatin immunoconjugate. *Clin Cancer Res* 2005 Jan;11(2 Pt 1):843–52.
26. Ross HJ, Hart LL, Swanson PM, Rarick MU, Figlin RA, Jacobs AD, McCune DE, Rosenberg AH, Baron AD, Grove LE, Thorn MD, Miller DM, Drachman JG, Rudin CM. A randomized, multicenter study to determine the safety and efficacy of the immunoconjugate SGN-15 plus docetaxel for the treatment of non-small cell lung carcinoma. *Lung Cancer* 2006 Oct;54(1):69–77.
27. Mandler R, Kobayashi H, Hinson ER, Brechbiel MW, Waldmann TA. Herceptin-geldanamycin immunoconjugates: pharmacokinetics, biodistribution, and enhanced antitumor activity. *Cancer Res* 2004 Feb;64(4):1460–67.
28. Bagshawe KD. Antibody directed enzymes revive anti-cancer prodrugs concept. *Br J Cancer* 1987 Nov;56(5):531–32.
29. Chester K, Pedley B, Tolner B, Violet J, Mayer A, Sharma S, Boxer G, Green A, Nagl S, Begent R. Engineering antibodies for clinical applications in cancer. *Tumour Biol* 2004 Jan;25(1–2):91–98.
30. Senter PD, Springer CJ. Selective activation of anticancer prodrugs by monoclonal antibody-enzyme conjugates. *Adv Drug Deliv Rev* 2001 Dec;53(3):247–64.
31. Sharma SK, Bagshawe KD, Begent RH. Advances in antibody-directed enzyme prodrug therapy. *Curr Opin Invest Drugs* 2005 Jun;6(6):611–15.
32. Francis RJ, Sharma SK, Springer C, Green AJ, Hope-Stone LD, Sena L, Martin J, Adamson KL, Robbins A, Gumbrell L, O'Malley D, Tsiompanou E, Shahbakhti H, Webley S, Hochhauser D, Hilson AJ, Blakey D, Begent RH. A phase I trial of antibody directed enzyme prodrug therapy (ADEPT) in patients with advanced colorectal carcinoma or other CEA producing tumours. *Br J Cancer* 2002 Sep;87(6):600–7.
33. Frankel AE. New anti-T cell immunotoxins for the clinic. *Leuk Res* 2005 Mar;29(3):249–51.
34. Frankel AE, Kreitman RJ, Sausville EA. Targeted toxins. *Clin Cancer Res* 2000 Feb;6(2):326–34.
35. Hexham JM, Dudas D, Hugo R, Thompson J, King V, Dowling C, Neville DM, Jr., Digan ME, Lake P. Influence of relative binding affinity on efficacy in a panel of anti-CD3 scFv immunotoxins. *Mol Immunol* 2001 Sep;38(5):397–408.
36. Posey JA, Khazaeli MB, Bookman MA, Nowrouzi A, Grizzle WE, Thornton J, Carey DE, Lorenz JM, Sing AP, Siegall CB, LoBuglio AF, Saleh MN. A phase I trial of the single-chain immunotoxin SGN-10 (BR96 sFv-PE40) in patients with advanced solid tumors. *Clin Cancer Res* 2002 Oct;8(10):3092–99.
37. Frankel AE. Reducing the immune response to immunotoxin. *Clin Cancer Res* 2004 Jan;10(1 Pt 1):13–15.
38. Youn YS, Na DH, Yoo SD, Song SC, Lee KC. Carbohydrate-specifically polyethylene glycol-modified ricin A-chain with improved therapeutic potential. *Int J Biochem Cell Biol* 2005 Jul;37(7):1525–33.
39. De Lorenzo C, Di Malta C, Cali G, Troise F, Nitsch L, D'Alessio G. Intracellular route and mechanism of action of ERB-hRNase, a human anti-ErbB2 anticancer immunoagent. *FEBS Lett* 2007 Jan;581(2):296–300.
40. Arndt MA, Krauss J, Vu BK, Newton DL, Rybak SM. A dimeric angiogenin immunofusion protein mediates selective toxicity toward CD22+ tumor cells. *J Immunother* 2005 May;28(3):245–51.
41. Van Pel A, Boon T. Protection against a nonimmunogenic mouse leukemia by an immunogenic variant obtained by mutagenesis. *Proc Natl Acad Sci USA* 1982 Aug;79(15):4718–22.

42. Manusama ER, Nooijen PT, Ten Hagen TL, Van Der Veen AH, De Vries MW, De Wilt JH, Van Ijken MG, Marquet RL, Eggermont AM. Tumor necrosis factor-alpha in isolated perfusion systems in the treatment of cancer: The Rotterdam preclinical-clinical program. *Semin Surg Oncol* 1998 Apr;14(3):232–37.
43. Keilholz U, Stoter G, Punt CJ, Scheibenbogen C, Lejeune F, Eggermont AM. Recombinant interleukin-2-based treatments for advanced melanoma: the experience of the European Organization for Research and Treatment of Cancer Melanoma Cooperative Group. *Cancer J Sci Am* 1997 Dec;3 Suppl 1:S22–S28.
44. Mattijssen V, De Mulder PH, De Graeff A, Hupperets PS, Joosten F, Ruiter DJ, Bier H, Palmer PA, Van den Broek BP. Intratumoral PEG-interleukin-2 therapy in patients with locoregionally recurrent head and neck squamous-cell carcinoma. *Ann Oncol* 1994 Dec;5(10):957–60.
45. Eggermont AM, De Wilt JH, Ten Hagen TL. Current uses of isolated limb perfusion in the clinic and a model system for new strategies. *Lancet Oncol* 2003 Jul;4(7):429–37.
46. Hoogenboom HR, Volckaert G, Raus JC. Construction and expression of antibody-tumor necrosis factor fusion proteins. *Mol Immunol* 1991 Sep;28(9):1027–37.
47. Gillies SD, Young D, Lo KM, Foley SF, Reisfeld RA. Expression of genetically engineered immunoconjugates of lymphotoxin and a chimeric anti-ganglioside GD2 antibody. *Hybridoma* 1991 Jun;10(3):347–56.
48. Becker JC, Pancook JD, Gillies SD, Furukawa K, Reisfeld RA. T cell-mediated eradication of murine metastatic melanoma induced by targeted interleukin 2 therapy. *J Exp Med* 1996 May;183(5):2361–66.
49. Schrama D, Voigt H, Eggert AO, Xiang R, Reisfeld RA, Becker JC. Therapeutic efficacy of tumor-targeted IL2 in LTalpha(–/–) mice depends on conditioned T cells. *Cancer Immunol Immunother* 2005 Sep;1–6.
50. Schrama D, Xiang R, Eggert AO, Andersen MH, Pedersen LS LO, Kampgen E, Schumacher TN, Reisfeld RR, Becker JC. Shift from systemic to site-specific memory by tumor-targeted IL-2. *J Immunol* 2004 May;172(10):5843–50.
51. Dela Cruz JS, Lau SY, Ramirez EM, De Giovanni C, Forni G, Morrison SL, Penichet ML. Protein vaccination with the HER2/neu extracellular domain plus anti-HER2/neu antibody-cytokine fusion proteins induces a protective anti-HER2/neu immune response in mice. *Vaccine* 2003 Mar;21(13-14):1317–26.
52. King DM, Albertini MR, Schalch H, Hank JA, Gan J, Surfus J, Mahvi D, Schiller JH, Warner T, Kim K, Eickhoff J, Kendra K, Reisfeld R, Gillies SD, Sondel P. Phase I clinical trial of the immunocytokine EMD 273063 in melanoma patients. *J Clin Oncol* 2004 Nov;22(22):4463–73.
53. Ko YJ, Bubley GJ, Weber R, Redfern C, Gold DP, Finke L, Kovar A, Dahl T, Gillies SD. Safety, pharmacokinetics, and biological pharmacodynamics of the immunocytokine EMD 273066 (huKS-IL2): Results of a phase I trial in patients with prostate cancer. *J Immunother* 2004 May;27(3):232–39.
54. Dela Cruz JS, Trinh KR, Morrison SL, Penichet ML. Recombinant anti-human HER2/neu IgG3-(GM-CSF) fusion protein retains antigen specificity and cytokine function and demonstrates antitumor activity. *J Immunol* 2000 Nov;165(9):5112–21.
55. Reisfeld RA, Gillies SD, Mendelsohn J, Varki NM, Becker JC. Involvement of B lymphocytes in the growth inhibition of human pulmonary melanoma metastases in athymic nu/nu mice by an antibody–lymphotoxin fusion protein. *Cancer Res* 1996 Apr;56(8):1707–12.
56. Shimizu M, Yoshimoto T, Nagata S, Matsuzawa A. A trial to kill tumor cells through Fas (CD95)-mediated apoptosis in vivo. *Biochem Biophys Res Commun* 1996 Nov;228(2):375–79.
57. Timmer T, de Vries EG, de Jong S. Fas receptor-mediated apoptosis: A clinical application? *J Pathol* 2002 Feb;196(2):125–34.

58. Samel D, Muller D, Gerspach J, Assohou-Luty C, Sass G, Tiegs G, Pfizenmaier K, Wajant H. Generation of a FasL-based proapoptotic fusion protein devoid of systemic toxicity due to cell-surface antigen-restricted activation. *J Biol Chem* 2003 Aug;278(34):32077–82.
59. Bremer E, ten Cate B, Samplonius DF, de Leij LF, Helfrich W. CD7-restricted activation of Fas-mediated apoptosis: A novel therapeutic approach for acute T-cell leukemia. *Blood* 2006 Apr;107(7):2863–70.
60. Bremer E, Samplonius DF, Peipp M, van Genne L, Kroesen BJ, Fey GH, Gramatzki M, de Leij LF, Helfrich W. Target cell-restricted apoptosis induction of acute leukemic T cells by a recombinant tumor necrosis factor-related apoptosis-inducing ligand fusion protein with specificity for human CD7. *Cancer Res* 2005 Apr;65(8):3380–88.
61. Stieglmaier J, Bremer E, Kellner C, Liebig TM, ten Cate B, Peipp M, Schulze-Koops H, Pfeiffer M, Buhring HJ, Greil J, Oduncu F, Emmerich B, Fey GH, Helfrich W. Selective induction of apoptosis in leukemic B-lymphoid cells by a CD19-specific TRAIL fusion protein. *Cancer Immunol Immunother* 2008 Feb;57(2):233–46.
62. Bremer E, Samplonius DF, van Genne L, Dijkstra MH, Kroesen BJ, de Leij LF, Helfrich W. Simultaneous inhibition of EGFR signaling and enhanced activation of TRAIL-R-mediated apoptosis induction by an scFv: sTRAIL fusion protein with specificity for human EGFR. *J Biol Chem* 2005 Mar;280(11):10025–33.
63. Ou X, Tan T, He L, Li Y, Li J, Kuang A. Antitumor effects of radioiodinated antisense oligonuclide mediated by VIP receptor. *Cancer Gene Ther* 2005 Mar;12(3):313–20.
64. Brignole C, Marimpietri D, Pagnan G, Paolo DD, Zancolli M, Pistoia V, Ponzoni M, Pastorino F. Neuroblastoma targeting by c-myb-selective antisense oligonucleotides entrapped in anti-GD(2) immunoliposome: immune cell-mediated anti-tumor activities. *Cancer Lett* 2005 Oct;228(1–2):181–86.
65. Wacheck V, Zangemeister-Wittke U. Antisense molecules for targeted cancer therapy. *Crit Rev Oncol Hematol* 2006 Jul;59(1):65–73.
66. Schiffelers RM, Ansari A, Xu J, Zhou Q, Tang Q, Storm G, Molema G, Lu PY, Scaria PV, Woodle MC. Cancer siRNA therapy by tumor selective delivery with ligand-targeted sterically stabilized nanoparticle. *Nucleic Acids Res* 2004;32(19):e149.
67. Moro M, Gasparri AM, Pagano S, Bellone M, Tornaghi P, Veglia F, Corti A, Casorati G, Dellabona P. Induction of therapeutic T-cell immunity by tumor targeting with soluble recombinant B7-immunoglobulin costimulatory molecules. *Cancer Res* 1999 Jun;59(11):2650–56.
68. Grosse-Hovest L, Hartlapp I, Marwan W, Brem G, Rammensee HG, Jung G. A recombinant bispecific single-chain antibody induces targeted, supra-agonistic CD28-stimulation and tumor cell killing. *Eur J Immunol* 2003 May;33(5):1334–40.
69. Bruenke J, Barbin K, Kunert S, Lang P, Pfeiffer M, Stieglmaier K, Niethammer D, Stockmeyer B, Peipp M, Repp R, Valerius T, Fey GH. Effective lysis of lymphoma cells with a stabilised bispecific single-chain Fv antibody against CD19 and FcgammaRIII (CD16). *Br J Haematol* 2005 Jul;130(2):218–28.
70. Kipriyanov SM, Cochlovius B, Schafer HJ, Moldenhauer G, Bahre A, Le Gall F, Knackmuss S, Little M. Synergistic antitumor effect of bispecific CD19 x CD3 and CD19 x CD16 diabodies in a preclinical model of non-Hodgkin's lymphoma. *J Immunol* 2002 Jul;169(1):137–44.
71. Gao Y, Xiong D, Yang M, Liu H, Peng H, Shao X, Xu Y, Xu C, Fan D, Qin L, Yang C, Zhu Z. Efficient inhibition of multidrug-resistant human tumors with a recombinant bispecific anti-P-glycoprotein x anti-CD3 diabody. *Leukemia* 2004 Mar;18(3):513–20.
72. Buhler P, Wolf P, Gierschner D, Schaber I, Katzenwadel A, Schultze-Seemann W, Wetterauer U, Tacke M, Swamy M, Schamel WW, Elsasser-Beile U. A bispecific diabody directed against prostate-specific membrane antigen and CD3 induces T-cell mediated lysis of prostate cancer cells. *Cancer Immunol Immunother* 2008 Jan;57(1):43–52.

73. Wang XB, Zhao BF, Zhao Q, Piao JH, Liu J, Lin Q, Huang HL. A new recombinant single chain trispecific antibody recruits T lymphocytes to kill CEA (carcinoma embryonic antigen) positive tumor cells in vitro efficiently. *J Biochem* (Tokyo) 2004 Apr;135(4):555–65.
74. Stohrer M, Boucher Y, Stangassinger M, Jain RK. Oncotic pressure in solid tumors is elevated. *Cancer Res* 2000 Aug;60(15):4251–55.
75. Yokota T, Milenic DE, Whitlow M, Schlom J. Rapid tumor penetration of a single-chain Fv and comparison with other immunoglobulin forms. *Cancer Res* 1992 Jun;52(12):3402–8.
76. Adams GP, Schier R, Marshall K, Wolf EJ, McCall AM, Marks JD, Weiner LM. Increased affinity leads to improved selective tumor delivery of single-chain Fv antibodies. *Cancer Res* 1998 Feb;58(3):485–90.
77. Hofmeister V, Vetter C, Schrama D, Brocker EB, Becker JC. Tumor stroma-associated antigens for anti-cancer immunotherapy. *Cancer Immunol Immunother* 2005 Oct;1–14.
78. Hassan R, Bullock S, Premkumar A, Kreitman RJ, Kindler H, Willingham MC, Pastan I. Phase I study of SS1P, a recombinant anti-mesothelin immunotoxin given as a bolus I.V. infusion to patients with mesothelin-expressing mesothelioma, ovarian, and pancreatic cancers. *Clin Cancer Res* 2007 Sep;13(17):5144–49.

15 Role of Vascular Targeting Agents in the Treatment of Solid Tumors

Current and Future Developments

Hans-Peter Gerber, Iqbal S. Grewal, and Ezogelin Oflazoglu

CONTENTS

KEY WORDS: vascular targeting agents, tumor vasculation, antibody–drug conjugates

ABSTRACT

Vascular targeting agents (VTAs), by definition, induce antitumor effects via rapid and selective destruction of blood vessels within established tumors, leading to collapse in tumor blood flow and tumor cell death due to ischemia and extensive hemorrhagic necrosis. VTAs can be broadly divided into two types: low-molecular-weight compounds and ligand-directed macromolecules. Both classes of compounds are designed to exploit pathophysiological differences between tumor and normal tissue endothelial cells to achieve selective targeting of tumor vessels. The clinical data generated with small molecule VTAs validated the safety and efficacy of this therapeutic strategy. Ligand-directed VTAs exploit differences in antigen expression to deliver a therapeutic agent selectively to the tumor vasculature. A variety of ligand-directed agents targeting antigens selectively expressed on tumor vasculature are being developed preclinically. Recently, the use of laser capture microdissection technology combined with gene expression profiling and subtractive proteomic mapping confirmed some of the vascular targets described earlier and identified novel, tumor vascular–specific antigens. In this chapter, we review the advantages of the vascular targeting approach and summarize the preclinical and clinical data generated with this class of compounds.

ADVANTAGES OF VASCULAR TARGETING AGENTS

Biologic Rationale

The recent Food and Drug Administration (FDA) approval of multiple therapeutic compounds interfering with vascular endothelial growth factor-A (VEGF-A)-regulated angiogenesis for the treatment of solid tumors validated the antiangiogenic approach clinically.[1] Given the therapeutic benefit obtained when blocking growth of tumor vasculature in a variety of carcinoma indications, future clinical and preclinical antiangiogenic research is likely to continue to evaluate the utility of novel vascular targeting agents (VTAs) for the treatment of solid tumors. The prerequisite for the success of the vascular targeting strategy is the presence of pathophysiological differences between tumor and normal tissue vessels that can be harnessed therapeutically. A variety of pathophysiological differences between normal and tumor vasculature were described, including the constant remodeling of tumor vessels, their reliance on an intact tubulin cytoskeletal network for their functional integrity, the lack of vessel-associated pericytes, and the increase in vascular permeability of the tumor vasculature.[2–6] Among the most significant differences between normal and tumor endothelial cells is the increased proliferation rate of the latter.[7,8]

A variety of VTAs are being developed, each designed to interfere with specific mechanisms involved in the regulation of tumor angiogenesis. Small molecule compounds targeting tubulin structures, which are critical for the cytoplasmic changes associated with proliferating cells, were shown to be efficacious as VTAs in preclinical and clinical studies. Among the most promising tubulin-interfering agents identified so far are derivatives of combretastatins, colchicines, and dolastatin 10.[9]

A growing body of evidence revealed the existence of genes or tumor-specific splice forms selectively expressed on pathologic tumor vasculature relative to normal, quiescent vessels (reviewed in References 5 and 10), representing unique therapeutic targets. More recently, gene expression profiling studies and proteomic mapping of endothelial cells isolated from vascular beds of experimental or patient tumors revealed distinct molecular signatures associated with tumor vasculature. Several of these newly identified genes are being pursued as putative targets for vascular targeting purposes.[11–13] Furthermore, the internalization and turnover rates of several membrane-associated genes are increased on endothelial cells exposed to tumor-like conditions *in vitro*.[14] Finally, the differences in the proteolytic environment and changes in the composition of the extracellular matrix within tumor vasculature can be exploited in the development of tumor-specific VTAs.[15,16] In sum, the magnitude and the variety of the molecular and cellular differences between the endothelium in tumors and normal tissues generate unique opportunities for the development of therapeutic compounds targeting tumor vasculature.

Pharmacological Rationale

In contrast to small-molecular-weight compounds, therapeutic antibodies and antibody–drug conjugates permeate tumors poorly and unevenly. Typically, only 0.001 to 0.01 percent of the injected antibody localizes to tumors in humans.[17] Such poor tumor distribution is caused by several interrelated factors. First, the dense packing of tumor cells and the fibrous tumor stroma constitute a physical barrier to macromolecular transport. Second, elevations in the interstitial fluid pressure within the solid tumor mass hinders extravasation and fluid convection.[18,19] Third, the distribution of antibodies with high binding affinities is often localized to perivascular regions, limiting the amounts of compounds reaching tumor cells at more distant sites and causing poor antitumor efficacy.[20]

Endothelial cells represent the first cell layer encountered by therapeutics administered intravenously. Because vascular endothelial cells are directly accessible to circulating agents, the limited tumor perfusion of high-molecular-weight compounds does not apply for VTAs. In addition, thousands of tumor cells are dependent on each tumor capillary for oxygen and nutrients, and therefore, even limited damage to the tumor vasculature is likely to cause death of multiple tumor cells.[21] From a drug development perspective, it is worth noting that the intravascular drug levels represent "on-target" concentrations. Such favorable exposure/efficacy relationship of vascular targeting agents provides an important advantage by potentially reducing undesired on- and/or off-target toxicities. In addition, the vascular targeting approach should be applicable to all solid tumors, because growth of most if not all carcinomas is dependent on intact tumor vasculature.[22] Importantly, endothelial cells are less likely to undergo somatic mutations and therefore are not prone to development of drug resistance, which is frequently associated with repeated administration of cytotoxic drugs targeting neoplastic cells.[23]

Because of the different identity of the cells targeted, improved antitumor effects when combining cytotoxic compounds targeting tumor cells and vascular targeting agents might be expected. In support of this notion, several VTAs, when combined with other antineoplastic agents, induced improved antitumor responses in preclinical studies.[24] Thus, combination of both types of compounds may be advantageous for the treatment of advanced-stage carcinomas. This may be of relevance for clinical development, as standard of care regimens frequently include chemotherapeutics. However, VTAs are designed to destroy the vasculature and therefore carry the potential to interfere negatively with drug delivery to the tumor cells, since the tumor distribution is dependent on a functional vasculature.[24] Therefore, the effects of combining VTAs with chemotherapy will need to be determined carefully for each indication and treatment regimen.

Technological Improvements Enabling the Identification of Novel Tumor Vascular-Specific Antigens

The recent advancements in genomic and proteomic technologies facilitate the identification of genes selectively upregulated on tumor vasculature. Several studies using these platforms confirmed previously identified targets and identified novel, tumor-specific vascular antigens. Employment of laser capture microdissection technology to study gene expression in endothelial cells in preclinical and/or patient tumors revealed several promising therapeutic targets, including angiopoietin-2.[12,25–27] In addition, several genes with tumor vasculature-specific splice variants were described,[28] and antibodies targeting such splice variants, including fibronectins and tenascin-C, are currently being evaluated in early-stage clinical trials.[5]

Subtractive proteomic mapping technology, including two-dimensional gel electrophoresis and multidimensional mass spectrometry, was also instrumental in the identification of novel tumor vascular targets. The strategies employed include *in vivo* coating of the vasculature of human and mouse tumors using colloidal silica.[29,30] Annexin A1 was among several proteins identified by this approach and ligand-based VTAs validated annexin A1 as a therapeutic target by inducing significant tumor growth delays in preclinical models.[29] Furthermore, *in vivo* chemical labeling of tumor vasculature with biotin led to the identification of 184 tumor vasculature-associated proteins expressed in tumor-bearing mice. Among them, periostin, annexin A4, melanoma-associated antigen MG50, and versican are significantly overexpressed on tumor relative to normal vessels.[31] Other groups employed *in vivo* phage display technology to identify peptides that bind to sites specifically expressed on tumor vasculature, but not normal vessels. This methodology is based on the perfusion of mouse or human tissues with targeting vehicles derived from phage display libraries. By using this approach, several specific peptides binding to NG2, MMP2/MMP-9, aminopeptidase N, and α_v integrins were identified.[32–34]

To monitor changes in gene expression on the mRNA level, serial analysis of gene expression (SAGE) was successfully applied and led to the identification of significant molecular differences between normal and tumor vasculature.[35] More recently, the same group studied gene expression patterns in endothelial cells derived from the blood vessels of normal, resting tissues, tumors, and regenerating livers using a model of syngeneic liver tumors in mice.[36] These studies identified multiple genes which became selectively upregulated on tumor vasculature of certain organs. Further analysis of CD276, the most promising cell surface tumor endothelial marker identified in this study, confirmed that this cell surface receptor is overexpressed during pathological, but not physiological angiogenesis in human tissues.[36] Finally, a subtractive algorithm was developed to analyze the gene expression information that is available within the public databases to identify novel endothelial-specific expressed genes.[37] Subsequent screening of the most promising "initial hits" by *in situ* hybridization identified magic roundabout (ROBO4) and disulfide isomerase (EndoPDI) as candidate tumor endothelial markers.[38,39] The cross section from overlapping gene expression data of endothelial cells exposed to tumor-like conditions *in vitro* and data from patient colon tumor samples resulted in the identification of Stanniocalcin (STC1) as a putative tumor vascular target.[40]

Despite the progress in the clinical testing of small-molecule compounds targeting tumor vasculature and the strong rationale for the development of ligand-directed VTAs, only a small number of ligand-directed VTAs are currently undergoing clinical development. The reasons for such delay in the clinical development of ligand-directed VTAs are unclear. However, the requirements for surrogate antibodies, recognizing the murine orthologs to investigate the antiangiogenic effects in preclinical tumor models in mice, represents a major disadvantage in the development of mouse hybridoma-derived antibodies for vascular targeting purposes. However, the availability of phage-derived, cross-species reactive vehicles may help to overcome these historic limitations.

VASCULAR TARGETING AGENTS IN CLINICAL DEVELOPMENT

Small-Molecule Vascular Targeting Agents

In the course of the clinical evaluation of several tubulin-binding agents such as colchicines and vinca alkaloids, it became apparent that this class of drugs induces vascular damage within tumors, but only at doses that were close to their maximum tolerated dose (MTD, reviewed in Reference 41). Currently, second generations of several low-molecular-weight tubulin-binding agents are undergoing clinical development (Table 15.1). A major advantage of these compounds compared to ligand-directed monoclonal antibodies is their cross-species reactivity, which allows for safety and efficacy assessments in the same species without the need to develop surrogate molecules. The most advanced compounds were shown to induce blood flow reductions in tumors when assessed by dynamic contrast-enhanced magnetic resonance imaging (DCE-MRI), and clinical antitumor activities were also reported.[42] The emerging target product profile for VTA derived from these early studies suggests that acute administration is preferable as opposed to chronic, because most VTAs induced more immediate effects, preferentially in the center of a solid tumor mass. In preclinical experiments, up to 95 percent of the tumor cells underwent apoptosis 2 hours posttreatment.[10] In summary, pronounced antitumor effects when

TABLE 15.1
Vascular Targeting Agents Undergoing Clinical Development

Drug	Drug Type	Company	Development Stage	Comments Refs.
L19 antibody and radio- and immunoconjugates	Splice form of fibronectin (EDB)	Schering AG	Phase II	IL-2 and TNFα conjugates, 87–91
F16 , G11, 81C2, BC-2, BC-4, ST2146 antibodies and radio- and immunoconjugates	Splice form of tenascin-C	Schering AG	Phase I+II	Immunocytokines, radioimaging and radioimmunotherapy, 49, 51, 92–95
3G4 Bavituximab	Phosphatidyl-serine	Peregrine Pharmaceuticals	Phase I	Naked mAb, 96–98
J591 MLN2704	PSMA	Millennium Pharmaceuticals, Inc.	Phase I/II	Immunoscintigraphic imaging and radioimmunotherapy in prostate cancer, 54, 99–101
LM985	Flavonoids		Phase I	46, 102, 103
DMXAA AS1404	Flavonoids	Antisoma	Phase I/II	47, 104
CA4P Prodrug Oxi4503	Tubulin-binding, prodrug of combretastatin A-4	Oxigen	Phase I/II	105
AVE8062A	Tubulin binding	Aventis	Phase I/II	Synthetic combretastatin, 104
ZD6126	Tubulin-binding, colchicine analogue	AstraZeneca	Phase I/II	104
ABT751	β-Tubulin inhibitor	Abbott	Phase I	Synthetic sulfonamide, 106
TZT-1027	Tubulin binding	Daiichi	Phase I	Synthetic dolastatin 10 derivative, also known as auristatin-PE, 104

testing small-molecular-weight VTAs against advanced solid tumors were observed, emphasizing the potential of this class of therapeutic compounds to prevent the progression of established tumors. Importantly, small-molecule VTAs can be combined with standard cytotoxics (reviewed in Reference 24) or experimental antiangiogenesis drugs[43] or radiotherapy,[44] resulting in increased efficacy without affecting toxicity in preclinical studies.

The clinically most advanced class of small-molecular-weight compounds are flavonoids (FAAs). Their mechanism of action remains mostly unknown. Induction of tumor necrosis factor alpha (TNFα), however, was shown to be antiangiogenic under certain experimental conditions and is considered a major contributor to their antiangiogenic activity.[45] Among the FAAs, LM985 and DMXAA are being developed in early-stage clinical trials (Table 15.1)[46,47] Tubulin-binding agents such as colchicines, vincristine, vinblastine, and combretastatins destabilize the tubulin cytoskeleton, leading to disruption of the tumor vasculature.[24] Several tubulin-binding agents were tested in clinical studies, including AVE8062A (combretastatin A-4 prodrug, formerly AC-7700), Oxi 4503 (combretastatin A-1 disodium phosphate), ZD6126 (phosphate prodrug of *N*-acetylcholine), ABT-751 (sulfonamide, binding to β-tubulin), and TZT-1027 (synthetic derivative of dolastatin 10, formerly Auristatin-PE) (Table 15.1). In general, these compounds induce vascular damage and a reduction in tumor blood perfusion. This is in marked contrast to conventional angiogenesis inhibitors, which are most active when administered during early stages of tumor progression.[10,24,48] The most advanced VTAs lack the traditional cytotoxic side effects of chemotherapeutic agents, including myelosuppression, stomatitis, mucositis, and alopecia. However, targeting the tumor vasculature can affect vascular functions in normal organs and cause undesirable effects, such as acute coronary syndromes and alterations in electrophysiological and hemodynamic parameters.[41]

Clinical Development of Ligand-Directed Vascular Targeting Agents

Ligand-directed VTAs bind to antigens that are selectively expressed and accessible on tumor vasculature. The two most advanced clinical programs include monoclonal antibodies targeting tumor-specific splice forms of fibronectin and tenascin. L19 is a monocolonal antibody interacting specifically with the EDB-containing splice form of fibronectin, which is selectively expressed on the tumor vasculature of a variety of solid tumors. A variety of therapeutic derivatives of the L19 antibody were generated and shown to be efficacious when tested in preclinical studies.[5]

Tenascins comprise a family of four extracellular matrix glycoproteins that are widely expressed in many different connective tissues. Detailed immunhistochemical analysis revealed that the C-terminal domain of tenascin-C is overexpressed in aggressive brain tumors and some lung tumors, with a prominent perivascular staining pattern.[49] Two high-affinity human antibodies, G11 and F16, were shown to bind to tumor vascular-specific isoforms and to induce antitumor effects when tested in preclinical experiments.[50,51]

Prostate specific stem cell antigen (PSMA) is a membrane glycoprotein that is predominantly expressed in the prostate, and serum concentrations are often increased in patients with prostate cancer.[52] Several studies reported overexpression of PSMA in the neovasculature of different solid tumors from patients.[53] The anti-PSMA antibody huJ591 has been conjugated to various toxins and radionucleotides and is currently being tested for therapeutic applications in prostate cancer and several solid tumor indications.[54]

PROMISING TARGETS FOR LIGAND-DIRECTED VASCULAR TARGETING AGENTS

Several therapeutic compounds targeting antigens expressed on the tumor endothelium, including gene therapy, monoclonal antibodies, and peptides, have been developed preclinically. The majority of ligand-directed VTAs tested in preclinical studies consist of function-blocking monocolonal antibodies or antibody–drug conjugates (ADCs). Here we review some of the most promising approaches and the efficacy data reported from preclinical studies testing ligand-directed vascular targeting compounds.

INTEGRINS

Integrin $\alpha_v\beta_3$ is among the most extensively studied tumor vascular-specific antigens. It is upregulated on a significant proportion of tumor vessels in colon, pancreas, lung, and breast carcinoma. However, $\alpha_v\beta_3$ is also expressed on normal vasculature in the colon (40 to 60 percent), pancreas (25 to 60 percent), lung (<35 percent), and breast (<25 percent).[55] On neoplastic cells, $\alpha_v\beta_3$ expression was reported on melanoma, glioblastoma multi forme (GBM), ovarian, breast, and prostate carcinomas.[56] The integrins $\alpha_5\beta_1$ and $\alpha_v\beta_5$ are expressed in a vascular bed-specific manner, and concomitant blocking of both integrins by antibodies was more efficacious compared to single-agent treatment. The integrins $\alpha_6\beta_4$, $\alpha_{IIb}\beta_3$ and $\alpha_5\beta_5$ were also found upregulated on the vasculature during tissue regeneration and inflammation, but are absent on normal vasculature (reviewed in Reference 56). Several RGD peptides binding to integrin $\alpha_v\beta_3$ were originally identified by using phage display technology. Peptide drug conjugates targeting $\alpha_v\beta_3$, such as conjugates of the tubulin-interfering auristatins, were successfully developed for imaging and therapy applications.[57,58] These experiments provided proof of concept validation of the utility of tubulin-interfering agents conjugated to vascular targeting vehicles to interfere with tumor angiogenesis.

ENDOGLIN (CD105)

Strong upregulation of CD105 on tumor vasculature was noticed when conducting immunohistochemical (IHC) analysis of sections derived from human breast and colon carcinomas. Moderate staining of the tumor vasculature in tumor lesions from patients with Hodgkin's disease (HD), lung, melanoma, ovarian, and osteosarcoma was also reported.[59] However, controversy exists regarding the expression of CD105 on normal vasculature. IHC analysis of normal tissues revealed CD105 expression on normal vasculature in all major organs, except breast.[60] Similarly, IHC analysis demonstrated expression of CD105 in the vasculature of normal adult adrenal gland, spleen, tonsil, thymus, and placenta.[60] In addition, significant levels of endoglin expression were found on normal brain endothelium, mesenchymal cells, arteries, and arteriols.[61] Monoclonal antibodies targeting endoglin were tested in biodistribution and imaging studies. Furthermore, ricin-A or radioimmunoconjugates were tested for therapy purposes. The latter were shown to induce moderate tumor growth delays in preclinical tumor xenograft experiments (reviewed in Reference 62).

Robo4

On human tumors, Robo4 was found to be exclusively expressed on the tumor vasculature.[38] When conducting *in situ* and immunohistochemical analysis of human tumor samples, Robo4 expression was noted on the vasculature of primary lung, liver, and kidney cancers and in their respective metastatic lesions.[63] Robo4 is expressed at relatively low levels in normal adult endothelial cells and other, not well defined cells in the lung, kidney, and liver.[63] Robo4 is involved in the regulation of axon guidance in the central nervous system (CNS) and is strongly upregulated in the developing embryo.[64] Recombinant Robo4 induced endothelial cell proliferation *in vitro*, demonstrating that it is has direct proangiogenic activities.[65] Robo4-directed therapy experiments have not been reported in the literature.

Aminopeptidase N (CD13)

By employing *in vivo* phage display technology, a peptide binding specifically to a form of aminopeptidase N, which is expressed predominantly on tumor vasculature, was identified.[66] Weak expression of CD13 on stromal and neoplastic cells in prostate cancer was also reported.[67] IHC analysis using conventional anti-CD13 monoclonal antibodies revealed vascular expression in renal, breast, and prostate tumors.[66] On normal cells, CD13 expression was found on most cells of myeloid origin, including monocytes, macrophages, and granulocytes. Abundant expression of CD13 was reported on normal renal proximal tubules, small intestine, prostatic epithelial cells, bile duct, mast cell, fibroblast, and smooth muscle cells. Moderate antitumor effects were observed for the anti-CD13 peptide (CNGRCC) fused to TNFα, when administered to mice implanted with syngeneic lymphomas.[67]

Delta-Like 4 (DLL4)

Selective upregulation of DLL4 on tumor vasculature was reported from *in situ* hybridization (ISH) studies conducted on sections from renal cell carcinomas, where a ninefold upregulation of DLL4 relative to normal kidneys was noticed.[68] Similarly, DLL4 upregulation in tumor vasculature was identified when analyzing 22 bladder cancers and 9 normal bladder tissues by ISH.[69] On normal cells, DLL4 protein is present on a subset of hematopoietic cells in the bone marrow. Recombinant DLL4 functionally suppresses the self-renewal capacity of acute myeloid leukemia (AML) cells and long-term growth via binding to the NOTCH receptors.[70] DLL4 is regulated by hypoxia, vascular endothelial growth factor (VEGF), and FGF, which provides a molecular and physiological explanation for the broad expression of DLL4 on the vasculature of most solid tumors in humans (reviewed in Reference 5). The heterozygous lethality observed in gene knockout experiments suggested that DLL4 plays a key role during embryonic vascular development.[71] Pharmacological blockade of DLL4 was associated with dysregulation of tumor angiogenesis and reduced tumor growth in preclinical studies.[72,73] However, efficacy and biodistribution studies with DLL4-directed VTAs have not been reported.

VEGFR-2 (CD 309)

VEGFR-2 was found upregulated on endothelial cells of most vascular beds in human tumors. In addition, VEGFR-2-positive neoplastic cells were found in several tumor types, including breast, lung, and ovarian.[74] In experimental tumor cell lines, VEGFR-2 is present on breast cancers and on lung and tumor cells present within lesions of HD.[75] On nonmalignant tissues in humans, low levels of VEGFR-2 expression were found on subsets of human hematopoietic stem cells. On normal tissues, VEGFR-2 expression is prominent on capillaries within lymph nodes and tonsils. The most significant expression of VEGFR-2 was found in the choriocapillary of the eye, in the capillaries in the duodenum, and in the appendix as revealed by IHC analysis. Tonic, low level of VEGFR-2 expression was noticed in the liver capillaries to the bile duct.[76] Despite the high interest in the VEGF-A/VEGFR-2 signal transduction pathway in oncology, only a few reports of ligand-directed VTAs targeting this pathway can be found in the literature. Among them, a VEGF-A diphtheria toxin conjugate induced significant antiangiogenic effects when tested in the chicken chorioallantoic membrane assay.[77] Similarly, a VEGF-A fusion protein of the toxin gelonin, a protein synthesis inhibitor, was effective in the inhibition of growth metastatic human breast tumors in mice.[78]

CXCR4 (CD184)

CXCR4 expression was found on neovessel endothelial cells in glioblastoma tumors[79] and is also upregulated on HUVEC cells exposed to hypoxia.[80] Among the members of the chemokine receptor family, CXCR4 is most commonly upregulated in human and murine cancer cell lines.[81] CXCR4-positive neoplastic cells are present in over 23 tumor cell lines, representing a variety of hematological malignancies and carcinomas, including breast, prostate, lung, kidney, and pancreas. On normal cells, CXCR4 is expressed on hematopoietic stem cells, T and B cells, dendritic cells, macrophages, neutrophils, and colonic epithelium.[81]

In breast and renal carcinoma lines *in vitro*, expression of CXCR-4 is regulated by VEGF and NFκB. CXCR4 is also involved in the regulation of homing of vascular progenitor cells (endothelial cell progenitor cells, EPCs) to tumor vasculature. Based on these observations, targeting of CXCR4 appears as a promising therapeutic strategy, as it may allow for targeting of both tumor vasculature and neoplastic tumor cells by the same compound.

Tie2 (CD202b)

IHC analysis of human tumor specimen revealed 80 to 98 percent of Tie2-positive vessels in melanoma, colorectal, pancreatic, breast, brain, prostate, and lung carcinomas.[82] Upregulation of Tie2 was also found on neoplastic cells in hepatocellular carcinoma, astrocytoma, and non-small cell lung carcinoma. Correlations between

Tie2 expression levels and the metastatic potential and tumor stage were identified, supporting an important role of Tie2 during tumor progression.[82] Several reports document moderate expression of Tie2 on blood vessels in the entire spectrum of the quiescent vasculature (arteries, veins, and capillaries) in a wide range of adult tissues. Importantly, a Tie2 form that was specifically immunoprecipitated from quiescent, adult tissues is tyrosine phosphorylated, suggesting a role in vascular maintenance functions in adults.[83] Tie2 expression was detected in the endothelium of normal breast tissues examined.[84] On neoplastic cells, Tie2 was present on prostate and gastric adenocarcinomas.[85] Tie2 appears not to be directly regulated by VEGF, but is induced by TNFα and hypoxia (reviewed in Reference 82).

FUTURE DIRECTIONS

The therapeutic utility of targeting tumor vasculature by using VTAs has been firmly established in a variety of experimental studies. Small-molecule VTAs induce rapid reductions in tumor blood flow and extensive necrosis in experimental tumors.[86] Clinical trials provided evidence of antitumor activity of VTAs when administered as monotherapy and preliminary evidence for tolerability in humans emerged. Only limited experience has been gathered with ligand-directed VTAs in the clinic. A handful of ligand-directed VTAs are undergoing clinical development (Table 15.1).[54,62,86] Thus, despite the promising preclinical antitumor activity of several antibody–drug conjugates directed toward cell surface antigens expressed on the tumor vasculature, only limited data regarding the effects of ligand-directed VTAs in clinical trials are available. The recent advances in genomics and proteomics will likely result in further improvements of the tumor selectivity of ligand-directed approaches to the endothelium (reviewed in Reference 5). The availability of phage display technology, enabling the selection of cross-species reactive compounds, holds great promise for accelerated preclinical and clinical development of VTAs. Finally, conjugation of ligand-directed vascular targeting vehicles with cytotoxic compounds shown to be effective VTAs, including tubulin inhibitors, is likely to result in ADCs with improved selectivity and potency to target tumor vasculature.

REFERENCES

1. Ferrara, N., et al., Targeting VEGF-A to treat cancer and age-related macular degeneration. *Annu Rev Med*, 2007. 58: 491–504.
2. Pluda, J.M., Tumor-associated angiogenesis: Mechanisms, clinical implications, and therapeutic strategies. *Semin Oncol*, 1997. 24(2): 203–18.
3. Jain, R.K., Molecular regulation of vessel maturation. *Nat Med*, 2003. 9(6): 685–93.
4. Jain, R.K., Normalization of tumor vasculature: An emerging concept in antiangiogenic therapy. *Science*, 2005. 307(5706): 58–62.
5. Neri, D. and R. Bicknell, Tumour vascular targeting. *Nat Rev Cancer*, 2005. 5(6): 436–46.
6. Darland, D.C. and P.A. D'Amore, Blood vessel maturation: vascular development comes of age. *J Clin Invest*, 1999. 103(2): 157–58.
7. Denekamp, J., Endothelial cell proliferation as a novel approach to targeting tumour therapy. *Br J Cancer*, 1982. 45(1): 136–39.

8. Denekamp, J. and B. Hobson, Endothelial-cell proliferation in experimental tumours. *Br J Cancer*, 1982. 46(5): 711–20.
9. Tozer, G.M., C. Kanthou, and B.C. Baguley, Disrupting tumour blood vessels. *Nat Rev Cancer*, 2005. 5(6): 423–35.
10. Thorpe, P.E., D.J. Chaplin, and D.C. Blakey, The first international conference on vascular targeting: Meeting overview. *Cancer Res*, 2003. 63(5): 1144–47.
11. Mittal, V. and D.J. Nolan, Genomics and proteomics approaches in understanding tumor angiogenesis. *Expert Rev Mol Diagn*, 2007. 7(2): 133–47.
12. Buckanovich, R.J., et al., Use of immuno-LCM to identify the *in situ* expression profile of cellular constituents of the tumor microenvironment. *Cancer Biol Ther*, 2006. 5(6): 635–42.
13. Pen, A., et al., Molecular markers of extracellular matrix remodeling in glioblastoma vessels: Microarray study of laser-captured glioblastoma vessels. *Glia*, 2007. 55(6): 559–72.
14. Lampugnani, M.G., et al., Vascular endothelial cadherin controls VEGFR-2 internalization and signaling from intracellular compartments. *J Cell Biol*, 2006. 174(4): 593–604.
15. Albig, A.R., et al., Transcriptome analysis of endothelial cell gene expression induced by growth on matrigel matrices: Identification and characterization of MAGP-2 and lumican as novel regulators of angiogenesis. *Angiogenesis*, 2007. 10(3): 197–216.
16. Sato, M., W. Arap, and R. Pasqualini, Molecular targets on blood vessels for cancer therapies in clinical trials. *Oncology* (Williston Park), 2007. 21(11): 1346–52; discussion 1354–55, 1367, 1370 passim.
17. Epenetos, A.A., et al., Limitations of radiolabeled monoclonal antibodies for localization of human neoplasms. *Cancer Res*, 1986. 46(6): 3183–91.
18. Baxter, L.T. and R.K. Jain, Transport of fluid and macromolecules in tumors. IV. A microscopic model of the perivascular distribution. *Microvasc Res*, 1991. 41(2): 252–72.
19. Jain, R.K., Vascular and interstitial barriers to delivery of therapeutic agents in tumors. *Cancer Metastasis Rev*, 1990. 9(3): 253–66.
20. Weinstein, J.N. and W. van Osdol, Early intervention in cancer using monoclonal antibodies and other biological ligands: Micropharmacology and the "binding site barrier." *Cancer Res*, 1992. 52(9 Suppl): 2747s–2751s.
21. Denekamp, J., Vascular attack as a therapeutic strategy for cancer. *Cancer Metastasis Rev*, 1990. 9(3): 267–82.
22. Folkman, J., Tumor angiogenesis. *Adv Cancer Res*, 1985. 43: 175–203.
23. Ferrara, N., H.P. Gerber, and J. LeCouter, The biology of VEGF and its receptors. *Nat Med*, 2003. 9(6): 669–76.
24. Horsman, M.R. and D.W. Siemann, Pathophysiologic effects of vascular-targeting agents and the implications for combination with conventional therapies. *Cancer Res*, 2006. 66(24): 11520–39.
25. Shen, Y., et al., A novel methodology to probe endothelial differential gene expression profile reveals novel genes. *Endothelium*, 2007. 14(6): 303–14.
26. Zhang, L., et al., Tumor-derived vascular endothelial growth factor up-regulates angiopoietin-2 in host endothelium and destabilizes host vasculature, supporting angiogenesis in ovarian cancer. *Cancer Res*, 2003. 63(12): 3403–12.
27. Espina, V., et al., Laser-capture microdissection. *Nat Protoc*, 2006. 1(2): 586–603.
28. Zheng, P.P., et al., Differential expression of splicing variants of the human caldesmon gene (CALD1) in glioma neovascularization versus normal brain microvasculature. *Am J Pathol*, 2004. 164(6): 2217–28.
29. Oh, P., et al., Subtractive proteomic mapping of the endothelial surface in lung and solid tumours for tissue-specific therapy. *Nature*, 2004. 429(6992): 629–35.

30. Scheurer, S.B., et al., Identification and relative quantification of membrane proteins by surface biotinylation and two-dimensional peptide mapping. *Proteomics*, 2005. 5(11): 2718–28.
31. Castronovo, V., et al., A chemical proteomics approach for the identification of accessible antigens expressed in human kidney cancer. *Mol Cell Proteomics*, 2006. 5(11): 2083–91.
32. Arap, W., et al., Steps toward mapping the human vasculature by phage display. *Nat Med*, 2002. 8(2): 121–27.
33. Rajotte, D., et al., Molecular heterogeneity of the vascular endothelium revealed by in vivo phage display. *J Clin Invest*, 1998. 102(2): 430–37.
34. Pasqualini, R. and E. Ruoslahti, Organ targeting in vivo using phage display peptide libraries. *Nature*, 1996. 380(6572): 364–66.
35. St Croix, B., et al., Genes expressed in human tumor endothelium. *Science*, 2000. 289(5482): 1197–202.
36. Seaman, S., et al., Genes that distinguish physiological and pathological angiogenesis. *Cancer Cell*, 2007. 11(6): 539–54.
37. Huminiecki, L. and R. Bicknell, In silico cloning of novel endothelial-specific genes. *Genome Res*, 2000. 10(11): 1796–806.
38. Huminiecki, L., et al., Magic roundabout is a new member of the roundabout receptor family that is endothelial specific and expressed at sites of active angiogenesis. *Genomics*, 2002. 79(4): 547–52.
39. Sullivan, D.C., et al., EndoPDI, a novel protein-disulfide isomerase-like protein that is preferentially expressed in endothelial cells acts as a stress survival factor. *J Biol Chem*, 2003. 278(47): 47079–88.
40. Gerritsen, M.E., et al., In silico data filtering to identify new angiogenesis targets from a large *in vitro* gene profiling data set. *Physiol Genomics*, 2002. 10(1): 13–20.
41. Cooney, M.M., et al., Drug insight: vascular disrupting agents and angiogenesis—Novel approaches for drug delivery. *Nat Clin Pract Oncol*, 2006. 3(12): 682–92.
42. Galbraith, S.M., et al., Reproducibility of dynamic contrast-enhanced MRI in human muscle and tumours: Comparison of quantitative and semi-quantitative analysis. *NMR Biomed*, 2002. 15(2): 132–42.
43. Siemann, D.W. and W. Shi, Efficacy of combined antiangiogenic and vascular disrupting agents in treatment of solid tumors. *Int J Radiat Oncol Biol Phys*, 2004. 60(4): 1233–40.
44. Siemann, D.W. and W. Shi, Targeting the tumor blood vessel network to enhance the efficacy of radiation therapy. *Semin Radiat Oncol*, 2003. 13(1): 53–61.
45. Pilat, M.J., J. McCormick, and P.M. LoRusso, Vascular targeting agents. *Curr Oncol Rep*, 2004. 6(2): 103–10.
46. Kerr, D.J., et al., Phase I and pharmacokinetic study of flavone acetic acid. *Cancer Res*, 1987. 47(24 Pt 1): 6776–81.
47. Jameson, M.B., et al., Clinical aspects of a phase I trial of 5,6-dimethylxanthenone-4-acetic acid (DMXAA), a novel antivascular agent. *Br J Cancer*, 2003. 88(12): 1844–50.
48. Gerber, H.P. and N. Ferrara, Pharmacology and pharmacodynamics of bevacizumab as monotherapy or in combination with cytotoxic therapy in preclinical studies. *Cancer Res*, 2005. 65(3): 671–80.
49. Carnemolla, B., et al., Identification of a glioblastoma-associated tenascin-C isoform by a high affinity recombinant antibody. *Am J Pathol*, 1999. 154(5): 1345–52.
50. Silverman, K.J., et al., Angiogenic activity of adipose tissue. *Biochem Biophys Res Commun*, 1988. 153(1): 347–52.
51. Brack, S.S., et al., Tumor-targeting properties of novel antibodies specific to the large isoform of tenascin-C. *Clin Cancer Res*, 2006. 12(10): 3200–8.

52. Bostwick, D.G., et al., Prognostic factors in prostate cancer. College of American Pathologists Consensus Statement 1999. *Arch Pathol Lab Med*, 2000. 124(7): 995–1000.
53. Liu, H., et al., Monoclonal antibodies to the extracellular domain of prostate-specific membrane antigen also react with tumor vascular endothelium. *Cancer Res*, 1997. 57(17): 3629–34.
54. Milowsky, M.I., et al., Vascular targeted therapy with anti-prostate-specific membrane antigen monoclonal antibody J591 in advanced solid tumors. *J Clin Oncol*, 2007. 25(5): 540–47.
55. Max, R., et al., Immunohistochemical analysis of integrin alpha vbeta3 expression on tumor-associated vessels of human carcinomas. *Int J Cancer*, 1997. 71(3): 320–24.
56. Cai, W. and X. Chen, Anti-angiogenic cancer therapy based on integrin alphavbeta3 antagonism. *Anticancer Agents Med Chem*, 2006. 6(5): 407–28.
57. Tarrus, M., et al., RGD-avidin-biotin pretargeting to alpha(v)beta (3) integrin enhances the proapoptotic activity of TNFalpha related apoptosis inducing ligand (TRAIL). *Apoptosis*, 2007.
58. Temming, K., et al., Evaluation of RGD-targeted albumin carriers for specific delivery of auristatin E to tumor blood vessels. *Bioconjug Chem*, 2006. 17(6): 1385–94.
59. Burrows, F.J., et al., Up-regulation of endoglin on vascular endothelial cells in human solid tumors: Implications for diagnosis and therapy. *Clin Cancer Res*, 1995. 1(12): 1623–34.
60. Balza, E., et al., Lack of specificity of endoglin expression for tumor blood vessels. *Int J Cancer*, 2001. 94(4): 579–85.
61. Matsubara, S., et al., Analysis of endoglin expression in normal brain tissue and in cerebral arteriovenous malformations. *Stroke*, 2000. 31(11): 2653–60.
62. Schliemann, C. and D. Neri, Antibody-based targeting of the tumor vasculature. *Biochim Biophys Acta*, 2007. 1776(2): 175–92.
63. Seth, P., et al., Magic roundabout, a tumor endothelial marker: Expression and signaling. *Biochem Biophys Res Commun*, 2005. 332(2): 533–41.
64. Park, K.W., et al., Robo4 is a vascular-specific receptor that inhibits endothelial migration. *Dev Biol*, 2003. 261(1): 251–67.
65. Bedell, V.M., et al., Roundabout4 is essential for angiogenesis *in vivo*. *Proc Natl Acad Sci USA*, 2005. 102(18): 6373–78.
66. Curnis, F., et al., Differential binding of drugs containing the NGR motif to CD13 isoforms in tumor vessels, epithelia, and myeloid cells. *Cancer Res*, 2002. 62(3): 867–74.
67. Curnis, F., et al., Enhancement of tumor necrosis factor alpha antitumor immunotherapeutic properties by targeted delivery to aminopeptidase N (CD13). *Nat Biotechnol*, 2000. 18(11): 1185–90.
68. Patel, N.S., et al., Up-regulation of delta-like 4 ligand in human tumor vasculature and the role of basal expression in endothelial cell function. *Cancer Res*, 2005. 65(19): 8690–97.
69. Patel, N.S., et al., Up-regulation of endothelial delta-like 4 expression correlates with vessel maturation in bladder cancer. *Clin Cancer Res*, 2006. 12(16): 4836–44.
70. Tohda, S., et al., Notch ligands, Delta-1 and Delta-4 suppress the self-renewal capacity and long-term growth of two myeloblastic leukemia cell lines. *Int J Oncol*, 2003. 22(5): 1073–79.
71. Gale, N.W., et al., Haploinsufficiency of delta-like 4 ligand results in embryonic lethality due to major defects in arterial and vascular development. *Proc Natl Acad Sci USA*, 2004. 101(45): 15949–54.
72. Noguera-Troise, I., et al., Blockade of Dll4 inhibits tumour growth by promoting non-productive angiogenesis. *Nature*, 2006. 444(7122): 1032–37.

73. Ridgway, J., et al., Inhibition of Dll4 signalling inhibits tumour growth by deregulating angiogenesis. *Nature*, 2006. 444(7122): 1083–87.
74. Gerber, H.P. and N. Ferrara, The role of VEGF in normal and neoplastic hematopoiesis. *J Mol Med*, 2003. 81(1): 20–31.
75. Ran, S., et al., Evaluation of novel antimouse VEGFR2 antibodies as potential antiangiogenic or vascular targeting agents for tumor therapy. *Neoplasia*, 2003. 5(4): 297–307.
76. Witmer, A.N., et al., Expression of vascular endothelial growth factor receptors 1, 2, and 3 in quiescent endothelia. *J Histochem Cytochem*, 2002. 50(6): 767–77.
77. Ramakrishnan, S., et al., Vascular endothelial growth factor-toxin conjugate specifically inhibits KDR/flk-1-positive endothelial cell proliferation *in vitro* and angiogenesis in *vivo*. *Cancer Res*, 1996. 56(6): 1324–30.
78. Ran, S., et al., The vascular-ablative agent VEGF(121)/rGel inhibits pulmonary metastases of MDA-MB-231 breast tumors. *Neoplasia*, 2005. 7(5): 486–96.
79. Rempel, S.A., et al., Identification and localization of the cytokine SDF1 and its receptor, CXC chemokine receptor 4, to regions of necrosis and angiogenesis in human glioblastoma. *Clin Cancer Res*, 2000. 6(1): 102–11.
80. Scheurer, S.B., et al., Modulation of gene expression by hypoxia in human umbilical cord vein endothelial cells: A transcriptomic and proteomic study. *Proteomics*, 2004. 4(6): 1737–60.
81. Balkwill, F., The significance of cancer cell expression of the chemokine receptor CXCR4. *Semin Cancer Biol*, 2004. 14(3): 171–79.
82. Fathers, K.E., et al., Heterogeneity of Tie2 expression in tumor microcirculation: influence of cancer type, implantation site, and response to therapy. *Am J Pathol*, 2005. 167(6): 1753–62.
83. Wong, A.L., et al., Tie2 expression and phosphorylation in angiogenic and quiescent adult tissues. *Circ Res*, 1997. 81(4): 567–74.
84. Peters, K.G., et al., Expression of Tie2/Tek in breast tumour vasculature provides a new marker for evaluation of tumour angiogenesis. *Br J Cancer*, 1998. 77(1): 51–56.
85. Moon, W.S., et al., Expression of angiopoietin 1, 2 and their common receptor Tie2 in human gastric carcinoma: Implication for angiogenesis. *J Korean Med Sci*, 2006. 21(2): 272–78.
86. Thorpe, P.E., Vascular targeting agents as cancer therapeutics. *Clin Cancer Res*, 2004. 10(2): 415–27.
87. Menrad, A. and H.D. Menssen, ED-B fibronectin as a target for antibody-based cancer treatments. *Expert Opin Ther Targets*, 2005. 9(3): 491–500.
88. Borsi, L., et al., Selective targeting of tumoral vasculature: Comparison of different formats of an antibody (L19) to the ED-B domain of fibronectin. *Int J Cancer*, 2002. 102(1): 75–85.
89. Borsi, L., et al., Selective targeted delivery of TNFalpha to tumor blood vessels. *Blood*, 2003. 102(13): 4384–92.
90. Birchler, M.T., et al., Immunoscintigraphy of patients with head and neck carcinomas, with an anti-angiogenetic antibody fragment. *Otolaryngol Head Neck Surg*, 2007. 136(4): 543–48.
91. Santimaria, M., et al., Immunoscintigraphic detection of the ED-B domain of fibronectin, a marker of angiogenesis, in patients with cancer. *Clin Cancer Res*, 2003. 9(2): 571–79.
92. Paganelli, G., et al., Antibody-guided three-step therapy for high grade glioma with yttrium-90 biotin. *Eur J Nucl Med*, 1999. 26(4): 348–57.
93. Riva, P., et al., Loco-regional radioimmunotherapy of high-grade malignant gliomas using specific monoclonal antibodies labeled with 90Y: A phase I study. *Clin Cancer Res*, 1999. 5(10 Suppl): 3275s–3280s.

94. Reardon, D.A., et al., Salvage radioimmunotherapy with murine iodine-131-labeled anti-tenascin monoclonal antibody 81C6 for patients with recurrent primary and metastatic malignant brain tumors: Phase II study results. *J Clin Oncol*, 2006. 24(1): 115–22.
95. Silacci, M., et al., Human monoclonal antibodies to domain C of tenascin-C selectively target solid tumors *in vivo*. *Protein Eng Des Sel*, 2006. 19(10): 471–78.
96. Beck, A.W., et al., Combination of a monoclonal anti-phosphatidylserine antibody with gemcitabine strongly inhibits the growth and metastasis of orthotopic pancreatic tumors in mice. *Int J Cancer*, 2006. 118(10): 2639–43.
97. Huang, X., M. Bennett, and P.E. Thorpe, A monoclonal antibody that binds anionic phospholipids on tumor blood vessels enhances the antitumor effect of docetaxel on human breast tumors in mice. *Cancer Res*, 2005. 65(10): 4408–16.
98. Ran, S., et al., Antitumor effects of a monoclonal antibody that binds anionic phospholipids on the surface of tumor blood vessels in mice. *Clin Cancer Res*, 2005. 11(4): 1551–62.
99. Milowsky, M.I., et al., Phase I trial of yttrium-90-labeled anti-prostate-specific membrane antigen monoclonal antibody J591 for androgen-independent prostate cancer. *J Clin Oncol*, 2004. 22(13): 2522–31.
100. Bander, N.H., et al., Targeted systemic therapy of prostate cancer with a monoclonal antibody to prostate-specific membrane antigen. *Semin Oncol*, 2003. 30(5): 667–76.
101. Bander, N.H., et al., Phase I trial of 177lutetium-labeled J591, a monoclonal antibody to prostate-specific membrane antigen, in patients with androgen-independent prostate cancer. *J Clin Oncol*, 2005. 23(21): 4591–601.
102. Galbraith, S.M., et al., Effects of 5,6-dimethylxanthenone-4-acetic acid on human tumor microcirculation assessed by dynamic contrast-enhanced magnetic resonance imaging. *J Clin Oncol*, 2002. 20(18): 3826–40.
103. Bibby, M.C. and J.A. Double, Flavone acetic acid—From laboratory to clinic and back. *Anticancer Drugs*, 1993. 4(1): 3–17.
104. Lippert, J.W., 3rd, Vascular disrupting agents. *Bioorg Med Chem*, 2007. 15(2): 605–15.
105. Young, S.L. and D.J. Chaplin, Combretastatin A4 phosphate: Background and current clinical status. *Expert Opin Investig Drugs*, 2004. 13(9): 1171–82.
106. Fox, E., et al., A phase 1 study of ABT-751, an orally bioavailable tubulin inhibitor, administered daily for 7 days every 21 days in pediatric patients with solid tumors. *Clin Cancer Res*, 2006. 12(16): 4882–87.

16 Recent Advancements in the Use of Antibody–Drug Conjugates for Cancer Therapy

Peter D. Senter and Dennis R. Benjamin

CONTENTS

KEY WORDS: monoclonal antibody, drug delivery, conjugate

ABSTRACT

A great deal of interest surrounds the use of monoclonal antibodies (mAbs) for the selective delivery of cytotoxic drugs to tumors. The premise behind this approach is that by targeting drugs selectively to tumors, it should be possible not only to improve therapeutic efficacy, but also to minimize the toxic side effects on noncancerous tissues. Early work in this field surrounded the use of antibody–drug conjugates (ADCs) comprising clinically approved drugs and mAbs that were highly immunogenic when administered in humans. The results of such studies have prompted investigation toward understanding the key parameters affecting activity and tolerability. These include the selection of the antigen target, the use of nonimmunogenic mAb carriers, the incorporation of highly potent drugs and novel conditionally stable linker technologies, and new conjugation technologies that minimally perturb the biologic characteristics of the mAb carrier and lead to well-characterized ADCs that can be manufactured reproducibly in high yields. Progress in this field has been significant, as evidenced from the pronounced clinical activities of SGN-35, an ADC directed against the CD30-positive malignancies, such as Hodgkin's disease and anaplastic large cell lymphoma, and trastuzumab-DM1 which has shown activity in metastatic breast carcinoma. This review details many of the technological advancements made and provides examples of promising ADCs that are progressing in clinical development.

INTRODUCTION

Monoclonal antibodies (mAbs) have demonstrated considerable utility in the clinical treatment of cancer.[1,2] Although the activities of unmodified mAbs such as Rituxan (rituximab) in non-Hodgkin's lymphoma, Erbitux (cetuximab) in colorectal carcinoma, Herceptin (trastuzumab) for metastatic breast cancer, and Avastin (bevacizumab) for colorectal and lung cancer are substantial, these agents are rarely curative. As a result, considerable attention has turned to enhancing antibody activity by appending cytotoxic drugs to them, thereby generating antibody–drug conjugates (ADCs) capable of site-selective drug delivery. The rationale for this approach is that by delivering cancer drugs to tumor cells, it may be possible both to enhance therapeutic efficacy and to spare normal tissues from chemotherapeutic damage. In the past few years, significant progress has been made in this field of research.[3–6] Key parameters for ADC development have been identified which include the choice of target antigen, the ability of the ADC to localize to target tissues, the fate of the antibody once bound to its cognate antigen, the stability of the linker used to attach the drug to the antibody both in the systemic circulation and inside the target cell, and the potency and mechanism of action of the released drug. Other important considerations include the methods used to generate ADCs, the composition and heterogeneity of the resulting product, and the effects that chemical modification have on such antibody properties as pharmacokinetics, biodistribution, antigen binding, and effector functions. This review describes several aspects of ADC development, with an emphasis on how addressing these parameters has led to promising agents that have advanced into clinical trials.

OPTIMIZATION OF ADC DRUG AND LINKER COMPONENTS

CONVENTIONAL DRUGS

Early work in antibody-mediated drug targeting surrounded the use of clinically approved drugs, since much was known about how they could be chemically manipulated and the agents were readily available. Drugs such as doxorubicin, methotrexate, mitomycin, 5-fluorouracil, and vinca alkaloids were typically used,[7] and little attention was paid to particular aspects of the mAb carrier, the mode and stoichiometry

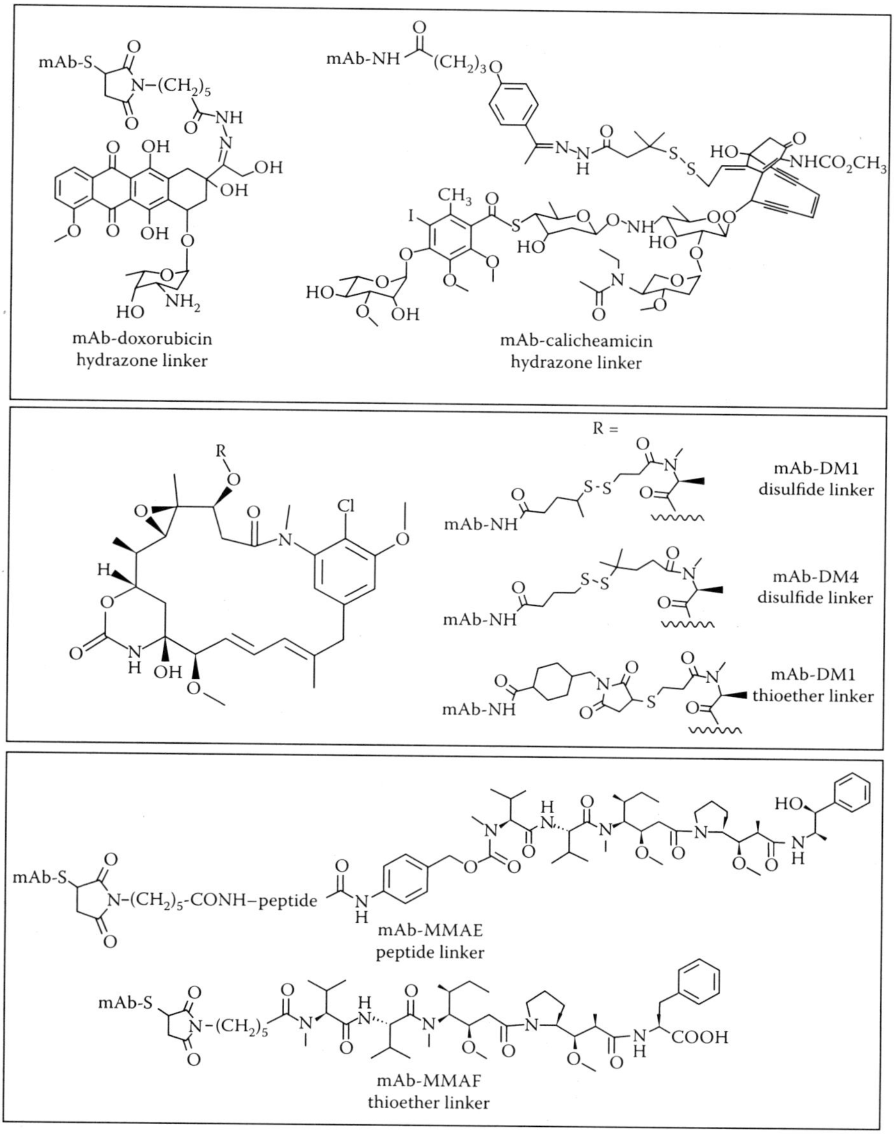

FIGURE 16.1 Structures of antibody drug conjugates for cancer therapy. The drugs used to prepare the conjugates range from doxorubicin to highly potent agents like the auristatin analogues MMAE and MMAF.

of drug attachment, and the mechanism of drug release. As a result, the activities of these agents were marginal, and few of them were extensively investigated.

An example of an earlier generation ADC is BR96-doxorubicin (Figure 16.1). This conjugate is directed against the LeY tetrasaccharide antigen on human carcinomas, and is composed of the clinically approved drug doxorubicin, which was known to be active in a broad array of tumors. The appeal of using such a drug was that although it lacked high potency, it was active in breast cancer, which was known to be LeY positive.[8,9] Once the ADC binds to cell surface antigens, it internalizes within acidic endosomal and lysosomal vesicles. As a result, an acid-labile hydrazone linker was developed for this ADC which was relatively stable at neutral pH, but unstable at pH 5, allowing for intracellular release of the active drug. BR96–doxorubicin contained eight drugs/mAb, which were attached to mAb thiol groups generated through intrachain disulfide bond reduction.

Treatment of tumor-bearing mice and rats with BR96–doxorubicin led to immunologically specific tumor cures, albeit at doses in excess of 100 mg/kg. The drug component of BR96–doxorubicin has low potency, and it is therefore not surprising that the ADC doses required for activity were very high. In a phase I clinical trial, the MTD of BR96–doxorubicin was approximately 700 mg/m^2, with dose-limiting toxicities of gastrointestinal origin.[10] A subsequent phase II trial revealed that the toxicities could be attributed to normal gut expression of LeY.[11] In both trials, little antitumor activity was obtained, and the measured half-life of systemic drug release was only 43 hours (Table 16.1). It became apparent that BR96–doxorubicin was significantly hampered by low drug potency component, the unstable linker used, and the presence of target antigen on highly sensitive nontumor cells.

Highly Potent Drugs

The results with low potency ADCs such as BR96–doxorubicin prompted significant efforts toward utilizing drugs with much higher potencies than doxorubicin for targeted delivery. The natural product calicheamicin has been the subject of extensive investigation for drug delivery, due to its ability to bind to the minor groove and effect cell kill at concentrations well below that of standard chemotherapeutic agents.[4,12] As with BR96–doxorubicin, an acid-labile hydrazone linker was used to link the drug to mAb carriers (Figure 16.1), with a similar half-life (Table 16.1) for drug release from the ADC in the range of 48 to 72 hours.[13] Pronounced antitumor activities have been obtained with mAb–calicheamicin ADCs on epithelial cancers at doses as low as approximately 2 mg (mAb component)/kg.[14–16] The enhanced potency of this ADC compared to the corresponding doxorubicin conjugate reflects the relative activities of the drug components. This is apparent from *in vitro* assays in which the ADCs differ in potency by as much as 1,000-fold.[14,17] Other promising ADCs using calicheamicin are directed against the CD22[18-20] and 5T4[15] antigens, recognizing B-cell lymphomas and epithelial tumors, respectively.

The sole clinically approved ADC to date, gemtuzumab ozogamicin (Mylotarg), consists of an anti-CD33 mAb linked to calicheamicin through an acid-labile hydrazone linker.[21,22] This ADC is clinically used for the treatment of acute myeloid leukemia at doses of 9 mg/m^2 given twice at 2 week intervals.[21,23] Complete remissions were obtained in approximately 30 percent of the patients treated with gemtuzumab ozogamicin, with side effects including myelosuppression and hepatic veno-occlusive disease.

TABLE 16.1
Representative ADCs for Cancer Therapy

ADC		Drug Component			Linker Component	
mAb–Drug	Antigen/Tumor	Drug Component	Mechanism of Activity	Free Drug Potency (M)	Cleavable Unit/ Mechanism for Drug Release	Leading Reference
BR96–doxorubicin	Le^Y/carcinomas	Doxorubicin	Topoisomerase II	10^{-7}	Hydrazone/low pH	9
mAb–calicheamicin	CD33/leukemia CD22/lymphomas Le^Y/carcinomas 5T4/carcinomas	Calicheamicin	DNA minor groove binder	10^{-10}–10^{-9}	Hydrazone/low pH	4
mAb–DM1	CD56/lung CanAg/colorectal CD33/leukemia PSMA/prostate CD79/lymphoma	Maytansine	Tubulin depolymerization	10^{-11}–10^{-9}	Disulfide/reduction	3
mAb–DM4	CanAg/colorecta Her2/neu α_v Integrin/carcinomas	Maytansine	Tubulin depolymerization	10^{-12}–10^{-11}	Disulfide/reduction	3
mAb–MMAE	CD30/Hodgkin's disease PSMA/prostate GPNMB/melanoma	Auristatin	Tubulin depolymerization	10^{-11}–10^{-9}	Peptide/enzymatic hydrolysis	27
mAb–MMAF	CD70/renal cell CD22/lymphoma CD79/lymphoma	Auristatin	Tubulin depolymerization	10^{-8}–10^{-7}	Peptide/enzymatic hydrolysis thioether/ mAb degradation	26

Although the effects of calicheamicin ADCs are in many cases antigen specific, it has been reported that gemtuzumab ozogamicin elicits potent antitumor activity in an antigen-independent manner on solid tumor *in vivo*.[24] The effects were attributed to passive targeting, and required the use of an acid-labile hydrazone linker. Therefore, it is possible that nonspecific drug release through linker instability contributes to the activities of gemtuzumab ozogamicin in these models. It is not clear that these effects would extend to acute myeloid leukemia (AML), although gemtuzumab ozogamicin has been shown to exhibit activity in patients with AML whose cancer cells do not apparently display the CD33 antigen.[24,25] To summarize, the results with mAb–calicheamicin conjugates demonstrate the importance of drug potency in developing ADCs with broad-spectrum activity. In addition, the approval of gemtuzumab ozogamicin represents a breakthrough in the field.

Auristatins[26,27] and maytansinoids[3] represent other classes of highly potent drugs that have been widely utilized for ADC development. These drugs are significantly more potent than most drugs used in cancer chemotherapy, and act by binding to tubulin at the same site as the vinca alkaloids and inhibiting tubulin polymerization. Maytansinoids are derived from a natural product, whereas the auristatins are totally synthetic.

Linker Technology

The methodology used for linking maytansinoids to antibody carriers involved the use of a disulfide bond in the linker (Figure 16.1). *In vivo* stability of disulfide bonds can be greatly enhanced through steric hindrance, as shown with immunotoxins.[28,29] The use of disulfide linkers exploits the observation that the intracellular concentration of thiols, such as glutathione and cysteine, are much higher than those in plasma. A series of maytansinoid–disulfide linker derivatives with varying degrees of steric hindrance were developed, from which DM1 and DM4 were selected as lead agents for antibody attachment.[30] ADCs were made by modifying antibody lysines with a bifunctional cross-linking reagent, and then adding thiol-containing DM1 or DM4. The resulting ADCs contained a mean of three to four drugs/antibody, and displayed high *in vitro* potency in an immunologically specific manner, as demonstrated with the humanized antibody, huC242, targeting CanAg, a tumor-selective carbohydrate epitope on MUC-1, a membrane-associated mucin.[5,31,32] As with the hydrazone-linked ADCs described earlier, the disulfide bond in huC242–DM1 is unstable, with a half-life of approximately 24 hours (Table 16.1). HuC242–DM4, having a more hindered disulfide, is more stable with a reported half-life of approximately 102 hours.[3] The more hindered disulfide ADCs have demonstrated superior antitumor activity in preclinical models.[3,30,33] HuC242–DM4 has demonstrated improved efficacy compared to HuC242–DM1 in some xenograft models. There are several other examples of maytansinoid ADCs that have shown promising preclinical activities.[34–39]

Use of peptide-based linker technologies may present significant advantages over use of those that are hydrolytically or reductively labile, since hydrolysis is enzymatic, and the enzymes can be selected for preferential expression within tumor cells or solid tumor masses. ADCs composed of drugs such as doxorubicin,[40] mitomycin C,[41] camptothecin,[42] talisomycin,[43] and auristatin family members[26,27,42,44] have been prepared using peptide sequences for intracellular drug release. Of these, the auristatins are of particular interest, because they are highly potent, totally synthetic, stable, and are amenable to chemical modification strategies to allow for linker attachment.

An auristatin derivative, monomethyl auristatin E (MMAE), was modified with dipeptide linkers, and the resulting drug derivatives were linked to solvent-accessible antibody thiol groups,[26,27] generating conjugates of uniform composition (Figure 16.1). The antibodies used were directed against the CD30 antigen on Hodgkin's disease and the Le^Y antigen on carcinomas. *In vitro* studies demonstrated that peptide-linked ADC were highly potent with 10- to 100-fold greater immunologically dependent cell kill compared to corresponding the hydrazone derived from auristatin E5-benzoylvaleric acid ester (AEVB) ADCs which contained a hydrazone-based linker system.[27] The *in vivo* half-lives of drug release from the peptide-based ADCs were 6 and 10 days in mice and cynomolgus macaques, respectively,[45] longer than previously described linkers, such as the hydrazone linker in AEVB ($t_{½}$ 1 to 2 days). Furthermore, the more stable peptide-linked MMAE ADCs were less toxic than corresponding hydrazone-linked ADCs.[27] The peptide that was selected for further studies was composed of valine-citrulline (Val-Cit), because it was shown to be stable in plasma, but labile in the presence of such enzymes as cathepsin B, which is a highly expressed lysosomal enzyme.[27]

mAb–Val-Cit-MMAE ADCs had pronounced antitumor activities *in vitro* (Figure 16.2A) due to the specific delivery of active drug within antigen-positive tumor cell lines (Figure 16.2B). Therapeutic efficacy studies were undertaken in mice with subcutaneous xenograft tumors, and therapy was initiated once the tumors were staged and rapidly growing. Cures and complete regressions were obtained at doses that were as low as 1/200th the maximum tolerated dose (Figure 16.3).[27] Since this work, auristatin ADC technology has been successfully applied to antibodies recognizing a wide range of tumor antigens, including BCMA,[46] CD19,[47] CD20,[48] CD70,[49] CD79,[37] E selectin,[50] EphB2,[51] glycoprotein NMB,[52] melanotransferrin/p97,[53] MUC16,[53] PSMA,[54] and TMEFF2.[55]

Because auristatins are totally synthetic, integral structural modifications can be made that significantly alter the properties of the drug. One such auristatin, MMAF, terminates with a negatively charged residue, phenylalanine, that impairs cell membrane permeability.[26] Consequently, ADCs containing MMAF that facilitate drug uptake by antigen-positive cells are >2,000-fold more potent than the free drug itself. MMAF linked to the cAC10 antibody through a peptide linker was active against anaplastic large cell lymphoma xenografts at well-tolerated doses.[26] Similar findings were obtained with anti-CD70 auristatin conjugates against renal cell carcinoma xenografts, and activity was demonstrated in multidrug-resistant cell lines.[49]

A finding that ran counter to expectations was that the cleavable peptide (Val-Cit) in MMAF ADCs could be entirely eliminated (Figure 16.1) without diminishing activity. ADCs in which the drug was directly attached to anti-CD30 and anti-Le^Y antibodies through thioether adducts retained activity.[26] Mass spectrometry showed that the released drug was a cysteine adduct of the linker-MMAF derivative, presumably resulting from antibody degradation within lysosomes. A closely related drug, MMAE, was not active when attached in this manner, indicating that ADCs requiring antibody degradation for drug release are highly dependent on the nature of the drug for activity. Similar results were obtained with a thioether adduct of

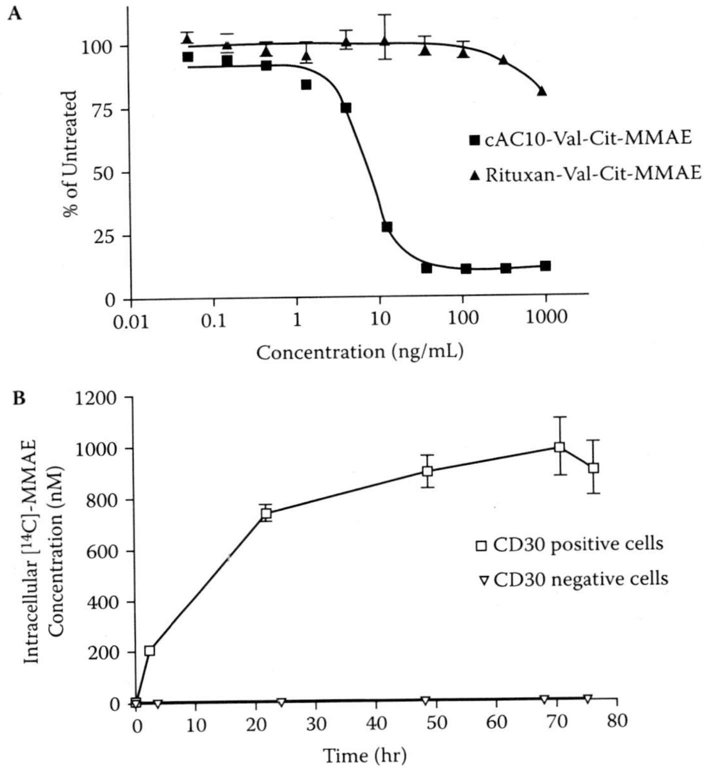

FIGURE 16.2 *In vitro* cytotoxic effects of MMAE ADCs. (A) CD30⁺ Karpas 299 cells were exposed to cAC10–Val-Cit-MMAE or the nonbinding Rituxan–Val-Cit-MMAE ADCs, and the cytotoxic effects were measured after 96 hours.[44] (B) The intracellular concentration of unconjugated MMAE was determined by exposing both antigen-positive (L540cy) and antigen-negative (WSU) cells to radiolabeled cAC10–Val-Cit-MMAE and measuring the amount of nonconjugated drug in cell extracts.

DM4 which was appended to lysine residues on the humanized antibody, huC242 (Figure 16.1). In this case, the released drug upon cell incubation was a lysine-DM1 derivative in which the lysine residue was the attachment residue on the antibody delivery vehicle.[56] The resulting ADC was active *in vitro*, but had little *in vivo* activity compared to the corresponding disulfide. It is not yet clear what the requirements are for effective cell kill using linker technologies that require antibody degradation for drug release, but it is apparent that activities are less predictable than those obtained with cleavable linker technologies.

The anti-CD70 ADC with a thioether and peptide linkages had similar half-lives of 7 days for drug release *in vivo*.[45] *Ex vivo* plasma stability studies revealed a shared mechanism involving fragmentation of the maleimide linkage to antibody thiols.[57] Thus, the weakest link in the antibody–MMAF thioether adduct is the thioether formed when the maleimido drug derivative is condensed with antibody thiols. A

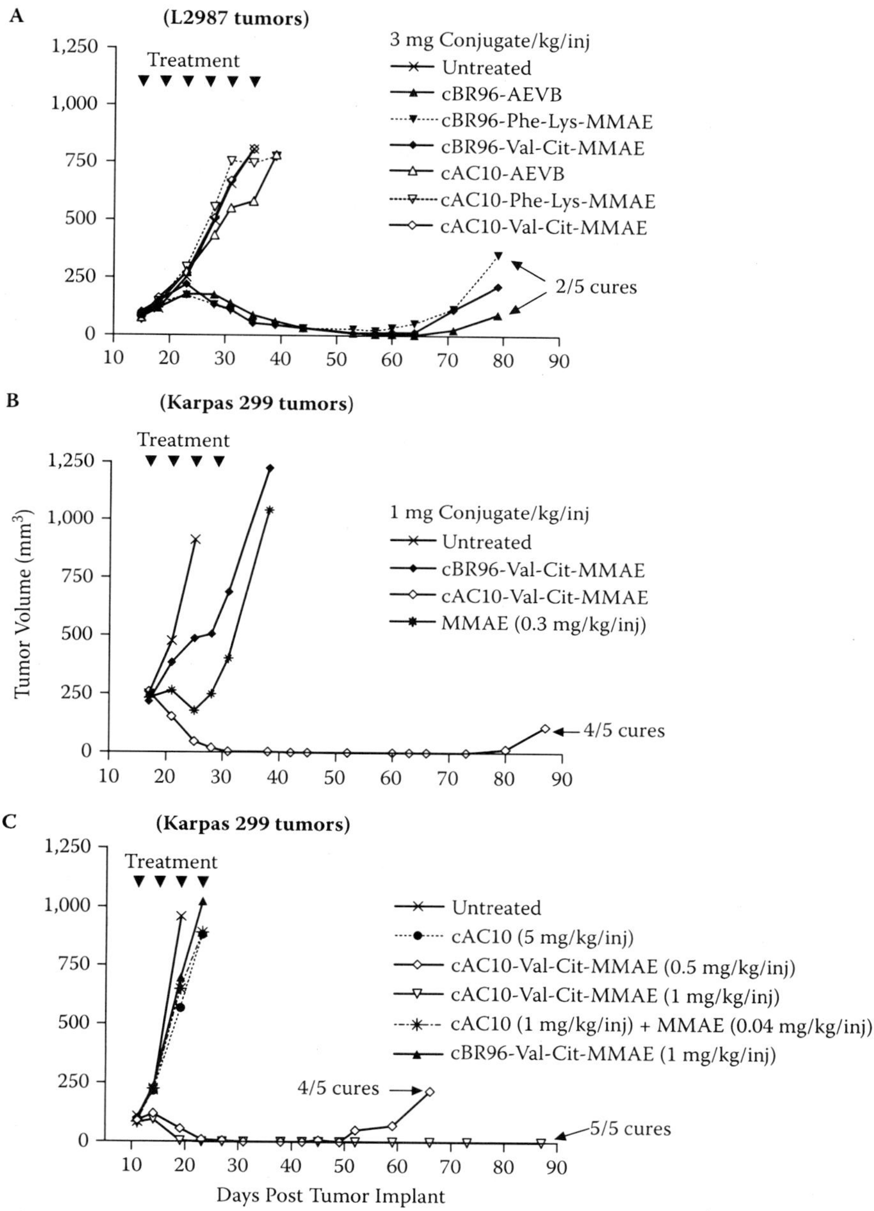

FIGURE 16.3 *In vivo* therapeutic effects of auristatin ADCs in mice with subcutaneous human tumor xenografts. (A) Comparison of various ADCs in mice with L2987 lung carcinoma tumors (BR96 antigen positive, cAC10 antigen negative). Animals were treated with either peptide-based (Val-Cit or Phe-Lys) or hydrazone-based (AEVB) ADCs according to the schedule shown by the arrows. (B and C) Effects of the Val-Cit ADCs on SCID mice with s.c. Karpas 299 tumors (cAC10 antigen positive, BR96 antigen negative). (Adapted from Reference 27.)

new generation of ADCs was subsequently made that used an acetamide-thiol adduct in place of the thioether (Figure 16.1). Circulating ADC did not lose any detectable drug over 2 weeks *in vivo*. The tolerability, efficacy, and intratumoral drug concentrations resulting from ADCs that were indefinitely stable *in vivo* were no different from the corresponding maleimido-thioether adducts. It is apparent that extending the stability of the ADCs beyond that of the 6 to 10 day half-life of the previously described peptide-based ADCs does not result in detectable improvements in the preclinical models tested. The effects of increasing drug retention half-life in this manner will undoubtedly require clinical testing in order to establish the impact on efficacy and tolerability.

OPTIMIZATION OF CONJUGATION TECHNOLOGY

METHODS FOR DRUG ATTACHMENT

The ADCs described thus far have been produced through the reaction of drugs or chemical cross-linking reagents with solvent-accessible reactive amino acids such as lysine and cysteine on the antibody. The process used to covalently attach the drug to the antibody delivery vehicle can have a major impact on activity, since antibody modification may compromise binding, aggregation state, pharmacokinetics, and biodistribution. For example, linker technologies have been developed to attach as many as 16 doxorubicin molecules/antibody, but many of the resulting ADCs were largely aggregated and had impaired antigen binding.[58] The other extreme involves the attachment of drugs that self-associate such as calicheamicin to antibodies. To avoid aggregation, the clinically approved ADC gemtuzumab ozogamicin was prepared with an average of two to three drugs/antibody, but half the antibody in the preparation contains no drug at all.[22] The heterogeneity of calicheamicin containing ADCs has recently been reduced by additives in the conjugation process.[59] Conjugation technology is a critical aspect in generating effective ADCs, and optimization strategies can vary with the drug, linker, and the antibody used.

DRUG TO ANTIBODY RATIOS

The first anti-CD30–Val-Cit-MMAE ADCs were formed by reducing the interchain disulfides on the cAC10 antibody and attaching the drugs to the resulting cysteine residues.[27,44] These ADCs contained eight drugs/antibody, and were highly uniform in composition, since there were only four reducible disulfides. Although the binding properties of the resulting ADC were well preserved, subsequent studies demonstrated that the pharmacokinetic properties of the antibody delivery vehicle were significant perturbed. The eight-loaded ADC cleared much more rapidly from the circulation than the unmodified antibody.[60] This led to the development of cAC10–Val-Cit-MMAE with an average of four drugs/antibody. Although the *in vitro* potency of the four-loaded ADC was half that of the eight-loaded version, the *in vivo* potencies of these agents were the same. To understand the basis for this, pharmacokinetic studies were undertaken, and it was found that the clearance rate of the eight-loaded ADC was approximately twice that of the four-loaded counterpart, and tumors therefore experienced comparatively longer exposure to ADCs with lower drug loading levels. Importantly, the maximum tolerated doses of the ADCs were dependent on drug loading, such that twice as much four-loaded ADC could be administered as eight-loaded. The results from this work demonstrated that drug loading was an important parameter for tolerability, and optimal therapeutic windows required careful attention to the amount of drug attached to the mAb. cAC10–Val-Cit-MMAE, with a mean loading of four drugs/antibody, is now known as SGN-35. This agent is currently in a dose-escalation phase 1 clinical trial for the treatment of Hodgkin's disease.[61,62]

Engineered mAbs for Drug Attachment

Recently, significant attention has been directed toward the use of recombinant strategies for the production of ADCs with predetermined sites for drug attachment. The first example of this approach involved engineered forms of cAC10 in which intrachain cysteines were replaced with serines.[63] The ADCs produced from these antibodies had two or four molecules of MMAE/antibody, and were produced in high yields with unprecedented levels of product uniformity.[63,64] The engineered ADCs were compared with those derived from the parent anti-CD30 immunoglobulin G_1 (IgG_1) for binding affinity, cytotoxic activity and for *in vivo* tolerability, pharmacokinetics, and efficacy. It was found that all the ADCs with four drugs/antibody were equally active and tolerated, independent of the site of drug substitution. These results provided evidence that the number of drugs/antibody influences the therapeutic window to a greater extent than the site to which the drugs are attached.

Another approach to generate engineered ADCs involved the incorporation of cysteines at defined sites available for drug conjugation.[65] The resulting mAbs, referred to as THIOMABs, contain two free cysteines in the mAb constant region, distant from both the antigen binding and effector function regions. ADCs were formed by condensing an anti-Muc16 THIOMAB with a maleimide derivative of Val-Cit-MMAE. The resulting THIOMAB–drug conjugate (TDC) had an average of approximately 1.6 drugs/mAb, with few other species present. One of the striking results of this work was that while the TDC was as potent as a chemically produced ADC with 3.1 drugs/mAb against human tumor xenografts in mice, the tolerability profile was improved, since it induced less neutropenia, weight loss, and liver toxicity. It is not clear from the work if the improved toxicity profile was due to the uniformity in product composition or to the lower drug/mAb ratio in the recombinant constructs. The latter possibility would be consistent with observations involving cAC10–Val-Cit-MMAE with varying drug loads.[60,63]

ADCs IN THE CLINIC

Gemtuzumab ozogamicin (Mylotarg) is the only clinically approved ADC. Since its approval in 2000 many other ADCs have entered clinical trials, and some of them are listed in Table 16.2. The ADCs used incorporate highly potent drugs with linkers ranging from disulfides, hydrazones, to dipeptides. SGN-35 and trastuzumab-DM1

TABLE 16.2
ADCs in the Clinic

Agent	Clinical Status	Target	Indication	Drug Class	Leading References
SGN-35	Phase I	CD30	Hodgkin's disease	Auristatin	26, 27, 57
CRO11–vc-MMAE	Phase II	GPNMB	Melanoma	Auristatin	52, 68, 69
Trastuzumab–DM1	Phase II	Her2/neu	Breast cancer	Maytansine	3, 66, 67
AVE9633	Phase I	CD33	Acute myelogenous leukemia	Maytansine	36
Gemtuzumab ozogamicin	Approved	CD33	Acute myelogenous leukemia	Calicheamicin	21, 22, 70
Inotuzumab ozogamicin	Phase III	CD22	Non-Hodgkin's lymphoma	Calicheamicin	18, 71
HuC242–DM4	Phase I	CanAg	Colorectal cancer	Maytansine	3, 72
HuN901–DM1	Phase I and II	CD57	Small cell lung cancer, multiple myeloma	Maytansine	39, 73, 74

are particularly noteworthy among these various agents, due to their pronounced levels of efficacy in phase I clinical trials.

As indicated earlier, SGN-35 is composed of the chimeric anti-CD30 mAb, cAC10, with an average of four MMAE molecules/mAb linked through a dipeptide linker. In a phase I clinical trial, patients were treated at doses starting at 0.1 mg/kg every 3 weeks, with 3.6 mg/kg as the highest dose administered.[61,62] The patients had an average of three prior chemotherapeutic regimens, and most had also received radiation therapy and autologous stem cell transplantation. Among 22 evaluable patients who were treated at 1.2 mg/kg and above, there were 23 percent complete regressions, 45 percent objective responses (complete and partial regressions), and 86 percent of the patients experienced clinical benefit from this single agent therapy. It was established that SGN-35 was minimally immunogenic, and many patients have received therapy for >6 months.

Another agent that has shown strong clinical potential is trastuzumab–DM1, an ADC that delivers a thioether-linked antimitotic maytansine derivative to Her2+ breast tumors. This ADC was designed to release drug upon complete degradation of the mAb in the lysosomes of target cells. A phase I clinical trial was carried out in patients with Her2+ metastatic breast cancer, in which patients were treated with escalating doses of trastuzumab–DM1 starting at 0.3 mg/kg and going up to 4.8 mg/kg every 3 weeks.[66,67] Partial responses were obtained in 25 percent of the patients, with dose-limiting toxicities of thrombocytopenia. These data prompted the initiation of a phase II clinical trial to further understand this agent.

CONCLUSION

Significant advancements have been made in ADC technology since the early clinical trials with low-potency, clinically approved anticancer drugs attached to mouse mAbs. One of the major developments surrounds the use of highly potent drugs such as calicheamicin, maytansinoids, and auristatins, attached to mAb carriers through conditionally stable linker technologies. In addition, the conjugation technologies used to make ADC have advanced to a point where both the site and stoichiometry of drug attachment can be controlled, and the resulting ADCs are well defined. ADCs that incorporate these improvements have demonstrated levels of activity and immunological specificity that were previously unattainable. Several such ADCs have advanced into clinical trials and are showing promising levels of activity. Although there are several unresolved issues surrounding ADC technology, such as the importance of antibody affinity, internalization rate, antibody isotype, position and stoichiometry of drug substitution, and the acceptability of normal tissue antigen expression, it is apparent from the progress in the past few years that ADCs will have an increasing impact in the field of cancer therapy.

REFERENCES

1. Reichert, J.M., et al., Monoclonal antibody successes in the clinic. *Nat Biotechnol*, 2005. 23(9): 1073–78.
2. Reichert, J.M. and V.E. Valge-Archer, Development trends for monoclonal antibody cancer therapeutics. *Nat Rev Drug Discov*, 2007. 6(5): 349–56.
3. Chari, R.V., Targeted cancer therapy: Conferring specificity to cytotoxic drugs. *Acc Chem Res*, 2008. 41(1): 98–107.
4. Damle, N.K., Tumour-targeted chemotherapy with immunoconjugates of calicheamicin. *Expert Opin Biol Ther*, 2004. 4(9): 1445–52.
5. Kovtun, Y.V. and V.S. Goldmacher, Cell killing by antibody-drug conjugates. *Cancer Lett*, 2007. 255(2): 232–40.
6. Wu, A.M. and P.D. Senter, Arming antibodies: prospects and challenges for immunoconjugates. *Nat Biotechnol*, 2005. 23(9): 1137–46.
7. Pietersz, G.A., et al., Chemoimmunoconjugates for the treatment of cancer. *Adv Immunol*, 1994. 56: 301–87.
8. Trail, P.A., H.D. King, and G.M. Dubowchik, Monoclonal antibody drug immunoconjugates for targeted treatment of cancer. *Cancer Immunol Immunother*, 2003. 52(5): 328–37.
9. Trail, P.A., et al., Cure of xenografted human carcinomas by BR96-doxorubicin immunoconjugates. *Science*, 1993. 261(5118): 212–15.
10. Saleh, M.N., et al., Phase I trial of the anti-Lewis Y drug immunoconjugate BR96-doxorubicin in patients with Lewis Y-expressing epithelial tumors. *J Clin Oncol*, 2000. 18(11): 2282–92.
11. Tolcher, A.W., et al., Randomized phase II study of BR96-doxorubicin conjugate in patients with metastatic breast cancer. *J Clin Oncol*, 1999. 17(2): 478–84.
12. Damle, N.K. and P. Frost, Antibody-targeted chemotherapy with immunoconjugates of calicheamicin. *Curr Opin Pharmacol*, 2003. 3(4): 386–90.
13. Boghaert, E.R., et al., Determination of pharmacokinetic values of calicheamicin-antibody conjugates in mice by plasmon resonance analysis of small (5 mul) blood samples. *Cancer Chemother Pharmacol*, 2008. 61: 1027–35.

14. Boghaert, E.R., et al., Antibody-targeted chemotherapy with the calicheamicin conjugate hu3S193-*N*-acetyl γ calicheamicin dimethyl hydrazide targets Lewis^Y and eliminates Lewis^Y-positive human carcinoma cells and xenografts. *Clin Cancer Res*, 2004. 10(13): 4538–49.
15. Boghaert, E.R., et al., The oncofetal protein, 5T4, is a suitable target for antibody-guided anti-cancer chemotherapy with calicheamicin. *Int J Oncol*, 2008. 32(1): 221–34.
16. Hamann, P.R., et al., An anti-MUC1 antibody-calicheamicin conjugate for treatment of solid tumors. Choice of linker and overcoming drug resistance. *Bioconjug Chem*, 2005. 16(2): 346–53.
17. Trail, P.A., et al., Effect of linker variation on the stability, potency, and efficacy of carcinoma-reactive BR64-doxorubicin immunoconjugates. *Cancer Res*, 1997. 57(1): 100–5.
18. DiJoseph, J.F., et al., Antibody-targeted chemotherapy with CMC-544: a CD22-targeted immunoconjugate of calicheamicin for the treatment of B-lymphoid malignancies. *Blood*, 2004. 103(5): 1807–14.
19. DiJoseph, J.F., et al., Potent and specific antitumor efficacy of CMC-544, a CD22-targeted immunoconjugate of calicheamicin, against systemically disseminated B-cell lymphoma. *Clin Cancer Res*, 2004. 10(24): 8620–29.
20. DiJoseph, J.F., et al., Antibody-targeted chemotherapy of B-cell lymphoma using calicheamicin conjugated to murine or humanized antibody against CD22. *Cancer Immunol Immunother*, 2005. 54(1): 11–24.
21. Bross, P.F., et al., Approval summary: Gemtuzumab ozogamicin in relapsed acute myeloid leukemia. *Clin Cancer Res*, 2001. 7(6): 1490–96.
22. Hamann, P.R., et al., Gemtuzumab ozogamicin, a potent and selective anti-CD33 antibody-calicheamicin conjugate for treatment of acute myeloid leukemia. *Bioconjug Chem*, 2002. 13(1): 47–58.
23. Larson, R.A., et al., Final report of the efficacy and safety of gemtuzumab ozogamicin (Mylotarg) in patients with CD33-positive acute myeloid leukemia in first recurrence. *Cancer*, 2005. 104(7): 1442–52.
24. Boghaert, E.R., et al., Tumoricidal effect of calicheamicin immuno-conjugates using a passive targeting strategy. *Int J Oncol*, 2006. 28(3): 675–84.
25. Jedema, I., et al., Internalization and cell cycle-dependent killing of leukemic cells by gemtuzumab ozogamicin: Rationale for efficacy in CD33-negative malignancies with endocytic capacity. *Leukemia*, 2004. 18(2): 316–25.
26. Doronina, S.O., et al., Enhanced activity of monomethylauristatin F through monoclonal antibody delivery: effects of linker technology on efficacy and toxicity. *Bioconjug Chem*, 2006. 17(1): 114–24.
27. Doronina, S.O., et al., Development of potent monoclonal antibody auristatin conjugates for cancer therapy. *Nat Biotechnol*, 2003. 21(7): 778–84.
28. Thorpe, P.E., et al., Improved antitumor effects of immunotoxins prepared with deglycosylated ricin A-chain and hindered disulfide linkages. *Cancer Res*, 1988. 48(22): 6396–403.
29. Thorpe, P.E., et al., New coupling agents for the synthesis of immunotoxins containing a hindered disulfide bond with improved stability *in vivo*. *Cancer Res*, 1987. 47(22): 5924–31.
30. Widdison, W.C., et al., Semisynthetic maytansine analogues for the targeted treatment of cancer. *J Med Chem*, 2006. 49(14): 4392–408.
31. Kovtun, Y.V., et al., Antibody-drug conjugates designed to eradicate tumors with homogeneous and heterogeneous expression of the target antigen. *Cancer Res*, 2006. 66(6): 3214–21.
32. Liu, C., et al., Eradication of large colon tumor xenografts by targeted delivery of maytansinoids. *Proc Natl Acad Sci USA*, 1996. 93(16): 8618–23.

33. Chen, Q., et al., α_v Integrin-targeted immunoconjugates regress established human tumors in xenograft models. *Clin Cancer Res*, 2007. 13(12): 3689–95.
34. Aboukameel, A., et al., Superior anti-tumor activity of the CD19-directed immunotoxin, SAR3419 to rituximab in non-Hodgkin's xenograft animal models: preclinical evaluation. *Blood*, 2007. 110: abstr 2339.
35. Henry, M.D., et al., A prostate-specific membrane antigen-targeted monoclonal antibody-chemotherapeutic conjugate designed for the treatment of prostate cancer. *Cancer Res*, 2004. 64(21): 7995–8001.
36. Legrand, O., et al., An open label, dose escalation study of AVE9633 administered as a single agent by intravenous (IV) infusion weekly for 2 weeks in 4-week cycle to patients with relapsed or refractory CD33-positive acute myeloid leukemia (AML). *Blood*, 2007. 110: abstr 1850.
37. Polson, A.G., et al., Antibody-drug conjugates targeted to CD79 for the treatment of non-Hodgkin lymphoma. *Blood*, 2007. 110(2): 616–23.
38. Tassone, P., et al., Cytotoxic activity of the maytansinoid immunoconjugate B-B4-DM1 against CD138+ multiple myeloma cells. *Blood*, 2004. 104(12): 3688–96.
39. Tassone, P., et al., *In vitro* and *in vivo* activity of the maytansinoid immunoconjugate huN901-N2′-deacetyl-N2′-(3-mercapto-1-oxopropyl)-maytansine against $CD56^+$ multiple myeloma cells. *Cancer Res*, 2004. 64(13): 4629–36.
40. Dubowchik, G.M. and R.A. Firestone, Cathepsin B-sensitive dipeptide prodrugs. 1. A model study of structural requirements for efficient release of doxorubicin. *Bioorg Med Chem Lett*, 1998. 8(23): 3341–46.
41. Dubowchik, G.M., et al., Cathepsin B-sensitive dipeptide prodrugs. 2. Models of anticancer drugs paclitaxel (Taxol), mitomycin C and doxorubicin. *Bioorg Med Chem Lett*, 1998. 8(23): 3347–52.
42. Walker, M.A., et al., Synthesis of an immunoconjugate of camptothecin. *Bioorg Med Chem Lett*, 2002. 12(2): 217–19.
43. Walker, M.A., et al., Monoclonal antibody mediated intracellular targeting of tallysomycin S(10b). *Bioorg Med Chem Lett*, 2004. 14(16): 4323–27.
44. Francisco, J.A., et al., cAC10-vcMMAE, an anti-CD30-monomethyl auristatin E conjugate with potent and selective antitumor activity. *Blood*, 2003. 102(4): 1458–65.
45. Sanderson, R.J., et al., *In vivo* drug-linker stability of an anti-CD30 dipeptide-linked auristatin immunoconjugate. *Clin Cancer Res*, 2005. 11(2 Pt 1): 843–52.
46. Ryan, M.C., et al., Antibody targeting of B-cell maturation antigen on malignant plasma cells. *Mol Cancer Ther*, 2007. 6(11): 3009–18.
47. Benjamin, D., et al., Humanized anti-CD19 auristatin antibody-drug conjugates display potent antitumor activity in preclinical models of B-cell malignancies. AACR-NCI-EORTC Conference on Molecular Targets and Cancer Therapeutics, 2007: abstr B60.
48. Law, C.L., et al., Efficient elimination of B-lineage lymphomas by anti-CD20-auristatin conjugates. *Clin Cancer Res*, 2004. 10(23): 7842–51.
49. Law, C.L., et al., Lymphocyte activation antigen CD70 expressed by renal cell carcinoma is a potential therapeutic target for anti-CD70 antibody-drug conjugates. *Cancer Res*, 2006. 66(4): 2328–37.
50. Bhaskar, V., et al., E-selectin up-regulation allows for targeted drug delivery in prostate cancer. *Cancer Res*, 2003. 63(19): 6387–94.
51. Mao, W., et al., EphB2 as a therapeutic antibody drug target for the treatment of colorectal cancer. *Cancer Res*, 2004. 64(3): 781–88.
52. Tse, K.F., et al., CR011, a fully human monoclonal antibody-auristatin E conjugate, for the treatment of melanoma. *Clin Cancer Res*, 2006. 12(4): 1373–82.
53. Chen, Y., et al., Armed antibodies targeting the mucin repeats of the ovarian cancer antigen, MUC16, are highly efficacious in animal tumor models. *Cancer Res*, 2007. 67(10): 4924–32.

54. Ma, D., et al., Potent antitumor activity of an auristatin-conjugated, fully human monoclonal antibody to prostate-specific membrane antigen. *Clin Cancer Res*, 2006. 12(8): 2591–96.
55. Afar, D.E., et al., Preclinical validation of anti-TMEFF2-auristatin E-conjugated antibodies in the treatment of prostate cancer. *Mol Cancer Ther*, 2004. 3(8): 921–32.
56. Erickson, H.K., et al., Antibody-maytansinoid conjugates are activated in targeted cancer cells by lysosomal degradation and linker-dependent intracellular processing. *Cancer Res*, 2006. 66(8): 4426–33.
57. Alley, S.C., et al., Contribution of linker stability to the activities of anticancer immunoconjugates. *Bioconjug Chem*, 2008. 19(3): 759–65.
58. King, H.D., et al., Monoclonal antibody conjugates of doxorubicin prepared with branched peptide linkers: Inhibition of aggregation by methoxytriethyleneglycol chains. *J Med Chem*, 2002. 45(19): 4336–43.
59. Hollander, I., A. Kunz, and P.R. Hamann, Selection of reaction additives used in the preparation of monomeric antibody-calicheamicin conjugates. *Bioconjug Chem*, 2008. 19(1): 358–61.
60. Hamblett, K.J., et al., Effects of drug loading on the antitumor activity of a monoclonal antibody drug conjugate. *Clin Cancer Res*, 2004. 10(20): 7063–70.
61. Younes, A., et al., A novel antibody–drug conjugate, SGN-35 (anti-CD30-auristatin), induces objective responses in patients with relapsed or refractory Hodgkin lymphoma, preliminary results of a phase I tolerability study. 7th International Symposium on Hodgkin Lymphoma, 2007: abstr PO99bis.
62. Younes, A., et al., Objective responses in a phase I dose-escalation study of SGN-35, a novel antibody-drug conjugate (ADC) targeting CD30, in patients with relapsed or refractory Hodgkin lymphoma. American Society of Clinical Oncology, 2008: abstr 8526.
63. McDonagh, C.F., et al., Engineered antibody–drug conjugates with defined sites and stoichiometries of drug attachment. *Protein Eng Des Sel*, 2006. 19(7): 299–307.
64. Sun, M.M., et al., Reduction-alkylation strategies for the modification of specific monoclonal antibody disulfides. *Bioconjug Chem*, 2005. 16(5): 1282–90.
65. Junutula, J.R., et al., Site-specific conjugation of a cytotoxic drug to an antibody improves the therapeutic index. *Nat Biotechnol*, 2008. 26: 925–32.
66. Beeram, M., et al., A phase I study of trastuzumab-MCC-DM1 (T-DM1), a first-in-class HER2 antibody-drug conjugate (ADC), in patients (pts) with HER2+ metastatic breast cancer (BC). *J Clin Oncol*, 2007. 25(18S): abstr 1042.
67. Krop, I.E., et al., A phase I study of trastuzumab-DM1, a first-in-class HER2 antibody-drug conjugate, in patients with advanced HER2+ breast cancer. 30th Annual San Antonio Breast Cancer Symposium, 2007: abstr 310.
68. Pollack, V.A., et al., Treatment parameters modulating regression of human melanoma xenografts by an antibody–drug conjugate (CR011-vcMMAE) targeting GPNMB. *Cancer Chemother Pharmacol*, 2007. 60(3): 423–35.
69. Sznol, M., et al., Phase I pharmacokinetic study of CR011-vcMMAE, an antibody toxin conjugate drug, in patients with unresectable stage III/IV melanoma. AACR-NCI-EORTC Conference on Molecular Targets and Cancer Therapeutics, 2007: abstr B47.
70. Sievers, E.L., Efficacy and safety of gemtuzumab ozogamicin in patients with CD33-positive acute myeloid leukaemia in first relapse. *Expert Opin Biol Ther*, 2001. 1(5): 893–901.
71. Dijoseph, J.F., et al., Therapeutic potential of CD22-specific antibody-targeted chemotherapy using inotuzumab ozogamicin (CMC-544) for the treatment of acute lymphoblastic leukemia. *Leukemia*, 2007. 21(11): 2240–45.

72. Mita, M.M., et al., A phase I study of a CanAg-targeted immunoconjugate, huC242-DM4, in patients with Can Ag-expressing solid tumors. *J Clin Oncol*, 2007. 25(18S): abstr 3062.
73. Chanan-Khan, A.A., et al., Phase I study of huN901-DM1 (BB-10901) in patients with relapsed and relapsed/refractory CD56-positive multiple myeloma. *Blood*, 2007. 110: abstr 1174.
74. McCann, J., et al., Phase II trial of huN901-DM1 in patients with relapsed small cell lung cancer (SCLC) and CD56-positive small cell carcinoma. *J Clin Oncol*, 2007. 25(18S): abstr 18084.

Index

B

C

D

H

I

J

K

L

M

N

O

P

R

S

T

U

V

Y